Reasoning and Sense-Making Activities for High School Mathematics

Selections from Mathematics Teacher

Edited by

Sarah Kasten
Northern Kentucky University, Highland Heights, Kentucky

Jill Newton
Purdue University, West Lafayette, Indiana

NATIONAL COUNCIL OF
TEACHERS OF MATHEMATICS

www.nctm.org/more4u
Access code: RSM13784

Library of Congress Cataloging-in-Publication Data

Reasoning and sense-making activities for high school mathematics : selections from
mathematics teacher / edited by Sarah Kasten, Jill Newton.
    p. cm.
  Includes bibliographical references.
  ISBN 978-0-87353-655-4
  1. Mathematics--Study and teaching (Secondary)--Activity programs. 2.
Reasoning. I. Kasten, Sarah, 1979- II. Newton, Jill, 1965-
  QA11.2.R433 2011
  510.71'2--dc23

ISBN 978-0-87353-655-4

The National Council of Teachers of Mathematics is a public voice of mathematics
education, supporting teachers to ensure equitable mathematics learning of the
highest quality for all students through vision, leadership, professional
development, and research.

Printed in the United States of America

Contents

Acknowledgments

The editors wish to extend heartfelt thanks to many individuals committed to secondary mathematics education who made this volume possible. Emre Gonulates, a graduate student at Michigan State University, contributed countless hours searching for articles, creating activity sheets, reading, writing, and using his masterful technology skills to solve many problems along the way. To him, we are incredibly grateful! We were lucky to have Dana Cox, Rachael Kenney, Sharon McCrone, and Mike Shaughnessy review chapters, and Glenda Lappan and Peg Smith review the overall book as it neared completion. We are indebted to them for sharing their time, which there never seems to be enough of, and expertise with us. We also would like to thank Anita Draper and Myrna Jacobs at NCTM for answering our questions—sometimes more than once—and guiding us patiently through the process.

Two other groups of people provided invaluable assistance:

Secondary mathematics educators who nominated articles from the *Mathematics Teacher* for inclusion in the volume—

John "Jed" Donovan	Lorraine Males	Mike Shaughnessy
Sherry Fraser	Joseph Malkeitch	Megan Staples
Christian Hirsch	Elizabeth Phillips	Robert Reys
Gary Kader	Adam Poetzel	Dawn Teuscher
Eric Knuth		

Mathematics Teacher authors who created student activity sheets to help teachers use the activities in their articles more easily—

Tom Ball	Karen Hansen	Todd Moyer
Kenneth Chelst	Gary Kader	Madhuri Mulekar
Susana Davidenko	Yukio Kobayashi	Mike Perry
Thomas Edwards	Geoffrey Lewis	Blake Peterson
Christine Franklin	Katherine McGivney	Maxine Pfannkuch
Marvin Gamble	Raymond McGivney	Nancy Powell
John Golzy	Jean McGivney-Burelle	Robert Quinn
Randall Groth	Deborah Moore-Russo	Wes White

Introduction

As a high school mathematics teacher, you are ever seeking new ways to engage your students in developing reasoning habits. This book is for you. We culled from the National Council of Teachers of Mathematics (NCTM) journal *Mathematics Teacher* articles and activities rooted in a wide variety of content areas and levels of mathematics (e.g., maximizing volume, linear regression). We hope that, when looking for mathematics activities to interest and engage your students, you turn to this compilation of activities and that it will become your reliable resource.

Focus in High School Mathematics: Reasoning and Sense Making (NCTM 2009) guided us and served as the screen through which we sifted our choices. We chose activities to highlight the reasoning habits in *Focus in High School Mathematics*. Reasoning is "the process of drawing conclusions on the basis of evidence or stated assumptions," and sense making means "developing understanding of a situation, context, or concept by connecting it with existing knowledge" (NCTM 2009, p. 4). The *Focus* authors emphasize developing these two mathematical processes for high school students: "At the high school level, reasoning and sense making are of particular importance, but historically 'reasoning' has been limited to very select areas of the high school curriculum, and sense making is in many instances not present at all. However, an emphasis on students' reasoning and sense making can help students organize their knowledge in ways that enhance the development of number sense, algebraic fluency, functional relationships, geometric reasoning, and statistical thinking" (NCTM 2009, p. 4). To this end, *Focus in High School Mathematics* presents four reasoning habits as productive ways to practice reasoning and sense making in the mathematics classroom:

1. Analyzing a problem (e.g., seeking patterns and relationships, considering special cases or simpler analogs)

2. Implementing a strategy (e.g., making purposeful use of procedures, making logical deductions)

3. Seeking and using connections (e.g., across mathematical domains, different representations)

4. Reflecting on a solution (e.g., considering the reasonableness of a solution, revisiting initial assumptions)

As mentioned, our goal here is to give you activities to engage your students in these reasoning habits. For example, Olson's (1991) activity in chapter 1 lets students *seek and use connections across mathematical domains* by studying the greatest common divisor in rectangular arrays.

By itself, though, a great activity does not ensure that students develop mathematical reasoning and sense-making habits. *Focus in High School Mathematics: Reasoning and Sense Making* offers several tips for you help your students develop these habits (NCTM 2009, p. 11):

- Provide tasks that require students to figure things out for themselves.

- Ask students to restate the problem in their own words, including any assumptions they have made.

- Give students time to analyze a problem intuitively, explore the problem further by using models, and then proceed to a more formal approach.

- Resist the urge to tell students how to solve a problem when they become frustrated; find other ways to support students as they think and work.

- Ask students questions that will prompt their thinking—for example, "Why does this work?" or "How do you know?"

- Provide adequate wait time after a question for students to formulate their own reasoning.

- Encourage students to ask probing questions of themselves and one another.

- Expect students to communicate their reasoning to their classmates and the teacher, orally and in writing, through using proper mathematical vocabulary.

- Highlight exemplary explanations, and have students reflect on what makes them effective.

- Establish a classroom climate in which students feel comfortable sharing their mathematical arguments and critiquing the arguments of others in a productive manner.

Finding appropriate activities involved several groups of people. We began by soliciting recommendations from secondary school mathematics educators, including teachers, teacher educators, curriculum developers,

TABLE 1

Student Activities Appearing in *Mathematics Teacher* by Content Area and Decade

Content area	No. of activities in decade beginning:								
	1920	1930	1940	1950	1960	1970	1980	1990	2000
Number and Measurement	2	3	3	7	3	23	34	30	31
Algebraic Symbols	1	5	3	1	3	6	13	18	3
Functions	2	1	1	2	1	10	13	40	55
Geometry	4	6	8	4	8	45	32	67	46
Statistics and Probability	0	0	1	0	0	7	17	24	34
Total	**9**	**15**	**16**	**14**	**15**	**91**	**109**	**179**	**169**

and researchers. At the same time we searched past volumes of *Mathematics Teacher* from 1908 to 2009 to identify all appropriate activities. From the recommendations we received and our own examination of the activities, we compiled 43 activities for this publication. When an activity did not include student activity sheets, we contacted the authors of the original activities and asked them to create one. or we created one ourselves. This collection of activities represents diverse reasoning habits, content areas, and years of publication, but it isn't meant to embody the "best" activities published in *Mathematics Teacher*; this book would not have had enough space to include such a scope.

The number of student activities published in *Mathematics Teacher* increased over the past century, and the content areas represented by the activities changed. Table 1, organized into the *Focus in High School Mathematics* content areas, summarizes these changes by decade beginning in 1920, when activities began to appear in earnest in the journal.

Student activities in *Mathematics Teacher* increased sixfold in the 1970s. For example, before 1970 only seven published activities addressed functions, whereas the number of function activities increased from 10 in the 1970s to 55 between 2000 and 2009. The increase in function activities in the 1990s may have been a result of the inclusion of a function standard in NCTM's 1989 *Curriculum and Evaluation Standards for School Mathematics*. Also notable is the recent inclusion of statistics and probability, with more than twice the activities in the 1990s and 2000s than were present from 1920 to 1980. The number of geometry activities increased markedly beginning in the 1970s, when almost half the activities published addressed geometry. Although not as strikingly as the other content areas, the number of published activities in algebraic symbols also increased beginning in the 1970s.

This book's chapters are similarly organized by the content areas in *Focus in High School Mathematics: Reasoning and Sense Making*: (1) Number and Measurement, (2) Algebraic Symbols (3) Functions, (4) Geometry, and (5) Statistics and Probability. Each chapter contains the following:

- an introduction;

- a table listing related articles and activities from Mathematics Teacher, summarizing their features; and

- the articles and activities themselves.

The articles and activities represent a range of difficulty level and mathematical focus. Additional articles and activities are available as PDF downloads through NCTM's More4U online resource center. Simply go to www.nctm.org/more4u and enter the access code that is on the title page of this book.

www.nctm.org/more4u

REFERENCES

National Council of Teachers of Mathematics (NCTM). *Curriculum and Evaluation Standards for School Mathematics*. Reston, Va.: NCTM, 1989.

———. *Focus in High School Mathematics: Reasoning and Sense Making*. Reston, Va.: NCTM, 2009.

Olson, Melfried. "A Geometric Look at Greatest Common Divisor." *Mathematics Teacher* 84 (March 1991): 202–8.

CHAPTER 1

Number and Measurement
Introduction

"Number and measurement, which receive substantial attention in kindergarten through grade 8, are foundational for high school mathematics; without reasoning skills in these areas, students will be limited in their reasoning in other areas of mathematics" (NCTM 2009, p. 21).

This chapter presents eight activities that relate to the content area of number and measurement. The activities range in level of difficulty, topic, and context. For example, one activity (Hall 2008) on More4U focuses on finding the best value for a box of popcorn and would be appropriate for students early in high school. Another (Herman, Milou, and Schiffman 2004) asks students to consider fractions and decimals in different bases and would be more engaging for students in the later years

of high school. The table below presents this chapter's number and measurement activities.

We chose the activities in this chapter because they exemplify activities that illustrate four essential elements of number and measurement and can help develop mathematical habits of mind. *Focus in High School Mathematics* (NCTM 2009) suggests four important elements of reasoning and sense making within number and measurement:

Number and Measurement Activities			
Author and title	Mathematical topic(s)	Context(s)	Materials
Albrecht (2001), "The Volume of a Pyramid: Low-Tech and High-Tech Approaches"	Volume, volume of a pyramid	Beginning with the volume of pyramids built with cubes and moving to the volume of pyramids	Cubic blocks; spreadsheet software (high-tech); hollow pyramids (low-tech); hollow prism (low-tech); water, sand, rice, or small pasta (low-tech); student activity sheets
Çağlayan (2006), "Visualizing Summation Formulas"	Summation, perfect squares	Relating the sum of consecutive odd integers to perfect squares	1-inch grid paper, 1-inch tiles, student activity sheets
Hansen and Lewis (2007), "Finding a Parking Spot for the Binomial Theorem"	Binomial theorem, Pascal's triangle	Finding a parking spot at a baseball game	Graphing calculator, student activity sheets
Herman, Milou, and Schiffman (2004), "Unit Fractions and Their 'Basimal' Representations: Exploring Patterns"	Decimal representations, working in bases other than 10	Looking at the period of repeating decimal representations in base 4 and base 10	Calculator, student activity sheets
Hill (2002), "Print-Shop Paper Cutting: Ratios in Algebra"	Ratios, remainders, scale drawings	Working in a print shop to minimize waste	5 × 7 inch paper, 17 × 22 inch paper, rulers, tape, staplers, student activity sheets
Olson (1991), "A Geometric Look at Greatest Common Divisor"	Greatest common divisor	Relating greatest common divisor to a geometric area model	Graph paper, scissors, student activity sheets
Slowbe (2007), "Pi Filling, Archimedes Style"	Area, π	Approximating π with polygons inscribed in a unit circle	Programmable calculator, student activity sheets
Willcutt (1973), "Paths on a Grid"	Pascal's triangle, permutations	Taxicab geometry	Student activity sheets

1. *Reasonableness of answers and measurements.* Judging whether a given answer or measurement has an appropriate order of magnitude and whether it is expressed in appropriate units

2. *Approximations and error.* Realizing that all real-world measurements are approximations and that unsuitably accurate values should not be used for real-world quantities; recognizing the role of error in subsequent computations with measurements

3. *Number systems.* Understanding number-system properties deeply; extending number system properties to algebraic situations

4. *Counting.* Recognizing when enumeration would be a productive approach to solving a problem and then using principles and techniques of counting to find a solution

Many of this chapter's eight articles address more than one key element, and all articles do so through engaging activities.

Hall (2008) addresses reasonableness of answers and measurements in the activity previously mentioned in which students compare the cost value of various brands of popcorn. A component of this key element is for students to judge whether an answer is given with appropriate units—an important aspect of this activity. "Students can improve their problem-solving skills if they understand that finding the unit of measure for the problem's answer is the *first* step" (Hall 2008, p. 609). Through participation in the activity, students take a close look at units related to the amount of popcorn produced per dollar. Wagner (2003), also on More4U, addresses this element by affording students the opportunity to work with square roots to solve problems about areas and side lengths of squares.

Slowbe (2007) also uses geometric figures to bring the idea of approximation and error in measurement to the forefront in an activity in which students generate the digits of π by using calculations of areas of regular polygons inscribed in a unit circle. "With sufficiently large *n*, we can obtain decimal approximations that become arbitrarily close to the exact value of π" (Slowbe 2007, p. 485).

The key element of number systems is the focus of the activity by Herman, Milou, and Schiffman (2004), in which students conjecture about the relationships between unit fractions and their decimal representations in both base ten and base four. Finally, the counting element appears in the activities by Hansen and Lewis (2007) and by Willcutt (1973). In the activity by Hansen and Lewis, students count the number of paths to a parking space and relate it to Pascal's triangle. In Willcutt's activity, students extend their initial ideas about the number of paths from one point to another on a grid to consider three-dimensional space. This counting activity "presents an excellent model for the need to simplify unwieldy mathematical problems, to look for patterns, and to formulate generalizations based on the results of the data collected for the simple cases" (Willcutt 1973, p. 303).

You can also find the reasoning that *Focus in High School Mathematics* (NCTM 2009) recommends in all the activities in this chapter. The activity by Albrecht (2001) develops the reasoning habit of *analyzing a problem*. Students use blocks and spreadsheets to find patterns to help them write the formula for the volume of a pyramid. Using the spreadsheet was integral for the teacher in this case: "My geometry class had not used spreadsheets before, and the students enjoyed the experience of using the efficiency of technology to compare hundreds—and even thousands—of shapes with ease" (Albrecht 2001, p. 58). In Çağlayan's (2006) activity, students *implement a strategy* to find that "every perfect square is the sum of consecutive odd integers" (p. 70). *Seeking and using connections* appears in Olson's (1991) activity, in which students "examine an arithmetic concept, greatest common divisor, from a geometric standpoint" (p. 202). Finally, Hill (2002) invites students to revisit initial assumptions as part of *reflecting on a solution* in an activity set in a print shop in which they are looking for the most efficient ways to make cuts from a large piece of paper.

REFERENCE

National Council of Teachers of Mathematics (NCTM). *Curriculum and Evaluation Standards for School Mathematics.* Reston, Va.: NCTM, 1989.

———. *Focus in High School Mathematics: Reasoning and Sense Making.* Reston, Va.: NCTM, 2009.

The Volume of a Pyramid:
Low-Tech and High-Tech Approaches

Masha Albrecht

This lesson came about spontaneously during a geometry unit on volume. I had used the lesson shown here in **activity sheet 1,** in which students use cubic blocks to rediscover the formulas for volumes of right prisms, that is, $V = Bh$ and $V = lwh$. This lesson was a simple review for my tenth-grade class, and they completed it easily before the end of the period. With the wooden cubes still on their desks, most of them used the remaining time to build towers and other objects. I noticed that many students piled the cubes into bumpy pyramidal shapes. Because the next day's lesson involved studying the volume of pyramids, I wondered whether these bumpy shapes could be useful for discovering the volume of a real pyramid with smooth sides. Students could compare the volumes of these "pyramids of cubes" with the volumes of corresponding right prisms and perhaps discover the ratio 1/3 to obtain the formula for the volume of a pyramid, $V = (1/3)Bh$. As it turns out, the ratio of 1/3 does not become evident right away. To my students' delight, we found that using a spreadsheet is an excellent way to investigate this problem. My geometry classes had not used spreadsheets before, and the students enjoyed the experience of using the efficiency of technology to compare hundreds—and even thousands—of shapes with ease.

Prerequisites: Students with only very basic mathematical knowledge can benefit from this lesson. Students should have some skill at describing a pattern with an algebraic equation and some familiarity with a spreadsheet. However, I used this lesson with students who had no previous spreadsheet experience.

Grade levels: Although I originally used this lesson with a regular tenth-grade geometry class, the lesson is appropriate for students at different levels and with different abilities. A prealgebra class could do the low-tech part of the lesson, in which students find patterns by using blocks, but they would need help with the formulas for the spreadsheet. Eleventh-grade or twelfth-grade students with more advanced algebra skills could be left on their own to find the spreadsheet formulas and could be given the difficult challenge of finding the closed formula for the volume in the "pyramid of cubes" column on **activity sheet 2.** A calculus class could find the limit of the ratio column as n goes to infinity before they check this limit on the spreadsheet.

Materials: The entire lesson works well in a two-hour block or in two successive fifty-minute lessons, with the low-tech lesson in the first hour and the high-tech spreadsheet lesson in the second. Cubic blocks are needed for the low-tech lesson. Because approximately forty blocks are needed for each group of four students, large classes will need many blocks. If you do not have enough blocks, groups can share. Simple wooden blocks work best; plastic linking cubes do not work as well, because their extruding joints can get in the way when students build the pyramids.

Spreadsheet software is needed for the high-tech lesson. If you are using a separate computer lab, sign out the lab for the second hour of this activity.

For the low-tech extension lesson, the following additional materials are needed: a hollow pyramid and prism with congruent bases and heights, as well as water, sand, rice, or small pasta.

TEACHING SUGGESTIONS

Sheet 1: Using cubic blocks—volume of prisms

This activity sheet is elementary, and more advanced students can skip it. Have students work in groups, with one set of blocks per group. Often one student quickly sees the answers without needing manipulatives, but the other group members are too shy to admit that they need to build the shapes. Require that each group build most of the solids, even if students protest that this activity seems easy.

Sheet 2: Using cubic blocks—volume of pyramids

Students may initially have difficulty understanding what the "pyramids of cubes" look like. Make sure that they build the one with side length 3 correctly. After using the blocks to build a few of the shapes, students recognize the patterns and start filling in the table without using the blocks. Calculating decimal answers for the last column of ratios instead of leaving answers in fraction form helps students look for patterns. Have a whole-class discussion about questions 4, 5, and 6 after students have had a chance to answer these questions in smaller groups, but do not reveal the answers to these questions. Students discover the answers when they continue the table on the spreadsheet.

The last row of the table, where students generalize the results for side length n, is optional. On the spreadsheet, students do not need the difficult closed formula for the second column. They can instead use the recursive formula, which is easier and more intuitive. The solutions include more explanation.

Sheet 3: Using a spreadsheet— volume of pyramids

This activity sheet is designed for students who have some spreadsheet knowledge. Having one pair of students work at each computer is useful if at least one student in each pair knows how to use computers and spreadsheets. For students who have no experience with spreadsheets, you can use this activity sheet as the basis for a whole-class discussion while demonstrating the process on an overhead-projection device. Do not bother photocopying **activity sheet 3** for students who are familiar with spreadsheets. Instead ask them to continue the table from **activity sheet 2,** and give them verbal directions as needed.

SOLUTIONS

Sheet 1, part 1:

1) Length	Width	Height	Volume
2 units	2 units	4 units	**16 cubic units**
1 unit	2 units	3 units	**6 cubic units**
2 units	2 units	**2 units**	8 cubic units
0.5 units	2 units	2 units	**2 cubic units**

2) $V = lwh$

Sheet 1, part 2

1)

Base	Area of the base	Height	Volume
	4 square units	2 units	**8 cubic units**
	3 square units	3 units	**9 cubic units**
	4 square units	3 units	12 cubic units
	1 1/2 square units	4 units	**6 cubic units**

2) $V = Bh$, where B is the area of the base.

Sheet 2:

1) 8 cubic units;

2) 5 cubic units

3)

Length of Side	Volume of Cubic Solid	Volume of "Pyramid of Cubes"	Volume of "Pyramid" Divided by Volume of Cubic Solid
1	1	1	1
2	8	5	5/8 = 0.625
3	27	14	14/27 0.518
4	64	30	30/64 0.469
5	125	55	55/125 = 0.440
6	216	91	91/216 0.421
7	343	140	140/343 0.408
8	512	204	204/512 0.398
9	729	285	285/729 0.391
10	1000	385	385/1000 = 0.385
n (if you can)	n^3	Volume of previous n + n^2 or $(1/3)n^3$ + $(1/2)n^2$ + $(1/6)n = (n/6) \cdot$ $(n+1)(2n+1)$	$[(n/6)(n+1) \cdot$ $(2n+1)]/n^3$

Some students may be interested in a derivation of the closed formula in the last cell of the "pyramid of cubes" column. One way to derive the formula from the information in the chart is to begin by establishing that the formula is a cubic function. Students who are familiar with the method of finite differences can see that the relationship is a cubic because the differences become constant after three iterations.

When students know that the formula is cubic, they know that they can write it in the form $f(n) = an^3 + bn^2 + cn + d$, where n is the side length. Because the four constants a, b, c, and d are unknown, they can be treated as variables for now. Students can use the first four rows of the data in the table to see that $f(1) = 1$, $f(2) = 5$, $f(3) = 14$, and $f(4) = 30$. They can write the following system of equations:

$$a(1)^3 + b(1)^2 + c(1) + d = 1$$
$$a(2)^3 + b(2)^2 + c(2) + d = 5$$
$$a(3)^3 + b(3)^2 + c(3) + d = 14$$
$$a(4)^3 + b(4)^2 + c(4) + d = 30$$

That system is equivalent to the following system:

$$a + b + c + d = 1$$
$$8a + 4b + 2c + d = 5$$
$$27a + 9b + 3c + d = 14$$
$$64a + 16b + 4c + d = 30$$

However students solve this system, they find that $a = 1/3$, $b = 1/2$, $c = 1/6$, and $d = 0$, from which students can obtain the formula shown in the chart. The system is actually not very difficult to solve by hand using linear combinations.

4) Accept any reasonable answer at this point. Such answers might be similar to, "The ratio gets smaller as the shapes get bigger." In fact, the ratio in the last column approaches 1/3, or 0.33333….

5) Again, accept any reasonable answer. The ratio approaches 1/3 because the pyramid of cubes becomes a closer approximation of an actual smooth-sided pyramid. The size of the cubic blocks does not change as the pyramids become larger, so bumps created by the edges of the blocks are less significant as the "pyramid" becomes larger. If students are familiar with the notion of a limit, you can discuss how the limit of these larger and larger shapes is an infinitely large pyramid with completely smooth sides.

6) Although the ratio in the last column keeps getting smaller, it never reaches 0. Let students discuss this result, but do not reveal the answer.

Sheet 3:

1) and 2)

Length of Side	Volume of Cubic Solid	Volume of "Pyramid"	Volume of "Pyramid" Divided by Volume of Cubic Solid
1	1	1	
2	8	5	

3) Although the formulas are displayed here, numbers should show in the cells on the students' spreadsheets.

	A	B	C	D
1	Length of Side	Volume of Cubic Solid	Volume of "Pyramid"	Volume of "Pyramid" Divided by Volume of Cubic Solid
2	1	1	1	=C2/B2
3	2	8	5	

4)

	A	B	C	D
1	Length of Side	Volume of Cubic Solid	Volume of "Pyramid"	Volume of "Pyramid" Divided by Volume of Cubic Solid
2	1	1	1	=C2/B2
3	2	8	5	=C3/B3

5)

	A	B	C	D
1	Length of Side	Volume of Cubic Solid	Volume of "Pyramid"	Volume of "Pyramid" Divided by Volume of Cubic Solid
2	1	1	1	=C2/B2
3	2	8	5	=C3/B3
4	=A3+1	=A4^3	=C3+A4^2	=C4/B4

6) A few sample rows are shown here.

Length of Side	Volume of Cubic Solid	Volume of "Pyramid"	Volume of "Pyramid" Divided by Volume of Cubic Solid
196	7 529 536	2 529 086	0.335888692
197	7 645 373	2 567 895	0.335875699
198	7 762 392	2 607 099	0.335862837
199	7 880 599	2 646 700	0.335850105
200	8 000 000	2 686 700	0.3358375

7) How students create this graph varies depending on the spreadsheet software and the platform. To select the side-length column and the nonadjacent ratio column, first select one column, then select the other while holding down the control key. Excel users should look for the Chart Wizard icon on the menu bar, click on this icon after selecting the side length and ratio column, and follow the menu choices until the appropriate graph appears.

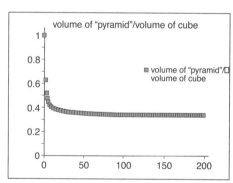

8) The numbers in the last column get closer and closer to 1/3.

9) No. The ratio will always be higher than 1/3.

10) $V = (1/3)Bh$.

POSSIBLE EXTENSIONS

My students enjoyed moving away from the computers for this low-tech finale. If you have a hollow pyramid-and-prism set that has congruent bases and congruent heights, have students use the pyramid as a measuring device to fill the prism with water, sand, rice, or pasta. They should find that three pyramids of water or sand fill the prism exactly to the brim.

I ended the lesson by giving students a picture of some Egyptian pyramids from a book on architecture. The caption to the picture includes measurements, so students can calculate the volume of one of the actual pyramids.

The pyramid of Cheops, the biggest of the three pyramids at Giza, measures 230.5 meters (756 feet) at its base and is 146 meters high. The slope is 51° 52'. At the center is the pyramid of Chephren. It is 215 meters (705 feet) at its base and 143 meters (470 feet) high. The pyramid of Mycerinus, in the foreground, is the smallest of the three. It measures 208 meters (354 feet) at its base and 62 meters (203 feet) in height, with a slope of 51°.

REFERENCE

Norwich, John Julius. *World Atlas of Architecture.* New York: Crescent Books, 1984.

Using Cubic Blocks—*Volume of Prisms*

Sheet 1

Part 1: Volume of a rectangular box

1. Construct each solid with your cubic blocks, and complete the chart. Use your imagination for the last answer.

Length	Width	Height	Volume
2 units	2 unit	4 units	
1 unit	2 unit	3 units	
2 units	2 unit		8 cubic units
0.5 units	2 unit	2 units	

2. Write a formula for the volume of a rectangular box. _____

Part 2: Volume of a right prism

1. Construct each solid with your cubic blocks, and complete the chart. Use your imagination for the last answer.

Base	Area of the Base	Height	Volume
		2 units	
		3 units	
			12 cubic units
		4 units	

2. Write a formula for the volume of any right prism. _____

Using Cubic Blocks—*Volume of Pyramids*   Sheet 2

Although we cannot build exact pyramids with cubes, we can approximate them by building "pyramids of cubes" such as the two pictured below. You will compare the volume of a "pyramid of cubes" with the volume of the prism having the same base and height.

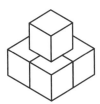

"Pyramid of cubes" with a
square base of side length 2
and height of 2

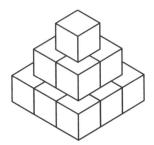

"Pyramid of cubes" with a
square base of side length 3
and height of 3

1. Find the volume of a cubic solid with a side of length 2. _____

2. Find the volume of the "pyramid of cubes" with a square base of side length 2 and a height of 2 (pictured above). _____

3. Complete the chart below. In the last column, compute the ratio of the number in the third column divided by the number in the second column.

Length of Side	Volume of Cubic Solid	Volume of "Pyramid of Cubes"	Volume of "Pyramid" Divided by Volume of Cubic Solid
1			
2			
3			
4			
5			
6			
7			
8			
9			
10			
n (if you can)			

4. What happens to the ratio in the last column as your solids become larger?

5. Why do you think that you obtain this result?

6. Does the ratio in the last column ever become 0?

Using a Spreadsheet—*Volume of Pyramids*   Sheet 3

As you can tell, finding the pattern in the last column of your table is difficult unless you continue the table. You can create a spreadsheet to do the work for you instead of doing the work by hand.

1. In a spreadsheet, type the headings for the four columns of your table, as shown. You may want to abbreviate the headings.

Length of Side	Volume of Cubic Solid	Volume of "Pyramid of Cubes"	Volume of "Pyramid" Divided by Volume of Cubic Solid

2. Enter the values for the first two rows into your spreadsheet. Do not enter numbers for the last column, because you will use a formula to cause the spreadsheet to calculate these values.

Length of Side	Volume of Cubic Solid	Volume of "Pyramid of Cubes"	Volume of "Pyramid" Divided by Volume of Cubic Solid
1	1	1	
2	8	5	

3. Enter a formula for ratio into the first empty cell in the last column. Remember that the formulas in a spreadsheet begin with an "=." Do not just type in the number 1.

4. Copy the ratio formula that you just wrote into the cell below it. Your spreadsheet should look something like the following:

Length of Side	Volume of Cubic Solid	Volume of "Pyramid of Cubes"	Volume of "Pyramid" Divided by Volume of Cubic Solid
1	1	1	1
2	8	5	0.625

5. The next row of your spreadsheet will contain only formulas. Enter all four appropriate formulas for the next row. For help, use the patterns that you noticed when you built the shapes with blocks. You can also work with other students.

6. Select the row of formulas that you just created, and copy them into the next row. Continue to copy down into more and more rows. Use any shortcut that your software allows, such as Fill Down, until your table is long enough that you are sure of a pattern in the last column.

7. Use the graphing feature of your spreadsheet to make a graph of the ratio numbers in the last column.

Use your spreadsheet to answer the following questions. Some of them are repeated from sheet 2.

8. What happens to the ratio in the last column as the solids become larger?

9. Will the ratio in this column ever be 0? Why or why not?

10. You can use your experience with the "bumpy" pyramids that you made with blocks to generalize the outcome for any pyramid. If a pyramid has a base of area B and a height of h, write a formula for its volume.

Visualizing Summation Formulas

Günhan Çağlayan

The use of colored tiles or paper cutouts as representational tools in teaching summation formulas provides students with opportunities to explore and discover algebraic connections between formulas and concrete operations. These activities provide teachers with an easily accessible concept-building activity for developing formulas for the summation of consecutive positive even and odd integers.

Sheet 1: Representing the Sum of Consecutive Positive Odd Integers

The goal of this activity is to have students recognize that every perfect square is the sum of consecutive odd integers. The first step is to show that every positive odd integer can be represented as a symmetric L-shaped figure composed of colored tiles. For instance, the symmetric L-shapes in **figure 1** represent the numbers 1, 3, 5, 7, 9, 11, 13, and 15, respectively.

After establishing the symmetrical L-shaped representations for positive odd integers, students will explore and discover the relationship between the visual/concrete representation and the more abstract formula

$$\sum_{i=1}^{n} 2i - 1 = n^2.$$

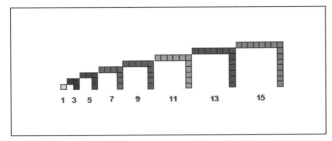

Fig. 1. Symmetric L-shapes representing the numbers 1, 3, 5, 7, 9, 11, 13, and 15

For this activity, 1-inch grid paper and 1-inch colored tiles or paper cutouts can be used to construct the L-shaped representations. The first stage of the activity establishes the relationship that holds when only one tile or paper cutout is used, namely, that the area of the square is $1 \times 1 = 1^2$, or $1 = 1^2$, as seen in **figure 2a**. The subsequent representations focus on building on this type of relationship. Placing the colored L-shaped figure for 3 on the 1 tile yields a representation of $1 + 3 = 4$, as shown in **figure 2b**. The area of the "big" square obtained from these two can be written as

$2 \times 2 = 2^2$. The connection between the area of the big square and the number of different colors in the representation can now be written as $1 + 3 = 2^2$. As shown in **figure 2c**, another colored L-shaped representation of 5 can be placed on the previous summation to yield a visual representation of $1 + 3 + 5 = 9$. With three different colors in the representation, once again the connection between the summation, $1 + 3 + 5$, and the number of terms, 3, yields the relationship $1 + 3 + 5 = 3^2$. Summations up to and including the first seven positive odd integers, as in **figure 2**, should follow in the same manner.

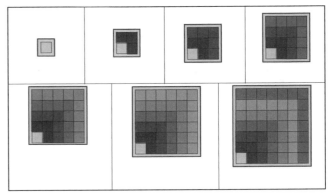

Fig. 2. Using L-shapes to show that the sum of the first n positive odd integers is n^2

Sheet 2: Representing the Sum of Consecutive Positive Even Integers

This activity is designed for students to recognize that every product of consecutive integers can be represented as the sum of consecutive even integers. The first step is to show that every positive even integer can be represented as a rectangle of width 2 and length half the integer. For instance, the shapes in **figure 3** represent the numbers 2, 4, 6, 8, 10, 12, 14, and 16, respectively.

$$\sum_{i=1}^{n} 2i = n(n + 1).$$

After establishing the rectangle representations for the even positive integers, students will explore and discover the relationship between the visual/concrete representation and the more abstract formula

A "building" strategy similar to the one used in the previous exploration will be used here. The initial step represents the equality $2 = 2$ by using a rectangular tile representation in the UP position, as shown in

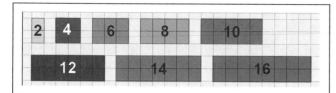

Fig. 3. Shapes representing the numbers 2, 4, 6, 8, 10, 12, 14, and 16

figure 4a. The area of this rectangle can be written as the product of length and width, $1 \times 2 = 2$. The subsequent representations use the same principles. The summation $2 + 4$ can be represented by placing the rectangle representing 4 horizontally to the RIGHT of the 2, as shown in **figure 4b**. The summation $2 + 4$ can also be written as the product of the length and width of the "big" rectangle, namely, $2 + 4 = 2 \times 3$, or $2 + 4 = 6$. The next summation for $2 + 4 = 6$ is built in the same manner by placing the rectangle representing 6 vertically DOWN from the previous representation, as shown in **figure 4c**. The representation yields the summation of $2 + 4 + 6 = 3 \times 4$, or $2 + 4 + 6 = 12$.

Complete the summation representations up to the first seven positive integers, being careful to note the UP-RIGHT-DOWN-LEFT positioning of each consecutive rectangle (see **fig. 4**).

TEACHER NOTES

In implementing this activity with students, it was established that most students prefer to manipulate the colored tiles or paper cutouts first, before transferring the representation to the 1-inch grid paper. Some students found that placing the cutouts directly onto 1-inch grid paper also aided in keeping the visual patterns intact. Possible extensions to this activity include altering the placement of the L-shaped odd integer representations onto the previous figures on **sheet 1**. However, alternate arrangements, although yielding the same results, are more difficult for students to interpret. In both activities, extension questions can include the development of alternate geometrical interpretations of both n^2 and $n(n + 1)$ and their significance in the visualization of summation formulas.

SOLUTIONS

Sheet 1:

1)a. square
  b. 4 in.^{2}

2) See table 1.

3) Each one is the next consecutive odd integer.

4) Summation "adds" up to the total area.

5)a. i is the stage; geometrically, each stage is represented by a different color or L-shaped figure.
  b. $2i - 1$ is the next consecutive odd integer added at stage i; geometrically, it is the number of tiles in the ith L-shaped figure.

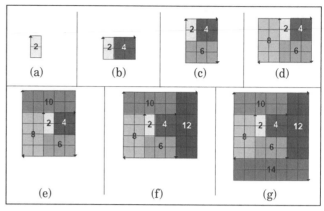

Fig. 4. Summation representations of the first seven positive even integers

  c. n is the total number of terms to add; geometrically, it is represented by the length of the side of a square or the total number of L-shaped figures.
  d. n^2 is the sum of the first n odd numbers; geometrically, it is represented by the total area of the square that is formed.

Sheet 2:

1)a. width = 2 in.
  b. length = 3 in.
  c. total area = 6 in.^{2}

2) See table 2.

3) Each one is the next consecutive even integer.

4) The summation "adds" up to the area of the rectangle.

5)a. i is the stage; geometrically, each stage is represente)d by a different color or rectangle.
  b. $2i$ is the next consecutive even integer at stage i—geometrically, it is represented by the number of tiles in or the area of the rectangle added at stage i.
  c. n is the total number of terms to add; geometrically, it is represented by the length of the shorter side of the rectangle that is formed.
  d. $n(n + 1)$ is the sum of the first n even numbers; geometrically, it is represented as the total area of the rectangle that is formed.

Table 1

Answer to Sheet 1, Question 2

Stage	Number of Squares Added	Summation	Length of Side of Square	Total Area
1	1	1	1	1
2	3	1 + 3	2	4
3	5	1 + 3 + 5	3	9
4	7	1 + 3 + 5 + 7	4	16
5	9	1 + 3 + 5 + 7 + 9	5	25
6	11	1 + 3 + 5 + 7 + 9 + 11	6	36
7	13	1 + 3 + 5 + 7 + 9 + 11 + 13	7	49
8	15	1 + 3 + 5 + 7 + 9 + 11 + 13 + 15	8	64

Table 2

Answer to Sheet 2, Question 2

Stage	Summation	Width of Rectangle	Length of Rectangle	Total Area of the Rectangle
1	2	2	1	2
2	2 + 4	2	3	6
3	2 + 4 + 6	4	3	12
4	2 + 4 + 6 + 8	4	5	20
5	2 + 4 + 6 + 8 + 10	6	5	30
6	2 + 4 + 6 + 8 + 10 + 12	6	7	42
7	2 + 4 + 6 + 8 + 10 + 12 + 14	8	7	56
8	2 + 4 + 6 + 8 + 10 + 12 + 14 + 16	8	9	72

Representing the Sum of Consecutive Positive Odd Integers

Sheet 1

Materials: colored pencils or crayons, 1-inch grid paper, 1-in.^{2} colored tiles or paper cutouts

Begin with one tile and color in the bottom left square on the grid paper. This is stage 1, representing a square with length = 1 unit, width = 1 unit, and area = 1 unit^2.

1. For stage 2, add an L-shaped figure using a different color "around" the previous arrangement (see **fig. a**).

    a. What type of new figure is formed?

    b. What is the area of this figure?

Figure a

Figure b

2. Complete the table by adding L-shaped figures to the previous arrangement (see **fig. b**).

Stage	Number of Squares Added	Summation	Length of Side of Square	Total Area
1	1	1	1	1
2	3	1 + 3	2	4

3. At each stage of the pattern you "add" another integer. Describe these integers.

4. What is the relationship between the summation and the total area at each stage?

5. Given the following summation formula,

$$\sum_{i=1}^{n} 2i - 1 = n^2,$$

    describe each of the following terms and provide a geometric description of each term based on your picture.

    a. What does i represent?

    b. What does $2i - 1$ represent?

    c. What does n represent?

    d. What does n^2 represent?

Representing the Sum of Consecutive Positive Even Integers

Materials: Colored pencils or crayons, 1-inch grid paper, 1-in.^{2} colored tiles or paper cutouts

All even integers can be represented by rectangles of width 2, e.g., for 2 and for 4.

Begin with the tile representing 2. This is stage 1. When transfering this representation to grid paper, begin in the middle of the page.

1. For stage 2, add a rectangle with 4 tiles to the RIGHT of the previous rectangle (see **fig. a**).

   a. What is the width of the new rectangle?

   b. What is the length of the new rectangle?

   c. What is the total area of the new rectangle?

Width / Length

Figure a

Figure b

2. Continue the pattern in an UP-RIGHT-DOWN-LEFT pattern (see **fig. b**) to represent the sums in the table.

Stage	Summation	Width of Rectangle	Length of Rectangle	Total Area of the Rectangle
1	2	2	1	2
2	2 + 4	2	3	6

3. At each stage of the pattern you "add" another integer. Describe these integers.

4. What is the relationship between the summation and the total area at each stage?

5. Given the following summation formula,

$$\sum_{i=1}^{n} 2i = n(n+1),$$

describe each of the following terms and provide a geometric description of each term based on your picture.

   a. What does i represent?

   b. What does $2i$ represent?

   c. What does n represent?

   d. What does $n(n+1)$ represent?

Finding a Parking Spot for the Binomial Theorem

Karen M. Hansen and Geoffrey J. Lewis

Oftentimes when learning the binomial theorem, students can apply the theorem's formula, but it has little meaning for them. Introducing the binomial theorem using an engaging application gives the theorem life and meaning. This activity relates the coefficients in the formula to finding a parking spot. First, the students discover Pascal's triangle using a parking spot simulation. Then the activity relates Pascal's triangle to the coefficients in the binomial theorem. Graphing calculators and combinations are used to verify the binomial expansion. This activity supports a variety of learning styles in an enjoyable application.

THE ACTIVITY

To begin the activity, distribute a worksheet like the one shown in **figure 1** and discuss the following scenario, as illustrated in the worksheet. A car pulls into the parking lot at Soldier Field in Chicago. Members of the Chicago Police Department direct traffic. There are four rows of police officers arranged in a triangular pattern, such that one officer is in the first row, two officers are in the second row, three officers are in the third row, and four officers are in the fourth row. They are directing traffic into five parking spots, labeled A through E in **figure 1**. Given that cars must proceed according to the traffic arrows, how many different paths are there into each of the parking spots?

Challenge students to try the problem on paper. Start by asking, "How many different paths can be taken from the entrance to each of the five parking spots?"

After about five minutes, have students simulate the activity. Create a parking lot in the classroom by clearing a small area and using desks as parking spots. Start with one driver, one police officer, and two parking spots as shown in **figure 2**. Next ask, "How many different paths can be taken from the entrance to each of the two parking spots?" Students easily realize that there is only one way to parking spot A and one way to parking spot B.

Next, repeat the activity using three police officers and three parking spots, A, B, and C, arranged according to **figure 3**. Again ask, "How many different paths can be taken from the entrance to each of the three parking spots?" Initially, students might think that there is only one way to get to each of the three parking spots because of the previous simulation. When student drivers travel the paths, however, they see that the pattern from the previous simulation does not apply in this

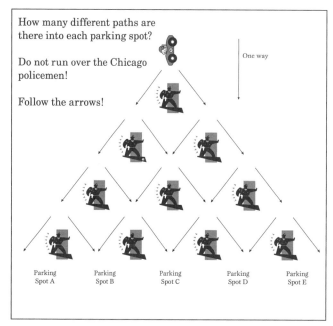

Fig. 1. Ten police officers direct cars to the five available parking spots.

situation. Students find that there is one way to each of parking spots A and C but that there are two ways to parking spot B, as shown in **figure 4**.

Repeat the activity again, but use six police officers and four parking spots as shown in **figure 5**. Students may be uncertain how to proceed with so many paths and parking spots. Ask the class to suggest a plan to find the number of paths to each spot systematically. Usually, a student suggests finding the paths to parking spot A first, then finding the paths to each of the other parking spots one at a time. With the student driver traveling the paths, students will realize that there is

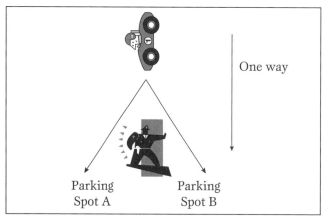

Fig. 2. Looking at a simpler problem

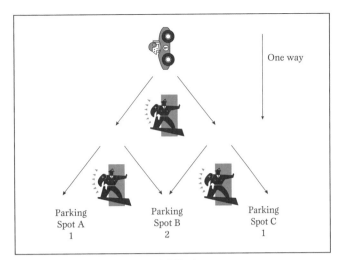

Fig. 3. An intermediate level of difficulty

Ask the class to identify patterns in the coefficients of the binomial expansion, how the pattern relates to the rows of Pascal's triangle, and how it indicates the number of paths to a parking spot. Students should find that the degree of the polynomial indicates the appropriate row of Pascal's triangle and that the numbers in the row correspond to both the coefficients of the polynomial's terms and the number of paths to a parking spot. For example, in the expansion of $(x + 1)^3 = 1x^3 + 3x^2 + 3x + 1$, the degree of 3 indicates the need to apply the third row of Pascal's triangle. The numbers in the third row, 1, 3, 3, 1, correspond to the coefficients

one way to parking spots A and D and three ways each to parking spots B and C.

Students should now be able to act out and draw on paper the results of the original problem with ten police officers and five parking spots.

CONNECTIONS

Summarize the results showing Pascal's triangle (**fig. 6**). Some students may recognize this as the triangle named after Blaise Pascal but actually discovered centuries earlier in China and the Middle East. Students immediately notice that the numbers in the triangle relate to the number of paths to a parking spot. The pattern in Pascal's triangle can now be used to introduce the pattern of the coefficients in the expansion of a binomial. Start with a simple binomial such as $(x + 1)^2$. Expand it by multiplying

$$(x + 1) \cdot (x + 1) = 1x^2 + 2x + 1.$$

Also expand

$$(x + 1)^3 = (x + 1) \cdot (x^2 + 2x + 1)$$
$$= 1x^3 + 3x^2 + 3x + 1$$

and

$$(x + 1)^4 = (x^2 + 2x + 1) \cdot (x^2 + 2x + 1)$$
$$= 1x^4 + 4x^3 + 6x^2 + 4x + 1.$$

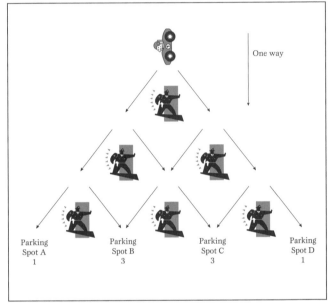

Fig. 5. The level of complexity increases when a third row of police officers is added.

of the four terms in the expansion of the polynomial as well as the number of paths to each parking spot when there are four parking spots. Ask students to relate the expansion of $(x + 1)^0$ to Pascal's triangle and the implied number of parking spots. The discussion concludes that $(x + 1)^0 = 1$, implying that there is one parking spot with one path to it and zero police officers. The class can also discuss the patterns in the exponents of the binomial expansion. Through the process of relating

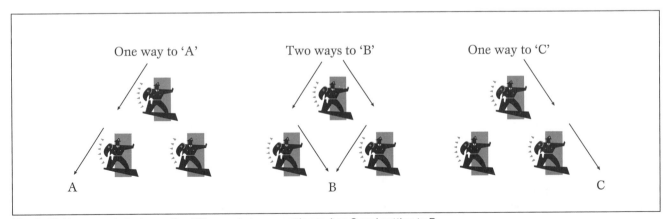

Fig. 4. Making sure students see the difference between getting to A or C and getting to B

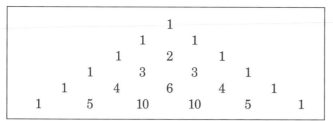

Fig. 6. Rows 0 through 6 of Pascal's triangle

the coefficients to a meaningful activity, students easily see the relationship between the coefficients in the binomial expansion, the rows in Pascal's triangle, the number of paths, and the number of parking spots. It deepens their understanding of the coefficients in the binomial expansion.

Graphing calculators can be used to check the results of the binomial expansion. Verify that graphs of $(x + 1)^2$ and $x^2 + 2x + 1$ are equal by having students enter the equations into the graphing calculator's "y=" function. One equation is entered in y_1 using a line graph style, while the other equation is entered in y_2 using a path graph style. The Graph Style is the icon immediately to the left of the equation labels y_1 and y_2. Positioning the cursor on the Graph Style icon and pressing Enter until the calculator reveals the appropriate icon changes the style of the graph. **Figure 7** shows the screen capture of these equations as well as the resultant graph. The two different graph styles help students distinguish the two graphs. The path graph style used in the second equation nicely animates the second graph on the display. Students always appreciate the animation. This technique is particularly useful on the larger degree functions such as $(x + 1)^5 = x^5 + 5x^4 + 10x^3 + 10x^2 + 5x + 1$.

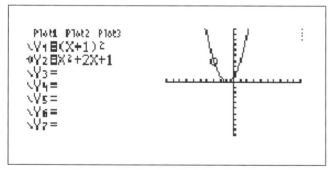

Fig. 7. Using the calculator to show that $(x + 1)^2 = x^2 + 2x + 1$

If students have been introduced to combinations, they can verify the relationships among the number of paths to $n + 1$ parking spots, the nth row of Pascal's triangle, and the number of combinations $_nC_r$, where r represents the $(r + 1)$th position in the nth row as r goes from 0 to n. For example, the fourth row of the triangle corresponds as follows:

$$1, \quad 4, \quad 6, \quad 4, \quad 1$$
$$_4C_0 \quad _4C_1 \quad _4C_2 \quad _4C_3 \quad _4C_4$$

At this point, students understand the generalized expansion of a binomial using the binomial theorem (see **fig. 8**).

$$(a+b)^n = {}_nC_0 a^n b^0 + {}_nC_1 a^{n-1} b^1 + {}_nC_2 a^{n-2} b^2 + \cdots + {}_nC_{n-1} a^1 b^{n-1} + {}_nC_n a^0 b^n$$
$$= \sum_{r=0}^{n} {}_nC_r a^{n-r} b^r$$
where $n \geq 1$

Fig. 8. Binomial theorem

After trying some examples, like $(x + y)^3$ or $(2x + y)^4$ or $(x - 3y)^5$, have students write a summary of the relationship between Pascal's triangle and the binomial theorem. The summary should include an explanation of the relationship between the parking spot simulation and Pascal's triangle, followed by expansion of a binomial such as $(x + y)^4$. Writing in their own words helps students to develop and summarize their thoughts.

Finding a Parking Spot for the Binomial Theorem   Sheet 1

1. Complete the first few rows of Pascal's triangle.

```
                        1
                    1       1
                1       2       1
            1       3      [ ]      1
        1       4      [ ]     [ ]      1
    1       5      [ ]     [ ]     [ ]      1
1      [ ]     [ ]     [ ]     [ ]     [ ]      1
```

2. Complete as you act out the parking spot activity by completing the paths to the parking places. State the number of routes to each spot. Remember the rule to not cross any paths. Parts A and B are done as an example.

    a.

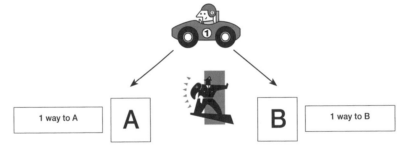

| 1 way to A | A | | B | 1 way to B |

    b.

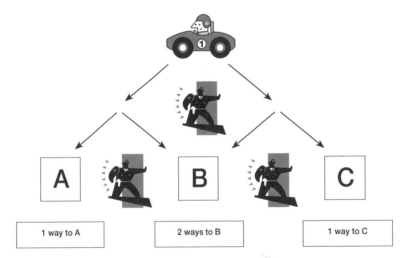

| 1 way to A | 2 ways to B | 1 way to C |

Finding a Parking Spot for the Binomial Theorem   Sheet 2

c.   Now you show the paths and state the total number of ways to get to each parking spot.

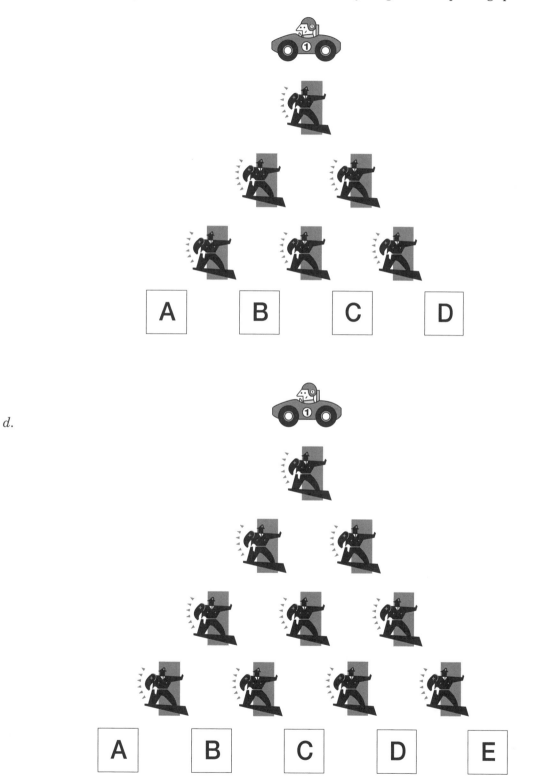

d.

Finding a Parking Spot for the Binomial Theorem   Sheet 3

e.

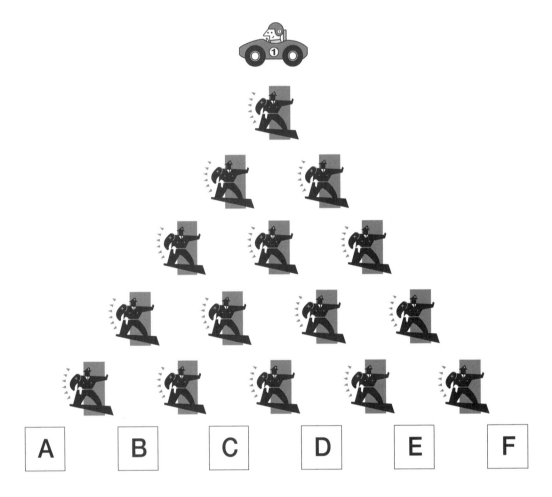

| A | B | C | D | E | F |

What patterns do you see?

3. Expand the following binomials by using Pascal's triangle.

    a. $(x + 1)^2$ *b.* $(x + 1)^3$

    c. $(x + 1)^4$ *d.* $(x + 1)^3$

    e. $(x + 1)^5$ *f.* $(x + 2)^2$

    g. $(x - 3)^2$ *h.* $(x - 3)^3$

    i. $(x + a)^2$ *j.* $(x + a)^3$

    k. $(x + 2a)^2$ *l.* $(x + 2a)^3$

4. Use a graphing calculator to expand $(x + 1)^9$.

5. Ben Wallace plays in the NBA. He makes about 25% of his free throws. What is the probability that he will make 2 out of 3 free throws?

Unit Fractions and Their "Basimal" Representations: Exploring Patterns

Marlena Herman, Eric Milou, and Jay Schiffman

Major foci of secondary mathematics include understanding numbers, ways of representing numbers, and relationships among numbers (NCTM 2000). This article considers different representations of rational numbers and leads students through activities that explore patterns in base ten, as well as in other bases. These activities encourage students to solve problems and investigate situations designed to foster flexible thinking about rational numbers. Preservice teachers in a college-level mathematics course carried out these activities. Their conjectures and ideas are incorporated throughout this article.

BASE-TEN INVESTIGATIONS

The set of rational numbers is often formally defined as follows:

$Q = \{ a/b: a$ and b are integers and $b \neq 0 \}$.

This definition uses a fractional representation of rational numbers. However, a rational number can also be represented as a terminating decimal or as a repeating decimal. When they study rational numbers, students are often asked to convert fractional representations to decimal representations, and vice versa. The connections that they find can suggest interesting problem-solving situations. This article focuses on investigations into the reasons that some rational numbers can be represented as terminating decimals while others repeat, as well as various properties of repeating decimals.

For simplicity's sake, only rational numbers whose fractional representations have a numerator of 1 are investigated in this article. Such a number is called a *unit fraction*, which is defined as a fraction of the form $1/n$ where n is a natural number greater than 1. Studying numbers whose fractional representations have numerators other than 1 is not necessary, since such numbers can be expressed as multiples of unit fractions (for example, $5/6 = 5 \cdot (1/6)$); hence, patterns that hold for unit fractions also hold for other fractions.

Consider the first ten unit fractions. As a beginning activity, students can find the decimal representations by doing simple division on a calculator, as shown in **table 1**. Separating the findings into terminating and repeating sets yields

Terminating: 1/2, 1/4, 1/5, 1/8, 1/10

Repeating: 1/3, 1/6, 1/7, 1/9, 1/11

Interestingly, the first ten unit fractions are split equally between terminating and repeating decimals. Students can conjecture which unit fractions terminate and which ones repeat. The preservice teachers who completed this article's activities, for example, quickly saw that all the unit fractions that terminate, except for 1/5, have even denominators, whereas all the unit fractions that repeat, except for 1/6, have odd denominators. Their insights prompted them to look at the next ten unit fractions, which are shown in **table 2**. Some of those decimal representations can be found

TABLE 1		
First Set of Ten Unit Fractions and Their Decimal Representations		
Unit Fraction	Decimal Representation	Terminating or Repeating
1/2	0.5	Terminating decimal
1/3	$0.333\ldots = 0.\overline{3}$	Repeating decimal
1/4	0.25	Terminating decimal
1/5	0.2	Terminating decimal
1/6	$0.1666\ldots = 0.1\overline{6}$	Repeating decimal
1/7	$0.142857142857\ldots = 0.\overline{142857}$	Repeating decimal
1/8	0.125	Terminating decimal
1/9	$0.111\ldots = 0.\overline{1}$	Repeating decimal
1/10	0.1	Terminating decimal
1/11	$0.090909\ldots = 0.\overline{09}$	Repeating decimal

TABLE 2		
Second Set of Ten Unit Fractions and Their Decimal Representations		
Unit Fraction	Decimal Representation	Terminating or Repeating
1/12	$0.08333\ldots = 0.08\overline{3}$	Repeating decimal
1/13	$0.076923076923\ldots = 0.\overline{076923}$	Repeating decimal
1/14	$0.0714285714285\ldots = 0.0\overline{714285}$	Repeating decimal
1/15	$0.0666\ldots = 0.0\overline{6}$	Repeating decimal
1/16	0.0625	Terminating decimal
1/17	$0.\overline{0588235294117647}$	Repeating decimal
1/18	$0.0555\ldots = 0.0\overline{5}$	Repeating decimal
1/19	$0.\overline{052631578947368421}$	Repeating decimal
1/20	0.05	Terminating decimal
1/21	$0.047619047619\ldots = 0.\overline{047619}$	Repeating decimal

by doing simple division on a calculator; however, others extend farther than the number of digits that a calculator can display before revealing a terminating or repeating pattern. Decimal representations of those unit fractions can be found by using such computer software as Mathematica. For this exploration, a teacher could simply provide the decimal representations of 1/17 and 1/19 to students. Summarizing to this point, terminating fractions are 1/2, 1/4, 1/5, 1/8, 1/10, 1/16, 1/20; and repeating fractions are 1/3, 1/6, 1/7, 1/9, 1/11, 1/12, 1/13, 1/14, 1/15, 1/17, 1/18, 1/19, 1/21.

The first twenty unit fractions are not split equally between terminating and repeating decimals. The preservice teachers noted that many more unit fractions seemed to repeat than to terminate, yielding many more rational numbers that can be represented as repeating decimals. "Many more" was the students' term; in fact, both sets have an infinite number of elements.

After students consider the first twenty unit fractions, they can test their first conjectures and conjecture which unit fractions terminate and which ones repeat. The preservice teachers still thought that for a unit fraction to terminate, it must have an even denominator (except for 1/5), whereas they thought that all the unit fractions that repeat have odd denominators, except for 1/6, 1/12, and 1/18. Some wondered whether all unit fractions with a denominator that is a multiple of 6 repeat. Others built from this conjecture, adding the thought that any unit fraction with a denominator that is a multiple of 3 or 7 repeats.

When students were prompted to look also at possible patterns with the terminating decimals, some of them noticed that unit fractions with denominators that are powers of 2 or powers of 5 terminate. This idea placed 1/2, 1/4, 1/5, 1/8, and 1/16 (but not 1/10 or 1/20) in the list of unit fractions that correspond to terminating decimals. A few preservice teachers noticed, however,

that the prime factorizations of 10 and 20 contain only 2s and 5s. In other words, they found an intriguing pattern in the unit fractions that terminate: the denominators can always be factored into powers of 2, 5, or both. All unit fractions with denominators that cannot be factored into only powers of 2, 5, or both seemed to repeat. This pattern accounted for everything that the preservice teachers had noticed. For example, multiples of 3 (including multiples of 6) and 7 contain powers of prime numbers other than 2 and 5. Thus, unit fractions with denominators that are multiples of 3 or 7 do indeed repeat.

The preservice teachers had, in fact, discovered a correct conjecture. Formally stated, a rational number a/b in simplest form can be written as a terminating decimal if and only if the prime factorization of the denominator contains no primes other than 2 and 5. Using the conjecture on the next four unit fractions gives 1/22 as a repeating decimal, 1/23 as a repeating decimal, 1/24 as a repeating decimal, and 1/25 as a terminating decimal. These results can be verified by division using a calculator or computer software.

Many preservice teachers became interested in the unit fractions 1/17 and 1/19 because of the large number of digits before the decimal repeats. The number of places before a decimal repeats is known as the *period*. When the preservice teachers noticed that 17 and 19 are prime numbers, that 1/17 has a period of 16, and that 1/19 has a period of 18, they wondered whether the next prime number, 23, would have a period of 22. Surely enough, it does. They conjectured that a unit fraction with denominator p, where p is any prime other than 2 or 5, repeats with a period of length $(p - 1)$.

To test this conjecture, students can build a chart like **table 3** of unit fractions with prime denominators and look at the period of each. The preservice teachers quickly noticed that counterexamples to the conjecture arise, since, for example, 1/13 has a period of 6 rather

		TABLE 3		
Lengths of Periods of Decimal Representations of Unit Fractions with Prime Denominators				
n	Unit Fraction	Decimal Representation	Length of Period
2	1/2	0.5	N/A
3	1/3	$0.0333\ldots = 0.0\overline{3}$	1
5	1/5	0.2	N/A
7	1/7	$0.142857142857\ldots = 0.\overline{142857}$	6
11	1/11	$0.090909\ldots = 0.\overline{09}$	2
13	1/13	$0.076923076923\ldots = 0.\overline{076923}$	6
17	1/17	$0.\overline{0588235294117647}$	16
19	1/19	$0.\overline{052631578947368421}$	18
23	1/23	$0.\overline{0434782608695652173913}$	22

than 12. However, they were determined not to give up on finding a pattern. They were then very interested in securing a conjecture with respect to a unit fraction's denominator and the length of its period. Thus, they modified the conjecture to state that a unit fraction with denominator p, where p is any prime other than 2 or 5, repeats with a period of at most $(p - 1)$. Further, they noticed that if p is prime and if its period is not of length $(p - 1)$, then the length of the period is a factor of $(p - 1)$. Returning to the example 1/13, although the period is not of length 12, the period's length of 6 is a factor of 12.

Several preservice teachers were still interested in investigating other properties of the repeating decimals. They noticed that some decimal expansions have repeating portions that begin immediately to the right of the decimal point, whereas others have repeating portions that do not. The purely periodic decimals, whose decimal expansions have repeating portions that begin immediately to the right of the decimal point, can be seen in the decimal representations of 1/3, 1/7, 1/9, 1/11, 1/13, 1/17, 1/19, 1/21, and 1/23. The delayed periodic decimals, whose decimal expansions have repeating portions that begin after a delay to the right of the decimal point, include 1/6, 1/12, 1/14, 1/15, 1/18, 1/22, and 1/24. For example, $1/6 = 0.1\overline{6}$ has an infinitely repeating part that begins one place to the right of the decimal point, yielding a delay of length 1. Similarly, $1/12 = 0.08\overline{3}$ has an infinitely repeating part that begins two places to the right of the decimal point, yielding a delay of length 2. **Table 4** summarizes the lengths of delay of the unit fractions considered thus far.

The preservice teachers remembered that they could use the prime factorization of the denominator of a unit fraction to determine whether its decimal representation terminates or repeats; they suspected that the length of delay of a repeating decimal can be determined by appealing to prime factorization. They decided to consider the repeating decimals with delay lengths that are greater than zero (that is, the delayed periodic decimals) and determine whether any patterns involved 2s and 5s again. The work for this investigation is shown in **table 5**.

One preservice teacher noted immediately that when the prime factorization involved 2s, the length of the delay matched the power of 2. Others noticed that when no 2s were in the prime factorization (such as with 1/15), they could look at 5s; the length of the delay matched the power of the 5. Most wanted to test these ideas with other cases and particularly wondered what would happen if both 2s and 5s appeared within a prime factorization. Clearly, they remembered that the denominator of the unit fraction must have at least one prime factor other than 2 or 5 for the decimal to repeat.

Cases such as the ones shown in **table 6** can be tested.

After testing some other cases, most preservice teachers were comfortable with the conjectures about factors involving 2s and 5s. They simply added that if both 2s and 5s appeared within a prime factorization, the length of delay is the highest power of either the 2 or the 5. Again, the preservice teachers had arrived at a correct hypothesis. That is, the length of delay of the decimal representation of a given rational number corresponds to the highest exponent on either the 2 or the 5 in the prime factorization of the denominator of the fractional representation of the rational number. It follows that a rational number is purely periodic if and only if the denominator has no factors of 2 and 5 (that is, the denominator is relatively prime to 10). It also follows that any rational number with a prime denominator greater than 5 must be purely periodic.

TABLE 4

Length of Delay of Unit Fractions Having Decimal Representations That Repeat

Unit Fraction	Decimal Representation	Length of Delay
1/3	$0.333\ldots = 0.\overline{3}$	0 (purely periodic)
1/6	$0.1666\ldots = 0.1\overline{6}$	1
1/7	$0.142857142857\ldots = 0.\overline{142857}$	0 (purely periodic)
1/9	$0.111\ldots = 0.\overline{1}$	0 (purely periodic)
1/11	$0.090909\ldots = 0.\overline{09}$	0 (purely periodic)
1/12	$0.08333\ldots = 0.08\overline{3}$	2
1/13	$0.076923076923\ldots = 0.\overline{076923}$	0 (purely periodic)
1/14	$0.0714285714285\ldots = 0.0\overline{714285}$	1
1/15	$0.0666\ldots = 0.0\overline{6}$	1
1/17	$0.\overline{0588235294117647}$	0 (purely periodic)
1/18	$0.0555\ldots = 0.0\overline{5}$	1
1/19	$0.\overline{052631578947368421}$	0 (purely periodic)
1/21	$0.047619047619\ldots = 0.\overline{047619}$	0 (purely periodic)
1/22	$0.0454545\ldots = 0.0\overline{45}$	1
1/23	$0.\overline{0434782608695652173913}$	0 (purely periodic)
1/24	$0.041666\ldots = 0.041\overline{6}$	3

TABLE 5

Prime Factorization of Denominators of Unit Fractions Having Delayed Periodic Decimal Representations

Unit Fraction	Decimal Representation	Length of Delay	Prime Factorization of Denominator
1/6	$0.1666\ldots = 0.1\overline{6}$	1	$2 \cdot 3$
1/12	$0.08333\ldots = 0.08\overline{3}$	2	$2^2 \cdot 3$
1/14	$0.0714285714285\ldots = 0.0\overline{714285}$	1	$2 \cdot 7$
1/15	$0.0666\ldots = 0.0\overline{6}$	1	$3 \cdot 5$
1/18	$0.0555\ldots = 0.0\overline{5}$	1	$2 \cdot 3^2$
1/22	$0.0454545\ldots = 0.0\overline{45}$	1	$2 \cdot 11$
1/24	$0.041666\ldots = 0.041\overline{6}$	3	$2^3 \cdot 3$

EXTENSION: BASE-FOUR INVESTIGATIONS

Learning whether the patterns found in base ten also work in other base systems is an interesting activity. For example, one might wonder how to tell whether a unit fraction in base four has a "basimal" representation that terminates or one that repeats. (Note that the word *basimal* replaces the word *decimal*, since *deci* refers specifically to base ten only, whereas *basi* refers to any base. We chose the term *basimal representation* because some readers may not be familiar with the term radix representation. Radix refers to the number base of a numeral system. An Internet search on the word basimal led to results that may be more applicable for use in the high school classroom than the word

radix, if a reader wants to further pursue the ideas presented in the article.)

Before we consider base-four basimals, we briefly review operations in base four. The base-ten system uses ten digits (0, 1, 2, 3, 4, 5, 6, 7, 8, and 9) and place value to denote sets of ten; the base-four system uses four digits (0, 1, 2, and 3) and place value to denote sets of four. Counting in base ten involves 0, 1, 2, 3, 4, 5, 6, 7, 8, and 9, followed by 10 to represent one set of ten. As the numbers increase, every new set of ten creates a new place-value position. For example, 100 represents ten sets of ten. Similarly, base-four counting begins with 0_{four}, 1_{four}, 2_{four}, and 3_{four}, and then continues with 10_{four} to represent one set of four. As the

TABLE 6

More Cases of Prime Factorization of Denominators of Unit Fractions Having Delayed Periodic Decimal Representations

Unit Fraction	Decimal Representation	Length of Delay	Prime Factorization of Denominator
1/30	$0.0\overline{3}$	1	$2 \cdot 3 \cdot 5$
1/48	$0.0208\overline{3}$	4	$2^4 \cdot 3$
1/56	$0.017\overline{857142}$	3	$2^3 \cdot 7$
1/60	$0.01\overline{6}$	2	$2^2 \cdot 3 \cdot 5$
1/75	$0.01\overline{3}$	2	$3 \cdot 5^2$
1/150	$0.00\overline{6}$	2	$2 \cdot 3 \cdot 5^2$
1/1200	$0.00008\overline{3}$	4	$2^4 \cdot 3 \cdot 5^2$
1/30,000	$0.00000\overline{3}$	4	$2^4 \cdot 3 \cdot 5^4$

numbers increase, every new set of four creates a new place-value position. In this manner, the numbers up to 100_{four} are 0_{four}, 1_{four}, 2_{four}, 3_{four}, 10_{four}, 11_{four}, 12_{four}, 13_{four}, 20_{four}, 21_{four}, 22_{four}, 23_{four}, 30_{four}, 31_{four}, 32_{four}, 33_{four}, and 100_{four}. Here, 100_{four} represents four sets of four, which is equivalent to 16 in base ten.

Addition, subtraction, multiplication, and division can be performed in base four with the same algorithms used in base ten. Addition can be completed by adding the digits in given place-value positions and carrying sets of four. Adding 213_{four} and 1220_{four}, for example, yields 2033_{four}. The 3 in the rightmost place-value position comes from adding 3_{four} and 0_{four}, and the 3 in the second place-value position (moving left) comes from adding 1_{four} and 2_{four}. The 0 in the third place-value position (continuing to move left) comes from adding 2_{four} and 2_{four}, getting 10_{four}, recording the 0, and carrying the 1 into the next place-value position (moving left). This carried 1_{four} is added to the 1_{four} from 1220_{four} to yield 2_{four} in the fourth position.

Much like addition, subtraction can be completed by subtracting the digits in given place-value positions and borrowing sets of four. Standard algorithms for multiplication and division incorporate addition and subtraction facts. The preservice teachers who completed the activities in this article practiced many base-four operations before doing a basimal investigation. While working on practice problems, they built a table, shown as **table 7**, of base-four multiplication facts to enable

them to complete their work more quickly. Some completed the table using repeated addition, for example,

$$31_{four} \times 3_{four} = 31_{four} + 31_{four} + 31_{four}$$
$$= 122_{four} + 31_{four}$$
$$= 213_{four},$$

whereas others noticed that quicker base-ten methods worked in base four (for example, $31_{four} \times 3_{four} = 213_{four}$, since $1_{four} \times 3_{four} = 3_{four}$ in the ones place-value column and since $3_{four} \times 3_{four} = 21_{four}$ in the next column, moving left).

A slightly more complicated multiplication example is $31_{four} \times 23_{four}$. When the standard algorithm for multiplication is used, the problem can be set up vertically and completed as in the steps in **figure 1**.

The multiplication facts in **table 7** are useful not only for more complicated multiplication problems, but also for long division. **Figure 2** shows the steps for completing $130_{four} \div 2_{four}$ by using multiplication and subtraction facts.

As **table 8** shows, students can use long division to complete work for a set of unit fractions. The first seventeen numbers in base four are 0_{four}, 1_{four}, 2_{four}, 3_{four}, 10_{four}, 11_{four}, 12_{four}, 13_{four}, 20_{four}, 21_{four}, 22_{four}, 23_{four}, 30_{four}, 31_{four}, 32_{four}, 33_{four}, and 100_{four}, so the first fifteen unit fractions can be compiled with these values (except for 0_{four} and 1_{four}) in denominators. Each long-division problem involves replacing the numerator of 1 with 1.0000.

TABLE 7

Base-Four Product Chart

×	0	1	2	3	10	11	12	13	20	21	22	23	30	31	32	33	100
0	0	0	0	0	0	0	0	0	0	0	0	0	0	0	0	0	0
1	0	1	2	3	10	11	12	13	20	21	22	23	30	31	32	33	100
2	0	2	10	12	20	22	30	32	100	102	110	112	120	122	130	132	200
3	0	3	12	21	30	33	102	111	120	123	132	201	210	213	222	231	300

$3\ 1_{\text{four}}$ $\times\ \ \ 2\ 3_{\text{four}}$ $2\ 1\ 3$ $\underline{1\ 2\ 2\ \ \ }$	Set up problem vertically. Multiply digits using standard algorithm. Place results in stair-step pattern ($31_{\text{four}} \times 3_{\text{four}} = 213_{\text{four}}$ written in the first row and $31_{\text{four}} \times 2_{\text{four}} = 122_{\text{four}}$ written in the second row one space from the rightmost place-value position).
$3\ 1_{\text{four}}$ $\times\ \ \ 2\ 3_{\text{four}}$ $2\ 1\ 3$ $\underline{1\ 2\ 2\ \ \ }$ $2\ 0\ 3\ 3_{\text{four}}$	Add digits in given place-value columns, carrying sets of four. Note: The 0 in the third place-value position (from the right) comes from adding 2_{four} and 2_{four}, getting 10_{four}, recording the 0, and carrying the 1 into the next place-value position (moving left). This 1 then gets added in the next (fourth) column.

Fig. 1. Steps for solving $31_{\text{four}} \times 23_{\text{four}}$

using as many 0s for place-holders as needed when carrying out the division process and looking for repeating patterns in the result. For example, the unit fraction $1/31_{\text{four}}$ requires an extension of seven 0s, the first six of which are needed to find the repeating portion of the basimal and the seventh of which is needed to discover when the repeating portion begins a new cycle. See **figure 3**.

The preservice teachers who completed **table 8** found it useful to see some long-division examples in base four and then completed the remaining problems. They then set out to find patterns using the denominators of the unit fractions that may determine whether the basimal representation terminates or repeats. Recalling that terminating decimals in base ten correspond to a prime factorization of the denominator containing no primes other than 2 and 5, the students guessed that a similar pattern may hold true in base four. Since the numeral 5 does not exist in base four, the preservice teachers knew that the exact pattern from base-ten computations could not hold true in base-four work. One student thought that the 5 could just be "thrown away" or ignored, leaving a pattern involving prime factorization of denominators containing no primes other than 2_{four}. Others thought that the 5 should be converted into base-four notation (that is, 11_{four}), yielding a pattern involving prime factorization of denominators containing no primes other than 2_{four} and 11_{four}.

To test these conjectures, the preservice teachers decided to stick with the idea of prime factorizations, and they used base-four multiplication facts to factor the denominators of the unit fractions in question.

$2_{\text{four}}\sqrt{130_{\text{four}}}$	Set up long division. Consider how many times 2_{four} goes into 1_{four}. Since 2_{four} does not go into 1_{four}, consider how many times 2_{four} goes into 13_{four}. Using multiplication facts, $$2_{\text{four}} \times 1_{\text{four}} = 2_{\text{four}},$$ $$2_{\text{four}} \times 2_{\text{four}} = 10_{\text{four}},$$ $$2_{\text{four}} \times 3_{\text{four}} = 12_{\text{four}},$$ and $$2_{\text{four}} \times 10_{\text{four}} = 20_{\text{four}}.$$ Since $2_{\text{four}} \times 3_{\text{four}}$ is less than 13_{four} and since $2_{\text{four}} \times 10_{\text{four}}$ is greater than 13_{four}, 2_{four} goes into 13_{four} at most 3_{four} times.
$2_{\text{four}}\sqrt{\begin{array}{l}3\\ \overline{130}_{\text{four}}\\ \underline{12}\\ 10\end{array}}$	Record the 3_{four} above the problem setup and the product 12_{four} below the problem setup. Subtract 12_{four} from 13_{four}, and bring down the next digit. This work yields a remainder of 10_{four}.
$2_{\text{four}}\sqrt{\begin{array}{l}32\\ \overline{130}_{\text{four}}\\ \underline{12}\\ 10\\ \underline{10}\\ 0\end{array}}$	Consider how many times 2_{four} goes into 10_{four}. Since $$2_{\text{four}} \times 2_{\text{four}} = 10_{\text{four}},$$ 2_{four} goes into 10_{four} exactly 2_{four} times. Record the 2_{four} above the problem setup and record the product 10_{four} below the problem setup. Subtract 10_{four} from 10_{four}. This work yields a remainder of 0_{four}. Since no other digits can be brought down, the problem is finished.
$130_{\text{four}} \div 2_{\text{four}}$ $= 32_{\text{four}}$	The result can be checked with multiplication: $$32_{\text{four}} \times 2_{\text{four}} = 130_{\text{four}}$$

Fig. 2. Steps for solving $130_{\text{four}} \div 2_{\text{four}}$

$$31_{\text{four}}\sqrt{\begin{array}{l}0.0103230\ldots_{\text{four}}\\ \overline{1.0000000}\ldots_{\text{four}}\\ \underline{31}\\ 300\\ \underline{213}\\ 210\\ \underline{122}\\ 220\\ \underline{213}\\ 100\\ \underline{31\ldots}\text{ (problem repeats from beginning)}\end{array}}$$

Fig. 3. With long division, the unit fraction $1/31_{\text{four}}$ requires an extension of seven zeros.

TABLE 8

Base-Four Unit Fractions and Their Basimal Representations

Unit Fraction	Decimal Representation	Terminating or Repeating
1/2	0.2	Terminating basimal
1/3	$0.111\ldots = 0.\overline{1}$	Repeating basimal
1/10	0.1	Terminating basimal
1/11	$0.0303\ldots = 0.\overline{03}$	Repeating basimal
1/12	$0.0222\ldots = 0.0\overline{2}$	Repeating basimal
1/13	$0.021021\ldots = 0.\overline{021}$	Repeating basimal
1/20	0.02	Terminating basimal
1/21	$0.013013\ldots = 0.\overline{013}$	Repeating basimal
1/22	$0.012012\ldots = 0.\overline{012}$	Repeating basimal
1/23	$0.0113101131\ldots = 0.\overline{01131}$	Repeating basimal
1/30	$0.0111\ldots = 0.0\overline{1}$	Repeating basimal
1/31	$0.010323010323\ldots = 0.\overline{010323}$	Repeating basimal
1/32	$0.0102102\ldots = 0.0\overline{102}$	Repeating basimal
1/33	$0.0101\ldots = 0.\overline{01}$	Repeating basimal
1/100	0.01	Terminating basimal

Table 9 shows this work. Since counterexamples arise, comparing the results in **table 8** and **table 9** rule out the conjecture about a pattern involving prime factorization of denominators containing no primes other than 2_{four} and 11_{four}. However, the conjecture about a pattern involving prime factorization of denominators containing no primes other than 2_{four} seems to work. On further reflection, the preservice teachers were content with this idea, thinking that the 2 and 5 are "special" in base ten because they are the prime factors of ten (10_{ten}) and likewise that the 2_{four} is "special" in base four because it is the only prime factor of four (10_{four}).

Thus, the conjecture that a rational number a/b in simplest form in base four can be written as a terminating decimal if and only if the prime factorization of the denominator contains no primes other than 2_{four} was generally accepted. Two more cases are checked in **table 10**.

SUMMARY OF INVESTIGATIONS

In working on all the preceding activities, students are encouraged to look for patterns, make conjectures, test conjectures, discover number-theory facts, and investigate mathematics. Exploring serious mathematics challenges both students and teachers beyond the usual mundane pursuits of ordinary arithmetic. The conjectures made by the preservice teachers who worked through the activities brought out many fascinating ideas and led them to delve further into mathematics. The preservice teachers developed theorems about rational numbers through discovery methods, rather than by memorizing them from a textbook. Many of the

TABLE 9

Base-Four Prime Factorization

Unit Fraction	Prime Factorization of Denominator
1/2	2
1/3	3
1/10	2^2
1/11	11
1/12	$2 \cdot 3$
1/13	13
1/20	2^3
1/21	3^2
1/22	$2 \cdot 11$
1/23	23
1/30	$2^2 \cdot 3$
1/31	31
1/32	$2 \cdot 13$
1/33	$3 \cdot 13$
1/100	2^{10}

students were genuinely excited about their pursuits and accomplishments and gained a sense of pride from developing their own ideas.

Teachers and students can use **sheets 1** and **2** to complete the activities that the preservice teachers did in this article. Solutions to these activity sheets can be found throughout the article. Readers are further

TABLE 10			
Base-Four Unit Fractions, Basimal Representations, and Prime Factorization			
Unit Fraction	Basimal Representation	Terminating or Repeating	Prime Factorization of Denominator
1/133	$0.0020100201\ldots = 0.\overline{00201}$	Repeating basimal	133
1/200	0.002	Terminating basimal	2^{11}

encouraged to explore methods of determining the length of the period and the length of the delay of basimal representations that repeat in base four. In fact, readers can repeat the above activities in various other bases. Building on the ideas presented in the article, readers can develop general theorems about the basimal representations of unit fractions, given any base.

REFERENCE

National Council of Teachers of Mathematics (NCTM). *Principles and Standards for School Mathematics.* Reston, Va.: NCTM, 2000.

Investigating Unit Fractions and Decimals

Sheet 1

1. Use a calculator to fill in the following chart.

Unit Fraction	Decimal Representation	Classification (Terminating or Repeating)	Length of Period (Number of Places That the Decimal Repeats)	Length of Delay (Number of Places to the Right of the Decimal Point before First Period Begins)
1/2				
1/3				
1/4				
1/5				
1/6				
1/7				
1/8				
1/9				
1/10				
1/11				
1/12				
1/13				
1/14				
1/15				
1/16				
1/17				
1/18				
1/19				
1/20				
1/21				
1/22				
1/23				
1/24				
1/25				

2. Make conjectures about ways to determine whether the decimal representation of a given unit fraction terminates or repeats.

Investigating Unit Fractions and Decimals—Continued Sheet 1

3. Make conjectures about ways that you can determine the length of the period of a repeating decimal.

4. Make conjectures about ways that you can determine whether the decimal representation of a repeating decimal will have a delay.

5. Make conjectures about ways that you can determine the length of the delay for a repeating decimal that has a delay.

6. Use a calculator to test your conjectures with other examples.

Unit Fraction	Decimal Representation	Classification	Length of Period	Length of Delay
1/30				
1/32				
1/48				
1/56				
1/75				
1/160				

Investigating Unit Fractions and Basimals in Base Four

1. Use long division to find the basimal representations of the unit fractions to determine whether the basimal representation terminates or repeats. The first two unit fractions have been completed for you.

Base Four Unit Fraction	Base Four Basimal Representation	Classification: Terminates or Repeats
1/2	0.2	Terminates
1/3	0.111 . . . = 0.1	Repeats
1/10		
1/11		
1/12		
1/13		
1/20		
1/21		
1/22		
1/23		
1/30		
1/31		
1/32		
1/33		

2. Write a conjecture that indicates how you can tell whether the basimal representation of a fraction in base four terminates or repeats. (*Hint:* In base ten, terminating decimals have something to do with denominators having factors of 2s and 5s.)

3. Use your conjecture to determine whether the basimal representations of the following fractions terminate or repeat in base four.

Unit Fraction	Terminates or Repeats
1/133	
1/200	

Print-Shop Paper Cutting:
Ratios in Algebra

Carelyn Hill

"A major goal of high school mathematics is to equip students with knowledge and tools that enable them to formulate, approach, and solve problems beyond those that they have studied. High school students should have significant opportunities to develop a broad repertoire of problem-solving (or heuristic) strategies. They should have opportunities to formulate and refine problems because problems that occur in real settings do not often arrive neatly packaged. Students need experience in identifying problems and articulating them clearly enough to determine when they have arrived at solutions. The curriculum should include problems for which students know the goal to be achieved but for which they need to specify—or perhaps gather from other sources—the kinds of information needed to achieve it." (NCTM 2000, p. 335)

Students have a difficult time transferring mathematics learned from a textbook to an activity that uses real data or that simulates a real-world experience. Several years ago, a vocational teacher expressed his frustration that students were not able to do the activities in a paper-cutting chapter in the course that he taught. In response, I borrowed the textbook (Cogoli 1980) and wrote the following activity for algebra students. This paper-cutting activity was developed specifically to help students gain experience in transferring skills and information that they have learned to different situations. The content focus of the activity is using ratios for various jobs that an employee in a print shop might encounter in a day's work.

Objectives. This mathematics activity helps students see connections between mathematics in the classroom and mathematics in the real world. Specifically, students learn to set up and solve ratios in a new way. Students also learn how to deal with remainders in a nontraditional way. Depending on the application, students truncate ratios or round them up. In addition, students interpret scale drawings and create an accurate scale drawing.

Prerequisites. Students should be familiar with ratios, percents, and scale drawings.

Grade levels. 7–12

Materials. Activity sheets, twelve 5 × 7-inch sheets of paper per student group, 17 × 22-inch stock sheets for each group, rulers, tape, and staplers; optional supplies include a ream of paper for demonstration purposes,

22 1/2 × 28 1/2-inch colored posterboard, and six 8 1/2 × 11-inch sheets of white paper.

Time required. 45–60 minutes

DIRECTIONS

Students can do this activity individually, or they can work in groups of three or four students. During the brief presentation that I use to introduce the activity, I inform students that this activity allows them to practice a skill needed when working in a print shop and gives them practice in doing some out-of-the-ordinary mathematical computations. I tell them that they can master the skills by using the material that they have learned in class, by doing some thinking, and by using problem-solving strategies. To help with vocabulary, I show the class a ream of paper and set it on the chalk tray. I also put a sheet of colored posterboard, labeled "stock sheet," on the tray. I then have 8 1/2 × 11-inch sheets of paper labeled "press sheets" to show students options for cutting press sheets from a stock sheet.

Student instructions

The **student instructions** give information, formulas, diagrams, and an example of a print-shop application. The students need this information to solve problems in the activity. They should read the material carefully and ask other students or the instructor to help clarify any part that they do not understand.

Sheet 3

The first two questions are factual questions about the reading material. These questions help students begin doing the ratio computations themselves. Question 1 lends itself toward a discussion of when they must compute both ratios and when they need to compute only one ratio. Question 3 asks students to connect the skill that they use in this activity with a skill that they might need.

Sheet 4

Job 1. This problem takes the students through the preliminary work for a job at a print shop. This job draws on ideas learned in the **student instructions.**

Job 2. This activity extends job 1 by asking students to consider the cost of paper.

Job 3. This job takes the activity to a level of greater complexity. The students make an accurate scale drawing of the layout of press sheets on a 17 × 22-inch stock sheet of paper. If this size is not available, the teacher should determine an appropriate scale and use paper that is available. Using the same size sheet, the students staple or tape the precut 5 × 7-inch sheets to learn whether they can find a more efficient way to line up press sheets than the number that they computed by using ratios and their drawing.

When the students are done with the activity, we discuss it. I ask students when a more efficient way might exist to cut the stock paper than the results indicated by the ratios. If the remainders are large, a more efficient result might be possible. Students can line up the ten press sheets in three different ways. I show the various ways on the board. See **figures 1a, 1b**, and **1c**. I ask students which of the three layouts they would choose if they were actually cutting the press sheets. Then I ask the students why they selected that arrangement. **Figure 1a** and **1c** are simpler to measure, draw, and cut than **figure 1b**.

SOLUTIONS

Sheet 1:

1)

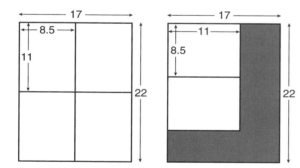

2) The number of letterheads that can be cut from two reams of 17 × 22-inch stock is (4 × 500 × 2), or 4000.

3) Answers will vary.

Sheet 2:

1) Job 1

$$\frac{17 \times 22}{8\,1/2 \times 11} = 4 \qquad \frac{17 \times 22}{11 \times 8\,1/2} = 2$$

The number of stock sheets required for the job is twenty-eight. The diagram should be the same as the one for **sheet 1**, problem 1.

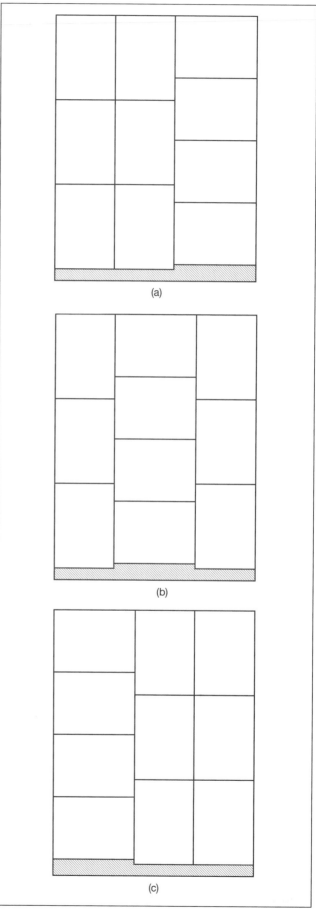

Fig. 1. Three ways to line up ten press sheets

2) Job 2

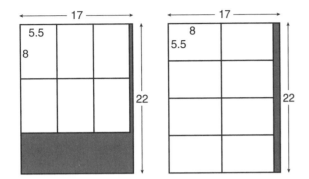

$$\underset{\uparrow\quad\uparrow}{\overset{3\quad2}{\frac{17\times22}{5\,1/2\times8}}}=6 \qquad \underset{\uparrow\quad\uparrow}{\overset{2\quad4}{\frac{17\times22}{5\,1/2\times8}}}=8$$

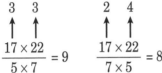

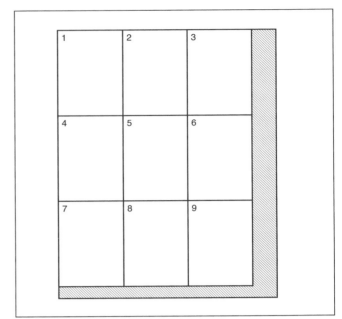

Fig. 2

a. You need 1500 stock sheets.
b. You need 3 reams of stock paper.
c. The cost of the stock paper is $61.50.

$$\underset{\uparrow\quad\uparrow}{\overset{3\quad3}{\frac{17\times22}{5\times7}}}=9 \qquad \underset{\uparrow\quad\uparrow}{\overset{2\quad4}{\frac{17\times22}{7\times5}}}=8$$

3) Job 3

You need sixty-three stock sheets.

a. See **figure 2**.

b. Answers will vary; see **figure 1** for some possible answers.

REFERENCES

Cogoli, John E. *Photo Offset Fundamentals*. Woodland
Hills, Calif.: Glencoe/McGraw-Hill, 1980.

National Council of Teachers of Mathematics (NCTM).
Principles and Standards for School Mathematics.
Reston, Va.: NCTM, 2000.

Paper-Cutting Activity: Student Instructions   Sheet 1

In this activity, you work in a print shop. The day's work includes several jobs, each of which first requires cutting sheets of paper to the proper size.

You need the following information before you can begin work.

A ream consists of 500 sheets of paper. Common papers are often packaged in this quantity.

To determine the number of 6×10-inch press sheets that can be obtained from a $22\ 1/2 \times 28\ 1/2$-inch stock sheet, calculate first as indicated in solution A, then change the orientation of the press sheets and calculate as indicated in solution B.

Divide the press-sheet size into the stock-sheet size. In solution B, switch the bottom numbers of your ratio.

Basic formula:

$$\frac{\text{Dimensions of the stock sheet}}{\text{Dimensions of the press sheet}} = \frac{\text{width}}{\text{width}} \cdot \frac{\text{height}}{\text{height}}$$

Solution A:

$$\frac{22\ \tfrac{1}{2}\text{ inches wide} \times 28\ \tfrac{1}{2}\text{ inches high}}{6\text{ inches wide} \times 10\text{ inches high}}$$

Solution B:

$$\frac{22\ \tfrac{1}{2}\text{ inches wide} \times 28\ \tfrac{1}{2}\text{ inches high}}{10\text{ inches wide} \times 6\text{ inches high}}$$

Divide the denominator into the numerator for each dimension. Round down to the nearest whole number.

A:

$$3 \quad \bullet \quad 2 \quad = 6$$
$$\uparrow \qquad\qquad \uparrow$$
$$\frac{22\ \tfrac{1}{2}\text{ inches wide} \times 28\ \tfrac{1}{2}\text{ inches high}}{6\text{ inches wide} \times 10\text{ inches high}}$$

B:

$$2 \quad \bullet \quad 4 \quad = 8$$
$$\uparrow \qquad\qquad \uparrow$$
$$\frac{22\ \tfrac{1}{2}\text{ inches wide} \times 28\ \tfrac{1}{2}\text{ inches high}}{10\text{ inches wide} \times 6\text{ inches high}}$$

Paper-Cutting Activity: Student Instructions

The following diagrams are visual representations of the two solutions:

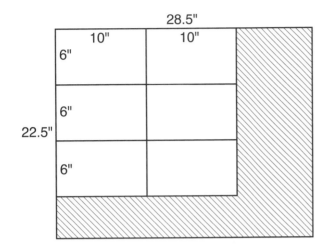 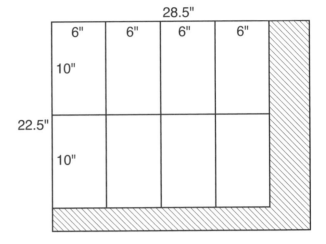

Solution B results in eight pieces, whereas solution A results in only six pieces. Because solution B wastes less paper than solution A, solution B is more efficient. However, you may be able to find a better solution. If you obtain large remainders when you perform the division, as in solutions A and B, you may be able to draw press sheets directly on the stock sheet and obtain a larger number of pieces per sheet. Refer to the diagram below.

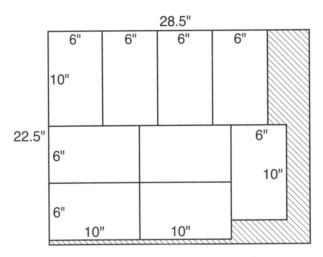

Diagram method of figuring stock

The diagram shows that nine 6 × 10-inch press sheets can be obtained from the same 22 1/2 × 28 1/2-inch stock sheet—one more press sheet per stock sheet than in solution B. Considerable savings could result if many sheets were to be cut.

Paper-Cutting Activity

Sheet 3

1. How many 8 1/2 × 11-inch letterheads can be cut from a 17 × 22-inch stock sheet? Show both ratios, and sketch the diagrams. Label the diagrams.

2. Use the larger number of letterheads that can be cut from a stock sheet in question 1, and determine the number of 8 1/2 × 11-inch letterheads that can be cut from two reams of 17 × 22-inch stock.

3. Make a connection between the paper-cutting skill used in this activity and some situation or occupation that uses this skill.

Paper-Cutting Activity

Sheet 4

1. **Job 1**

   The order asks for one hundred 8 ½-by-11-inch letterheads. The stock size is 17 inches by 22 inches. Use a 10 percent allowance for spoilage, that is, sheets ruined during printing. How many stock sheets of paper are required for the job? You must write ratios or draw diagrams.

2. **Job 2**

   This order asks for 12,000 pieces of notepaper, each measuring 5 1/2 inches by 8 inches. The stock size is 17 inches by 22 inches, and it sells for $20.50 a ream. Assume that no spoilage occurs. Write ratios, and draw diagrams.

   a. How many stock sheets do you need? _____

   b. How many reams of stock paper do you need? _____

   c. What is the cost of the stock paper? _____

3. **Job 3**

   You have reached the last job of the day. However, it is a big one, so take your time and get it right.

   The last order requests 500 5-by-7-inch report cards. How many stock sheets of paper do you need if the stock-sheet size is 17 inches by 22 inches? Allow 10 percent for spoilage.

   Use ratios to show the two options that indicate the number of 5-by-7-inch sheets that you can get from one stock sheet.

   a. Use a 17-by-22-inch stock sheet. Complete the diagram by showing how you would cut the paper to get the most efficient result from your ratios. Use a ruler. Shade in the waste.

   b. You again have a 17-by-22-inch stock sheet. For this diagram, use the cutout 5-by-7-inch sheets of paper. Determine whether your previous diagram is in fact the most efficient way to cut the stock sheet. Whatever way you decide to cut the stock sheet, place the 5-by-7-inch sheets in that order, and staple or tape them to the diagram.

A Geometric Look at Greatest Common Divisor

Melfried Olson

TEACHER'S GUIDE

Introduction: One of the four standards specified at all grade levels in NCTM's *Curriculum and Evaluation Standards for School Mathematics* (1989) is mathematical connections. The curriculum standards suggest that the study of mathematics should include opportunities to make connections so that students can link conceptual and procedural knowledge, relate various representations of concepts or procedures to one another, recognize relationships among different topics in mathematics, use a mathematical idea to further their understanding of other mathematical ideas, and recognize equivalent representations of the same concept. It is important to afford opportunities for connections so that students will not view mathematics as a collection of isolated topics nor think of mathematics as computation only.

The following activity examines an arithmetic concept, greatest common divisor, from a geometric standpoint. This activity could also be used to demonstrate conversions between units of area. The method presented here engages students intellectually and challenges them to apply prior knowledge of numerical relationships to explore the concept of greatest common factor from another slant. This activity uses the area model for multiplication and division in another attempt to emphasize connections between arithmetic and geometric representations. Problem solving and reasoning in the exploration of number theory and patterns are also included.

Grade levels: 5–10

Materials: Copies of sheets 1–4 for each student, scissors, extra sheets of graph paper

Objectives: To develop and enhance students' understanding of the idea of greatest common divisor

Prerequisites: Before beginning this activity, the teacher should make sure that the students understand multiplication and division from an area-model perspective. Ideally, students should be able not only to model 3×7 in rectangular form as but also to model the distributive property for 14×23 in rectangular form (see **fig. 1**.)

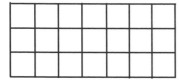

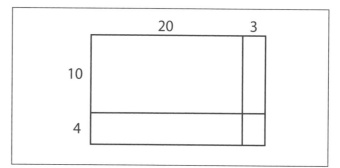

Fig. 1

Directions: This activity is probably best used over two or three days. This time can be shortened by not having all students cut each rectangle in problem 1 of sheet 2. Instead, the teacher can divide the class into three groups and have each group do two rectangles.

Sheet 1: Students should cut out the appropriate rectangle from sheet 4. Students may need some help in identifying the largest square that can be removed. The teacher should work through the first activity with the students to familiarize them with "removing the largest possible square from one of the rectangle's edges." Once this concept is clear, the students should cut out the second square as directed. Students should be careful to keep the remaining rectangle. The information regarding the size of the last square removed will be recorded on the chart on sheet 1, as well as later on sheet 2. The purpose of the cutting is to find the special square related to each rectangle. This special square is the size of the last square removed, and the length of a side of the square is the greatest common divisor of the dimensions of the rectangle. As can be seen on sheet 3, each rectangle can be partitioned into squares all of which are the size of the last square removed.

Sheet 2: Students must use their own grid paper for work on this sheet. They repeat the square-removal process from sheet 1 and record what they find. Students predict the size of the last square removed and check their prediction on other rectangles. This process yields the greatest common divisor of the dimensions of a rectangle because it geometrically pictures the steps used in the Euclidean algorithm to find the greatest common divisor of two numbers. For example, to find the greatest common divisor of 24 and 44 (problems 1–4, sheet 1) by using the Euclidean algorithm we calculate thus:

1.

$$24\overline{)44}$$ with quotient 1, 24 below, 20 remainder

One 24×24 square is removed; a 20×24 rectangle remains.

2.

$$20\overline{)24}$$ with quotient 1, 20 below, 4 remainder

One 20×20 square is removed; a 4×20 rectangle remains.

3.

$$4\overline{)20}$$ with quotient 5, 20 below, 0 remainder

A 0 remainder gives 4 as the greatest common divisor of 44 and 24.

Similarly, using the Euclidean algorithm for finding the greatest common divisor of 18 and 51 (problems 5 and 6, sheet 1), we write this:

1.

$$18\overline{)51}$$ with quotient 2, 36 below, 15 remainder

Two 18×18 squares are removed; a 15×18 rectangle remains.

2.

$$15\overline{)18}$$ with quotient 1, 15 below, 3 remainder

One 15×15 square is removed; a 3×15 rectangle remains.

3.

$$3\overline{)15}$$ with quotient 5, 15 below, 0 remainder

A 0 remainder gives 3 as the greatest common divisor of 51 and 18.

Sheet 3: Students reverse the process of removing squares and try to draw in and visualize how many squares can be found. After drawing in squares, the students should record the information in the chart in problem 3. The extension into least common multiple is probably expected but often not seen in the same manner as asked in question 4. That this strategy indeed does produce the least common multiple can be seen by one of the methods used to find the least common multiple of two numbers. For example, the least common multiple of 20 and 42 is found by factoring both numbers:

$$20 = 2 \cdot 10$$
$$42 = 2 \cdot 21$$

Therefore, the least common multiple of 20 and 42 is $2 \cdot (10 \cdot 21)$.

SOLUTIONS

Sheet 1: 1) a. 24×24, b. no, c. 20×24; 2) a. 20×20, b. No, c. 4×20; 3) a. 4×4, b. yes: 5, c. No rectangle remains; 5) a. First, two 18×18 squares are removed leaving a 15×18 rectangle; b. Next, one 15×15 square is removed, leaving a 3×15 rectangle; c) Finally, five 3×3 squares are removed so that no rectangle remains; 6) For the 24×44 rectangle: 4×4; for the 18×51 rectangle; 3×3.

Sheet 2: 1) 20×42—2×2, 16×36—4×4, 21×35—7×7, 17×30—1×1, 12×42—6×6, 18×45—9×9; 2) Among the observations are the following: The dimension of the last square removed is the greatest common divisor of the dimensions of the rectangle, the original rectangle can be divided into squares that are all the same size, a last square can usually be found—for most rectangles, that is not a 1×1 square; 3) a. 7×7, b. 8×8, c. 1×1, d. 3×3, e. 4×4; 4) See directions for sheet 2.

Sheet 3: 1) 66 squares; 2) 102 squares of size 3×3

3)

Dimensions of Rectangle	Number of Squares (N)	Size of Squares (S)
20×42	210	2×2
16×36	36	4×4
21×35	15	7×7
12×42	14	6×6
18×45	10	9×9
14×35	10	7×7
16×40	10	8×8
7×24	168	1×1
9×21	21	3×3
140×288	2520	4×4
$ab \times ac$	bc	$a \times a$

4) Students may need to be reminded of other uses, including "simplifying" fractions. Suppose we treat the length of the sides of a rectangle like the numerator and denominator of a fraction. When all the lines are drawn in the rectangle, as in **figure d** in problem 1 (see sheet 3), the common factor is revealed along each side, with each side being "divided" by the greatest common divisor; the "reduced" fraction can be seen by counting the number of divisions along each side. This representation emphasizes dividing both numerator and denominator by the same amount; 5) The product of the number in column N and the length of a side in column S gives the least common multiple of the dimensions of the rectangle.

REFERENCE

National Council of Teachers of Mathematics (NCTM), Commission on Standards for School Mathematics. *Curriculum and Evaluation Standards for School Mathematics.* Reston, Va.: NCTM, 1989.

Removing Squares

1. Cut out the 24×44 rectangle (rectangle A) on sheet 4. Remove the largest possible square "from one of the rectangle's edges."

    a. What size of the square is removed? _____

    b. Can more than one of these squares be removed? _____

    If yes, how many can be removed? _____

    c. What size of rectangle remains? _____

2. Repeat the process in step 1 with the 20×24 rectangle left after step 1c.

    a. What size of square is removed? _____

    b. Can more than one of these squares be removed? _____

    If yes, how many can be removed? _____

    c. What size of rectangle remains? _____

3. Repeat the process in step 1 with the 4×20 rectangle after step 2c.

    a. What size of square is removed? _____

    b. Can more than one square be removed? _____

    If yes, how many can be removed? _____

    c. What size rectangle remains? _____

4. Record in the table at the bottom of the page the size of the last square removed and the number of squares removed so that nothing remained.

5. Cut out one 18×51 rectangle (rectangle B) from sheet 4 and remove from it the largest possible square, as in step 1. Continue the process until the whole 18×51 rectangle has been depleted. Answer the following questions at each stage.

    a. What size of square is removed? _____

    b. Can more than one of these squares be removed? _____
    If so, how many? _____

    c. What size of rectangle is left? _____

6. When you get to the last step, record in the chart below the size of the last square removed and the number of squares removed so that nothing remained.

Rectangle	Size of Last Square Removed	Number of Squares Removed
24×44	_____	_____
18×51	_____	_____

Predicting Last Square Removed   Sheet 2

1. For the rectangles given, repeat the removing-squares process used on sheet 1. In the right-hand column of the chart, record the size of the last square removed. The information for the 24×44 and 18×51 rectangles from sheet 1 is recorded for you.

Rectangle	Size of Last Square Removed
24×44	4×4
18×51	3×3
20×42	
16×36	
21×35	
17×30	
12×42	
18×45	

2. What observations can you make about the relationship between the size of the original rectangle and the size of the last square removed? _____

   _____

   _____

   _____

   _____

3. Given the dimensions of a rectangle, can you predict the size of the last square to be removed? Try your predictions on the following rectangles:

   a. 14×35 _____

   b. 16×40 _____

   c. 7×24 _____

   d. 9×21_____

   e. 140×288 _____

4. Why does this square-removal process give us information needed to find the greatest common divisor of the dimensions of a rectangle? _____

   _____

   _____

   _____

How Many Squares?

1. The results of the work on question 1 on sheet 1 can be represented in the following manner:

   a.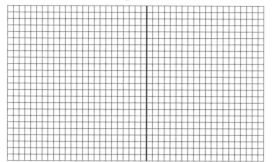

   b.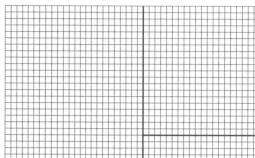

   c.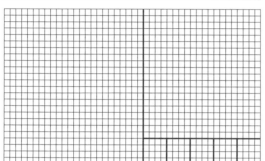

   d.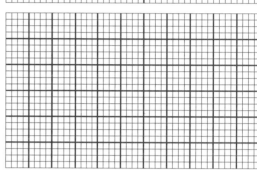

   By inserting heavy lines into figure c, we can obtain figure d, which contains squares of only size 4×4.
   How many 4×4 squares appear in figure d?

2. Repeat the process shown above for the remaining 18×51 rectangle (rectangle C) from sheet 4 to find how many 3×3 squares are contained in the rectangle.

3. Together with the process shown above, use the information from questions 1 and 2 on sheet 2 to complete the following chart.

Dimensions of Rectangle	Number of Squares (N)	Size of Squares (S)
20×42		
16×36		
21×35		
12×42		
18×45		
14×35		
16×40		
7×24		
9×21		
140×288		
$ab \times ac$*		

Where b and c have no common factors

4. How does this concept of greatest common divisor relate to other uses you have made of the concept of greatest common divisor?

5. How can the numbers in columns N and S be used to find the least common multiple of the dimensions of the rectangle?

Grids

A

B

C

Pi Filling, Archimedes Style

Jason Slowbe

A lesson on finding the area of regular polygons using the formula

$$A = (1/2) \text{ (apothem)(perimeter)}$$

can provide a rich exploration of the relationship between inscribed polygons and the development of an algorithm that systematically generates the digits of π.

In the spirit of Archimedes' method of approximating π, we will inscribe regular polygons in the unit circle, which is known to have area π. As the number of sides, n, in each regular polygon increases, the polygons fill more and more of the area inside the unit circle. Naturally, increasing n causes the areas of the n-gons to approach π. With sufficiently large n, we can obtain decimal approximations that become arbitrarily close to the exact value of π. Of course, a problem arises with large n, as the calculations quickly become unbearably tedious without the use of a calculator. Fortunately, this process can be automated very easily by writing a program to compute the area for any n-gon, thus providing a compelling exploration of a method for generating the never-ending digits of π.

CALCULATING AREAS

First, consider the case of a regular (equilateral) triangle inscribed in the unit circle. The distance from the center to any vertex is equal to the circle's radius. Using the well-known formula for the sum of the interior angles, S, of any n-gon, we have

$$S = (n - 2) \cdot 180.$$

Therefore, each angle of the regular triangle measures

$$\frac{(3 - 2) \cdot 180}{3} = 60°$$

See **figure 1**.

Now, consider the isosceles triangle formed by two radii and a side of the equilateral triangle. The radii bisect the angles to which they are drawn, so the base angles of the isosceles triangle measure

$$\frac{(n - 2) \cdot 180}{3n} = \frac{(3 - 2) \cdot 180}{2 \cdot 3}$$
$$= 30°.$$

See **figure 2**.

We label the apothem y and the bases of each right triangle x. Using trigonometry, we can find the lengths of the apothem, y, and the base, $2x$. The base will then be multiplied by n, the number of sides, to obtain the perimeter. Finally, we will be able to use $A = (1/2)$ (apothem)(perimeter) to find the area of the regular triangle. Solve for x and y, as shown in **figure 3**.

With the apothem $= \sin(30°)$ and the perimeter $= 3 \cdot 2 \cos(30°)$, we can calculate the area of the regular triangle in the following way:

$$
\begin{aligned}
\text{Area} &= (1/2) \text{ (apothem)(perimeter)} \\
&= [1/2][\sin(30°)][3 \cdot 2 \cos(30°)] \\
&= [1/2][\sin(30°)][3 \cdot 2 \cos(30°)] \\
&\approx 1.2990 \text{ sq. units} \qquad (1)
\end{aligned}
$$

Following this same process for a square, we get:

$$
\begin{aligned}
\text{Area} &= (1/2)\text{(apothem)(perimeter)} \\
&= [1/2][\sin(45°)][4 \cos(45°)] \\
&= [1/2][\sin(45°)][4 \cdot 2 \cos(45°)] \\
&= 2.0000 \text{ sq. units} \qquad (2)
\end{aligned}
$$

See **figure 4**.

In fact, we quickly notice that equations (1) and (2)

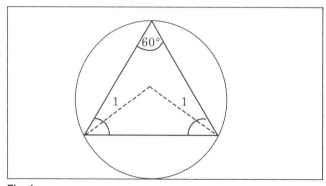

Fig. 1

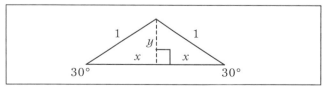

Fig. 2

45

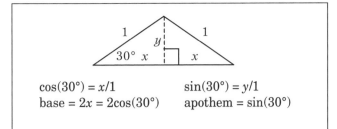

$$\cos(30°) = x/1 \qquad \sin(30°) = y/1$$
$$\text{base} = 2x = 2\cos(30°) \qquad \text{apothem} = \sin(30°)$$

Fig. 3

are remarkably similar. The number of sides n has changed, as have the degree measures of each base angle in the isosceles triangle. But those base angles are dependent on n. In particular, the degree measure of the base angles of the inner isosceles triangles can be found by finding the total measure of the interior angles of the regular n-gon, dividing this sum by n to get the measure of each angle, and dividing that angle by 2 to account for the bisection by the radius. So, we can generalize the area of the regular n-gons inscribed in the unit circle by the following:

$$\text{Area} = [1/2]\left[\sin\left(\frac{(n-2)\cdot 180}{2n}\right)\right]$$
$$\cdot \left[n \cdot 2\cos\left(\frac{(n-2)\cdot 180}{2n}\right)\right] \qquad (3)$$

We now want to program a graphing calculator to complete these calculations for any value of n specified by the user. Based on your students' programming experiences, you can provide varying levels of guidance to challenge each student appropriately. For purposes of this article, I used a TI-83, but any calculator with programming capabilities can be used as well.

One way to introduce and motivate programming for inexperienced students is to store a particular numeric value for the number of sides of the polygon into the calculator, then enter the area expression just as it appears in (3). The variable N is a "memory location" in the calculator, where numeric values can be stored and then substituted in subsequent calculations.

The downside to such an approach is the need to re-store new values of n for each polygon, then recall the long expression to calculate the area of each n-gon. A program can make this process a bit easier (**fig. 5**). The user inputs the desired number of sides n, the calculator computes the area and stores this value into the variable A, and the calculator displays the number of sides N and its corresponding area A.

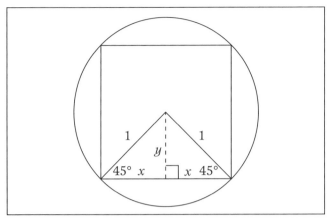

Fig. 4

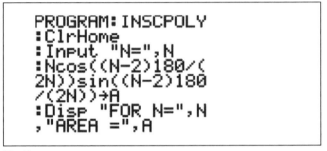

```
PROGRAM:INSCPOLY
:ClrHome
:Input "N=",N
:Ncos((N-2)180/(
2N))sin((N-2)180
/(2N))→A
:Disp "FOR N=",N
,"AREA =",A
```

Fig. 5

```
PROGRAM:INSPOLY1
:ClrHome
:3→N
:Lbl B
:Ncos((N-2)180/(
2N))sin((N-2)180
/(2N))→A
:Disp "FOR N=",N

,"AREA =",A
:Pause
:1+N→N
:Goto B
```

Fig. 6

GENERATING π

As n increases, the areas displayed by the program will approach 3.141592654 (the maximum number of digits displayed on the viewing screen; the last digit, 4, is rounded up from 3). With a minor accommodation, we can have the program automatically display areas for increasing values of n (**fig. 6**). The program starts with $n = 3$, displays the number of sides and the corresponding area, pauses to allow the user time to see the results on the screen, then adds 1 to n and returns to the Lbl B line, repeating the process for $n = 4, 5, 6,$ and so on.

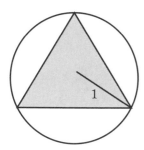

As we increase the number of sides in regular polygons inscribed in the unit circle, what happens to their areas?

Example: Triangle

(a) The goal is to find the area of the regular (equilateral) triangle, represented below.

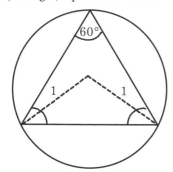

(b) Consider the isosceles triangle below, formed by drawing radii to two of the vertices of the triangle. Note that the base angles of the isosceles triangle are half the angles of the regular triangle. These base angles can be represented as

$$\frac{1}{2} \cdot \frac{(n-2)(180)}{n} = \frac{(n-2)(180)}{2n}$$

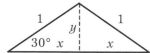

(c) Next, form a right triangle by drawing the apothem from the center of the circle perpendicular to the base (see figure below). Using sine and cosine, we can find the lengths of the apothem, y,

and the base, $2x$. Use degree mode.

$\cos(30°) = x/1 = x$ $\qquad$ $\sin(30°) = y/1 = y$

base $= 2x = 2\cos(30°)$ $\quad$ apothem $= \sin(30°)$

(d) We now have enough information to calculate the area of the regular triangle.

Area $= (1/2)(\text{perimeter})(\text{apothem})$

$\qquad = [1/2] [2\cos(30°) + 2\cos(30°) + 2\cos(30°)] [\sin(30°)]$

$\qquad = [1/2] [3 \cdot 2\cos(30°)][\sin(30°)]$

$\qquad \approx 1.2990$ sq. units

Fig. 7. Example: Triangle

An interesting question to explore is "After what value of n do the digits of π stabilize in the hundredths place value?" This does not occur until $n = 114$, indicating that this algorithm for generating the digits of π is not very efficient. To improve its efficiency, incrementing n by a value larger than 1 will allow the user to see the areas of large n-gons without needing to hit Enter many times. The second-to-last line of the program would be changed to ":10+N→N" in order to increment n by 10, for example. The calculator would then display the areas of the 3-gon, 13-gon, 23-gon, and so on.

CONCLUSION

Graphing calculators are widely used in high school mathematics classrooms, but their programming capabilities are underused. These capabilities can transform an ordinary geometry lesson into a rich mathematical exploration that is interesting and accessible to students with a wide range of abilities. One of the most beneficial aspects of this lesson is teaching students to extend and generalize mathematical results, which is an important skill for students to learn as they begin studying more abstract mathematical ideas. As well, training students to think more broadly and universally helps them develop critical thinking skills and build a strong foundation for future mathematical learning. The lesson helps to satisfy student curiosity regarding the irrational nature of π, making use of the graphing calculator to develop an infinite algorithm that makes the "never-ending" nature of π more apparent.

TABLE 1											
Answer to Question 1											
n	3	4	5	6	7	8	9	10	...	50	...	200
Area	1.2990	2.0000	2.3776	2.5981	2.7364	2.8284	2.8925	2.9389	...	3.1333	...	3.1411

SOLUTIONS

Figure 7 shows a complete solution for a regular inscribed triangle. The worksheet itself provides response questions for students for the case of the triangle and more advanced cases as well. Depending on your students, you may want to give them the triangle solution so they can reference it for other n-gons. Or, to leave the problem open-ended, you can copy only the worksheet for students, then provide whatever guidance or assistance is most appropriate for the needs of your students.

1 See **table 1**; **2** The areas will approach the area of the unit circle, $\pi(1)^2 = \pi$; **3** Circumscribing regular polygons around the unit circle will overestimate the value of π. Therefore, using these methods with circumscribed polygons will generate areas that approach π from above; **4** See **figure 5** for code that allows users to input any specific value for n. See **figure 6** for code that starts with $n = 3$ and increments n by a determined value each time the user hits Enter.

Pi Filling, Archimedes Style

Sheet 1

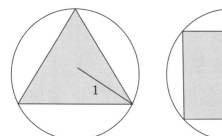 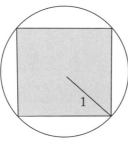 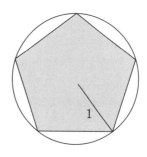 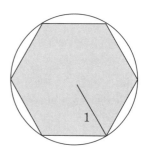

As the number of sides in a regular polygon inscribed in the unit circle increases, what happens to the area of the polygon?

1. Calculate the area of each regular n-gon inscribed in the unit circle rounded to the nearest ten-thousandth. Use calculator values of sine and cosine throughout the problem, and round only in the last step. The areas of the regular triangle and quadrilateral are provided to check your work.

n	3	4	5	6	7	8	9	10	...	50	...	200
Area	1.2990	2.0000							...		...	

2. As n increases, what number should the areas approach? Explain.

3. What would be different about the areas if the regular polygons were circumscribed around the unit circle? Explain.

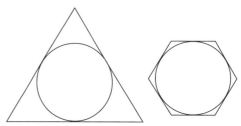

4. *(Challenge)* Write a calculator program to generalize your results. That is, with input n, the program should output the area of the n-gon.

Paths on a Grid

Robert Willcutt

The classic street problem can be used to present new challenges for your students. My classes have always appreciated the number patterns and relationships that are found in its solution (Pólya 1962; Johnson and Rising 1967). This problem presents an excellent model for the need to simplify unwieldy mathematical problems, to look for patterns, and to formulate generalizations based on the results of the data collected for the simple cases. The extension of the basic problem to city layouts that include lakes, parks, or one-way streets provides a good check on a student's understanding of the basic solution process. An extension to three dimensions provides a setting for discovering several new generalizations.

THE BASIC PROBLEM

The basic problem begins with a grid that represents the streets and avenues of a town, as shown in **figure 1**.

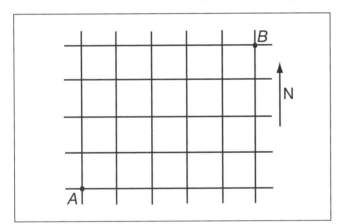

Fig. 1

Bill lives in a house on the corner marked *A*, and Betty lives in a house on the corner marked *B*. Bill and Betty decide to see each other after school every day at Betty's house. Bill decides he will continue to see Betty as long as he can travel a different route from his house (*A*) to Betty's house (*B*) each day. By "different," Bill means that at least one block of the trip will be different from any previous trip. Bill also decides not to backtrack—that is, he will always travel toward Betty's house in either a northerly or easterly direction. Two possible different routes are shown in **figure 2**. How many trips will Bill make to Betty's house?

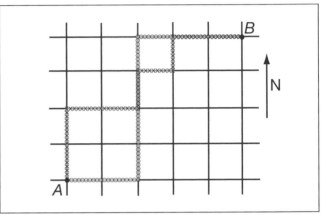

Fig. 2

One way to guide the student's exploration of the problem is to use questions such as these:

1. What can you do to make the problem simpler?

2. What location of Betty's house would make the simplest problem?

3. If you change the location of Betty's house, what pattern seems to evolve?

4. If you do find a number pattern, how can you describe it with a mathematical expression?

The emergence of Pascal's triangle as a solution for this problem is usually a pleasant surprise. Many students use their common sense along with some logical thinking to solve the problem quickly and in so doing, create the theory behind Pascal's triangle. For example, after looking at a few simple cases, one student determined the data shown in **figure 3**. The numbers at each intersection

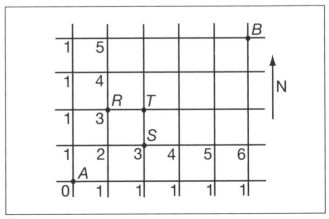

Fig. 3

represent the total number of different routes possible to that intersection from point A. Next, the student considered point T. How many different routes are there between point A and point T? The model is still simple enough so that the routes can be traced and counted, but we can avoid that process if we are thinking logically. How does Bill get to point T? Where does he come from? Through what intersections must he pass just before he arrives at point T? (He must pass through the intersections at points R or S just before he arrives at point T.) We know there are three different routes from point A to point R and three different routes from point A to point S; therefore, there must be six different routes from point A to point T. A student who sees this reasoning need only determine the data for the basic skeleton of the two boundary streets and then he can proceed to complete Pascal's triangle.

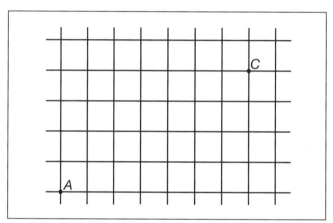

Fig. 4

EXTENSIONS OF THE PROBLEM

1. Can you solve the original problem if Betty had moved into the house at point C, as shown in **figure 4**?

2. If Betty lives at point D in the town shown in **figure 5**, what is the number of paths from A to D?

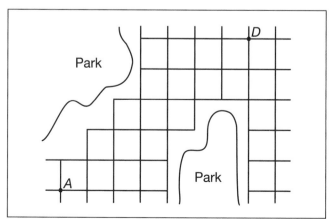

Fig. 5

3. Another town has the map shown in **figure 6**. How many paths from A to F?

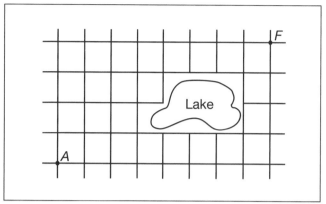

Fig. 6

4. Next extend the problem to three dimensions. Suppose Bill and Betty live in a modern, high-rise apartment building. This building is laid out in groups of apartments called villages. Each village is connected to all other villages by a series of halls and stairways. A model is shown in **figure 7** in which a total of 24 villages is represented. If Bill lives in an apartment located at point A and Betty lives in an apartment located at point G, how many different routes can Bill take to Betty's house? (Assume the exterior lines of the model represent outdoor open passageways and therefore can be used in the trips from point A to point G.) The interior passageways are represented by dotted lines.

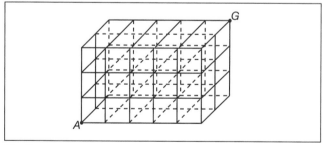

Fig. 7

To find a solution, students collect data, find patterns, and form generalizations. The challenge is to find a generalization that will tell how many different routes exist between point A and any other point H. For example, if Bill lives at point A and Betty lives at point H in the model shown in **figure 8**, how many different routes are possible?

Data for this particular problem appear in **figure 9**.

The patterns seem to be everywhere! Notice Pascal's triangle emerges for the first floor as well as the front face and side face of the building. Each floor has its own pattern, and, of course, all the different patterns can be related. Students often refer to the second floor

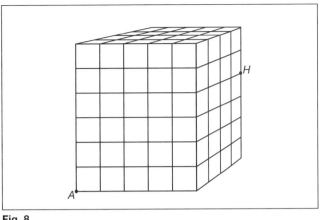

Fig. 8

as Pascal's triangle-2d floor, the third floor patterns as Pascal's triangle-3d floor, and so forth.

GENERALIZING THE PATTERN

How can any one of these patterns be generalized? Is there a total generalization for the entire model?

If you remember that one way to generate Pascal's triangle is the following, where n stands for the row and r represents the position of an element in that row, then the generalizations shown in **figure 10** emerge for each floor.

$$\frac{(n-1)!}{(r-1)!(n-r)!}$$

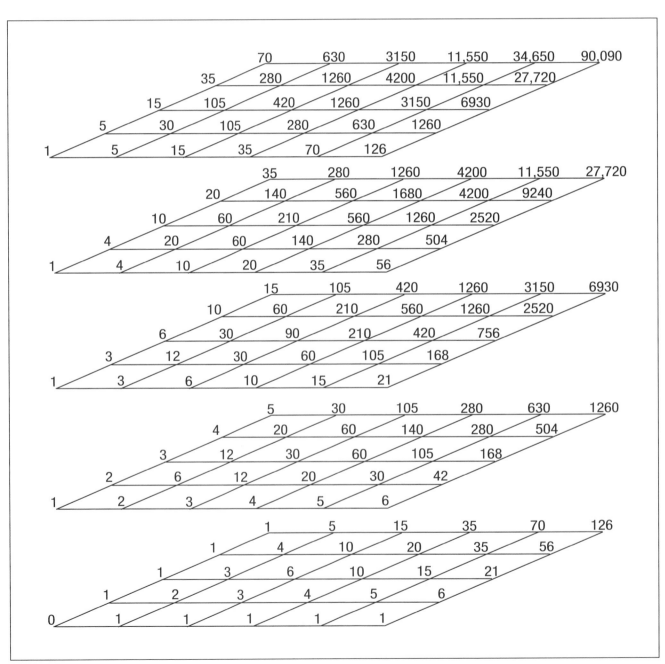

Fig. 9

The pattern within this set of generalizations begins to emerge. A mathematical notation that stands for the generalization is

$$\frac{(n-1)!}{(r-1)!(n-r)!} \prod_{l=2}^{k} \frac{n+(k-l)}{(k-l)+1}$$

where k represents the floor number, n represents the row, and r the position in the row. Thus for the sixth floor:

$$\frac{(n-1)!}{(r-1)!(n-r)!} \prod_{l=2}^{k} \left(\frac{n+(k-l)}{(k-l)+1} \right)$$

$$= \frac{(n-1)!}{(r-1)!(n-r)!}$$

$$\cdot \left(\frac{n}{1} \right)\left(\frac{n+1}{2} \right)\left(\frac{n+2}{3} \right)\left(\frac{n+3}{4} \right)\left(\frac{n+4}{5} \right)$$

We consider the ground floor as floor number one and we make the condition that if $k = 1$, then

$$\prod_{l=2}^{k} \left(\frac{n+(k-l)}{(k-l)+1} \right) = 1$$

Now, to find how many different routes are possible if Bill lives at point A and Betty lives at point H (see **fig. 8**), $k = 5$, $n = 10$, and $r = 5$. Using these values and our generalization, the number of different routes equals

$$\frac{(10-1)!}{(5-1)!(10-5)!}$$

$$\cdot \left(\frac{10}{1} \right)\left(\frac{10+1}{2} \right)\left(\frac{10+2}{3} \right)\left(\frac{10+3}{4} \right)$$

$$= 90,090.$$

Now, we can determine the number of different routes from any point A to any other point H in the building. The solution process that has been described is only one way of attacking this problem. With so many different number patterns involved, students can find many other ways of solving the problem. In fact, the solution process we have described may not be the most efficient one. But that is not the important issue. What remains is a problem full of challenges, one that can be extended from its basic form to greater depths, and one in which students can find many different solution processes.

Floor	Generalization
1st	$\dfrac{(n-1)!}{(r-1)!(n-r)!}$
2d	$\dfrac{(n-1)!}{(r-1)!(n-r)!} \left(\dfrac{n}{1} \right)$
3d	$\dfrac{(n-1)!}{(r-1)!(n-r)!} \left(\dfrac{n}{1} \right)\left(\dfrac{n+1}{2} \right)$
4th	$\dfrac{(n-1)!}{(r-1)!(n-r)!} \left(\dfrac{n}{1} \right)\left(\dfrac{n+1}{2} \right)\left(\dfrac{n+2}{3} \right)$
5th	$\dfrac{(n-1)!}{(r-1)!(n-r)!} \left(\dfrac{n}{1} \right)\left(\dfrac{n+1}{2} \right)\left(\dfrac{n+2}{3} \right)\left(\dfrac{n+3}{4} \right)$

Fig. 10

SOLUTIONS

Sheet 1: $\dfrac{7!}{4!3!} = 35$

Sheet 2: 1) $\dfrac{11!}{4!7!} = 330$

2) 434

3) 218

Sheet 3: 4) $\dfrac{(3+2+5)!}{3!2!4!} = 2520$

5) Look at activity for answer.

REFERENCES

Apostol, Tom M. *Calculus.* New York: Blaisdell Publishing Co., 1964.

Johnson, Donovan, and Gerald Rising. *Guidelines for Teaching Mathematics.* Belmont, Calif.: Wadsworth Publishing Co., 1967.

Pólya, George. *Mathematical Discovery.* Vol. 1. New York: John Wiley & Sons, 1962.

Paths on a Grid

Bill lives in a house on the corner marked A, and Betty lives in a house on the corner marked B (see **fig. 1**). Bill and Betty decide to see each other after school every day at Betty's house. Bill decides he will continue to see Betty as long as he can travel a different route from his house (A) to Betty's house (B) each day. By "different," Bill means that at least one block of the trip will be different from any previous trip. Bill also decides not to backtrack—that is, he will always travel toward Betty's house in either a northerly or easterly direction. Two possible different routes are shown in **figure 2**. How many trips will Bill make to Betty's house?

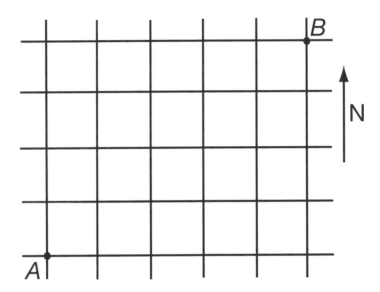

Fig. 1

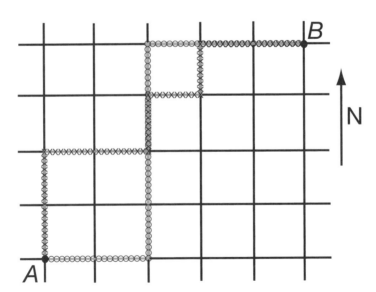

Fig. 2

Paths on a Grid

Sheet 2

1. Can you solve the original problem if Betty had moved into the house at point *C*, as shown in **figure 4**?

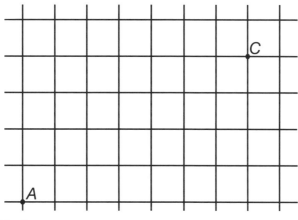

Fig. 4

2. If Betty lives at point *D* in the town shown in **figure 6**, what is the number of paths from *A* to *D*?

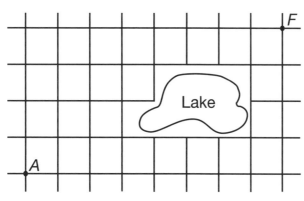

Fig. 6

3. Another town has the map shown in **figure 5**. How many paths from *A* to *F*?

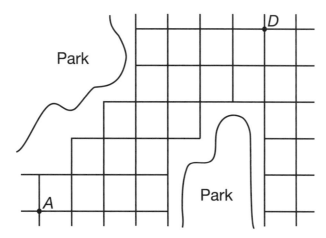

Fig. 5

Paths on a Grid

4. Next extend the problem to three dimensions. Suppose Bill and Betty live in a modern, high-rise apartment building. This building is laid out in groups of apartments called villages. Each village is connected to all other villages by a series of halls and stairways. A model is shown in **figure 6** in which a total of 24 villages is represented. If Bill lives in an apartment located at point A and Betty lives in an apartment located at point G, how many different routes can Bill take to Betty's house? (Assume the exterior line of the model represent outdoor open passageways and therefore can be used in the trips from point A to point G.) The interior passageways are represented by dotted lines.

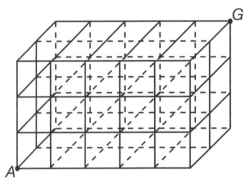

Fig. 6

5. Find a general solution that will tell how many different routes exist between point A and point H. For example, if Bill lives at point A and Betty lives at point H in the model shown in **figure 7**, how many different routes are possible?

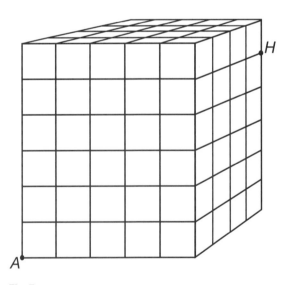

Fig. 7

CHAPTER 2

Algebraic Symbols

Introduction

"The algebraic notation we use today is a major accomplishment of humankind, allowing for the compact representation of complex calculations and problems (Fey 1984; Radford and Puig 2007). However, that very compactness can be a barrier to sense making (Radford and Puig 2007; Saul 2001). A basic task for teachers of algebra is to help students reason their way through that barrier" (NCTM 2009, p. 31).

This chapter presents eight activities that pertain directly to using algebraic symbols. We chose these activities to help you and your students accomplish the task articulated by *Focus in High School Mathematics: Reasoning and Sense Making* (NCTM 2009). As one example, Uth's (1955) activity engages students in creating models of the expressions $(a + b)$ and $(a - b)$ and using these models to reason about why $(a + b)(a - b) = a^2 - b^2$. **Table 2.1** presents the algebraic symbols activities, along with a description of their key elements, found in this chapter.

Focus in High School Mathematics (NCTM 2009) suggests five key elements of reasoning and sense making within algebraic symbols:

1. *Meaningful use of symbols.* Choosing variables and constructing expressions and equations in context; interpreting the form of expressions and equations; manipulating expressions so that interesting interpretations can be made

2. *Mindful manipulation.* Connecting manipulation with the laws of arithmetic; anticipating the results of manipulations; choosing procedures purposefully in context; picturing calculations mentally

3. *Reasoned solving.* Seeing solution steps as logical deductions about equality; interpreting solutions in context

4. *Connecting algebra with geometry.* Representing geometric situations algebraically and algebraic situations geometrically; using connections in solving problems

5. *Linking expressions and functions.* Using multiple algebraic representations to understand functions; working with function notation

This chapter presents eight articles spanning more than fifty years of activities from *Mathematics Teacher* that address these key elements as well as the reasoning habits that NCTM (2009) puts forth.

TABLE 2.1

Algebraic Symbols Activities

Author and title	Mathematical topic(s)	Context(s)	Materials
Gamble (2005), "Teaching Logarithms, Day One"	Logarithms	Using rules of exponents to understand logarithms	Calculator, student activity sheets
House (1987), "An Electrifying Introduction to Algebra"	Variable, associative, commutative, and distributive properties	Electrical circuits	Representations of the circuit diagrams and tables, student activity sheets
Johnson (1986), "Making $-x$ Meaningful"	$-x$	Number line	Student activity sheets
Kinach (1985), "Solving Linear Equations Physically"	Solving linear equations	Balance pan	Scissors, cardboard, tagboard, student activity sheets
Kobayashi (2006), "Relations among Powers of 2, Combinations, and Symbolic Algebra"	Summation, combinations	Using playing card suits to develop a formula	Student activity sheets
Leiva (1980), "Math Magic"	Variable	Magic trick	Student activity sheets
Uth (1955), "Teaching Aid for Developing $(a + b)(a - b)$"	Multiplying binomials	Cutting paper to create geometric models	$8 1/2 \times 11$ inch paper, scissors, student activity sheets
Vandyk (1990), "Expressions, Equations, and Inequalities"	Root, domain, range, inequality	Studying functions in a table	Student activity sheets

We found the *meaningful use of symbols* key element in many of the articles we included in this chapter. As an example, the activity by Johnson (1986) asks students to consider what $-x$ means when x takes on a value that is positive, negative, or zero. "Students are so inured to the use of the word *negative* and the definition of a negative number [that] they do not appreciate the meaning of the expression $-x$" (Johnson 1986, p. 507). This activity is included as the meaningful use of symbols because it clearly calls on students to interpret the form of the expression $-x$.

The *mindful manipulation* key element is present in many of the activities. In one such activity presented by House (1987), students use models of electrical circuits to reason about the associative, commutative, and distributive properties for variables. Students have the opportunity in this activity to apply "the behavior of current in electrical circuits to create a symbolic algebra" (House 1987, p. 302). They then reflect on how the properties of circuits, modeled with symbols, relate to properties of numbers.

Reasoned solving appears in several activities, including one by Kinach (1985) in which models are again used. The models provided in the activity for the teachers and students to create are similar to several available commercial products. In this activity, students use models of variables, units, and balance pans to reason about and write algebraic expressions and then to solve linear equations.

Models are also used in Uth's (1955) activity, which falls in the key element of *connecting algebra and geometry*. The activity could help students explore the connection between a geometric area representation and the fact that $(a + b)(a - b) = a^2 - b^2$. Finally, the key element of *linking expressions and functions* is in the activity by Gamble (2005). By taking part in this activity, students have the opportunity to work with different representations of logarithmic functions to understand some of the laws of logarithms.

As with the key elements, the reasoning habits also appear in many of the activities. As one example, in the activity by Vandyk (1990), students consider the domain, range, and roots of functions represented as expressions. By seeking patterns and relationships and making preliminary deductions and conjectures, this activity affords students the opportunity to *analyze a problem*. In Leiva's (1980) activity, students *implement a strategy* to detect a magician's magic trick and then to create their own. *Seeking and using connections* appears in Kobayashi's (2006) activity, in which students "use algebra as a tool to think about counting problems" (p. 578).

REFERENCE

National Council of Teachers of Mathematics (NCTM). *Focus in High School Mathematics: Reasoning and Sense Making*. Reston, Va.: NCTM, 2009.

Teaching Logarithms, Day One

Marvin Gamble

Students being taught logarithms for the first time often memorize the exponent and logarithm properties and never understand the material. Teachers who have taught this concept have repeatedly told the students that *logarithms* are *exponents*. For some reason students hear the terms exponents and logarithms but often do not understand the relationship between them. I have had a great deal of success in helping students understand the relationship between exponents and logarithms in high school and college courses using the following procedure.

The procedure works best if all students have a calculator with either 10^x or y^x keys and they all use the same model of calculator, preferably a graphing calculator that shows the previous calculations. Students graph a few exponential functions, such as $y = 2^x$ and $y = 3^x$, using paper and pencil; then they check their work with the calculator. The students compare the graphs for similarities, such as the y-intercept and the range and domain of the graphs.

Then we consider the following exponent properties:

1. $x^a \cdot x^b = x^{a+b}$
2. $x^a/x^b = x^{a-b}$
3. $x^0 = 1$
4. $x^{a(b)} = x^{ab}$

After the students review the powers of base ten, particularly when the exponent is 0, 1, 2, and 3, I ask a student for a number between 10 and 100 and I discuss how to find a decimal approximation of the exponent x in 10^x that will be close to the number given by the student. The students use an estimate-and-check approach to determine the exponent, using base ten, to the nearest ten-thousandth. Students typically become so involved that stopping with four significant digits is not acceptable. If a student picked the number 54, for example, the other students wanted to see how close to 54 they could come. Using their graphing calculators, they found 54 is approximated by $10^{1.7323937598}$. Pointing out to students that this result is an approximation and not an equality is important.

Next, I ask a student to pick a number between 100 and 1000. For the following examples I will use 342. I then have the students find its exponent in base ten. The students will find that 342 is approximated by $10^{2.534026106}$. Once these exponents are found, students write the results on the board.

At this time, I review exponent property (1) of 10^{a+b}. Substituting into the property the two numbers whose base ten exponents we just found, they discover that

$$54(342) = 10^{1.7323937598}\,(10^{2.534026106})$$

or

$$10^{a+b} = 10^{4.2664198658}$$
$$= 18468$$
$$= 54(342).$$

The students seem to be amazed by their discovery.

Using the same two numbers with exponential property (2) $x^a/x^b = x^{a-b}$, the students again estimate the decimal answer and the value of the exponent. I have them find both 342/54 and 54/342.

Solving for 342/54, we have

$$342/54 = (10^{2.534026106})/(10^{1.7323937598})$$
$$= 10^{(2.534026106 - 1.7323937598)}$$
$$= 10^{0.8016323462}$$
$$= 6.33$$
$$\sim 342/54$$

Solving for 54/342, we have

$$54/342 = (10^{1.7323937598})/(10^{2.534026106})$$
$$= 10^{(1.7323937598 - 2.534026106)}$$
$$= 10^{-0.8016323462}$$
$$= 0.1578947369$$
$$\sim 54/342$$
$$= 0.1578947368.$$

To verify the third property, I use the following steps. We know that $54/54 = 10^a/10^a$, by property (1) of exponents, is 10^{a-a}, or 10^0. Therefore, by the transitive property, we conclude that $10^0 = 1$.

After reviewing the fourth exponent property, $x^{a(b)} = x^{ab}$, I ask a student to pick an integer between 2 and 5 for b. To help illustrate this, b will be 2. Using the original two numbers picked by the students, I take the smaller of the two numbers, 54 in this example, to illustrate the meaning of the property:

$$54^2 = 54(54)$$
$$= (10^{1.7323937598})(10^{1.7323937598})$$
$$= 10^{(1.7323937598 + 1.7323937598)}$$
$$= 10^{(1.7323937598) \cdot 2}$$
$$= 10^{3.4647875196}$$
$$= 2916$$
$$= 54^2$$

Continuing the concept, I have the students find log 54 and log 342 on their calculators. Students

see that log 54 = 1.7323937598 and log 342 = 2.534026106. I lead the students to discover that $10^{1.7323937598} = 54$ and log 54 = 1.7323937598, and that $10^{2.534026106} = 342$ and log 342 = 2.534026106. These problems demonstrate to students the similarities and relationships between the logarithm expression and the base ten expression. I show the students that $\log_{10}x = a$ is the same as $10^a = x$. If time permits we review the first three laws of logarithms:

If M and N are positive, $b > 0$, and b is not equal to 1, then

Law 1 $\log_b MN = \log_b M + \log_b N$

Law 2 $\log_b(M/N) = \log_b M - \log_b N$

Law 3 $\log_b(N^k) = k \log_b N$

Logarithms, Day One Worksheet

Sheet 1

Every number can be written as a power of 10. An example is $32 \approx 10^{1.50515}$.

1. Pick a number between 11 and 99:
   Find the exponent for your number such that 10^{exponent} will give the number within ± 0.00001 of your number.

2. Pick another number between 101 and 999:
   Find the exponent for this number within ± 0.00001 of your number.

3. The exponents that you found should have either a 1 or a 2 followed by a decimal point. Could you have predicted this?

4. If we picked a four-digit number, what would the first digit before the decimal point be?

   .

5. If we picked a ten-digit number, what would the number before the decimal point be?

   The exponent properties are as follows:

   (1) $x^a x^b = x^{a+b}$

   (2) $x^a / x^b = x^{a-b}$

   (3) $x^0 = 1$

   (4) $x^{a(b)} = x^{ab}$

6. Apply property (1) to the exponents that you found above; add the exponents together and raise 10 to that power. Record your answer.
   Multiply your two original numbers together; what is your answer?_____
   Compare the above two results.

Logarithms, Day One Worksheet

7. Apply property (2) to the two numbers. First subtract the smaller exponent from the larger exponent you found using your two numbers. Raise 10 to your result.
   Divide the smaller number into the larger number.
   How do the two answers compare?

8. Subtract the smaller exponent from the larger exponent. Raise 10 to this power.
   Divide the larger number by the smaller number.
   Compare your two results.

9. Using property (4) above, multiply the first exponent that you found by 4. Raise 10 to this power.
   Raise your first number you picked by the power of 4 and compare your answers.

10. Are your answers exact? _____ If not, why? _____
    What we are doing is working with logarithms. The definition of a logarithmic equation, of base 10, is as follows: $\log_{10} x = a$ is equivalent to $10^a = x$.

11. Take your original numbers and find the log of each number and compare it to the exponents you found for the numbers.

12. Write your original numbers, using the definition of logarithms.

    The first three laws of logarithms are as follows. If M and N are positive, $b > 0$ and $b \neq 1$, then

    Law 1 $\log_b MN = \log_b M + \log_b N$

    Law 2 $\log_b (M/N) = \log_b M - \log_b N$

    Law 3 $\log_b (N^k) = k \log_b N$

13. Fill in the blank. Logarithms are _____

14. Using a graphing calculator, graph $y = \log x$, setting the x range as $-10 \leq x \leq 10$ and the y range as $0 \leq y \leq 20$. What are some characteristics of the graph?

An Electrifying Introduction to Algebra

Peggy A. House

TEACHER'S GUIDE

Introduction: Teachers frequently have interpreted *An Agenda for Action's* (NCTM 1980) much-quoted call for problem solving to be the focus of school mathematics in a manner much narrower than that intended by the writers of the document. Problem solving, as it must occur in the mathematics classroom, is not limited to textbook story problems or to clever nonroutine problems and puzzles. Yet, in reality, many students have not experienced problem solving beyond these two types.

A more careful reading of *An Agenda for Action* (NCTM 1980, pp. 2–4) gives a description of a problem-solving classroom that is rich in a variety of activities that engage students in higher order thinking and in inquiry and discovery. Among the actions recommended in the Agenda (pp. 2–4) for problem solving, we read of the need for—

> mathematical methods that support the full range of problem solving, including ... the use of mathematical symbolism to describe real-world relationships, the use of deductive and inductive reasoning to draw conclusions about such relationships, and the geometrical notions so useful in representing them; opportunities for the student to confront problem situations in a greater variety of forms than the traditional verbal forms alone; for example, presentation through activities, graphic models, observation of phenomena, schematic diagrams, simulations of realistic situations, and interaction with computer programs; [illustrating] the enormous versatility of mathematics ... by presenting as diversified a collection of applications as possible at the given grade level.

This call for applications in the mathematics classroom has also been subject to overly narrow interpretation at times. In particular, applications are frequently dealt with in the most literal sense of the word as an "applying" or laying on of examples after a concept or relationship has been presented. For example, students are taught the Pythagorean theorem and then are expected to apply it to determining the heights of window sills above the ground by leaning a ladder against the sill.

But applications can also be approached in an a priori manner as the source for developing mathematical ideas and relationships. In this use, students are confronted with phenomena, usually derived from known physical relationships, and they use these phenomena to make observations, explain, predict, formulate and test hypotheses, and make decisions. This inquiry should produce generalizations that become the basis for new mathematical understanding. Also, this activity should promote the realization that mathematics is not a collection of arbitrary rules and formulas but a creation of the human mind in response to the need to solve problems arising from many sources.

The activity described here is an example of an application that draws on the behavior of current in electrical circuits to create a symbolic algebra. It is especially appropriate for secondary school students because it lends itself to concrete representation, it involves only a small finite system, it yields a mathematical structure (Boolean algebra) that is generally unfamiliar to the students, and it can be related to such other topics as set theory and symbolic logic. In an elementary algebra class it might be used to reinforce such concepts as the associative and distributive laws; in geometry courses it could serve to promote study of truth tables and logic; in advanced courses it could be the basis for studying a finite algebraic system or the application of Boolean algebra in computers.

Grade levels: 8–12

Materials: The activity can be implemented using only copies of the activity sheets for each student and, for group discussion, transparencies of the various circuit diagrams and tables. However, the enterprising teacher will find that wiring the circuit boards is a relatively easy task that will allow demonstration of the various examples. Suggestions for necessary hardware and wiring can be found in most introductory physical science textbooks. A science teacher in your school may already have the necessary materials and would be pleased to join you in this integration of science and mathematics. The use of such a board to illustrate and verify outcomes by actually lighting a bulb enhances the activity greatly.

Objectives: To enable students (1) to investigate a physical model of electrical circuits as a basis for establishing an algebraic system as a mathematical model; (2) to use mathematical symbolism to describe physical phenomena and to relate mathematical symbolism to physical embodiments; (3) to develop and test hypotheses derived from the physical situation; and (4) to encounter familiar concepts and relationships in an unfamiliar setting

Procedures: The activity should be conducted as a combination of total-class investigation and small, cooperative group activity. At the beginning, discussion with the whole class (at least for sheet 1 and the first part of sheet 2) should be employed to assure that students understand the workings of the electrical circuits and the symbols used in the circuit diagrams and tables. This approach is especially effective if an actual circuit board is available.

The worksheets should be distributed one at a time to each student, and opportunity should be afforded for class discussion of solutions following the completion of each sheet. In the early stages of the activity, students will need to construct the charts to represent the state of the bulb under all possible conditions of the switches. Later, some will be able to work abstractly with the symbols and others will continue to rely on completing the charts. Either approach is acceptable, and students should be allowed to proceed according to their individual needs. The activity can be expected to take several class periods.

Sheet 1: By way of introduction, explain that an electrical circuit is an unbroken path (usually a wire) through which electrons can move. Whenever electrons flow through such a path, an electrical current is produced. However, if the path is broken, no current can flow. **Figure 1** represents a simple circuit in which current flows in a wire from a voltage source, such as a battery (represented by ⊣⊢), through a light bulb, and back to the battery. As long as the circuit remains closed, the bulb will be lit. The first two figures on sheet 1 show the effect of adding a switch to a circuit.

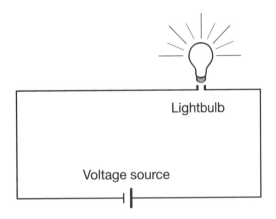

Lightbulb

Voltage source

Fig. 1

After the background, pupils should have little difficulty, with simple examples, in determining the state of the light bulb. Students should conclude that for current to flow in the series circuit, it is necessary that all the switches be closed (A *and* B) but that in a parallel circuit it is sufficient for one of the branches to be a closed path (A *or* B). This qualitative observation is then given the binary symbolism of 0 (switch open; light off) or 1 (switch closed; light on).

A familiar concrete example of the difference between the two types of circuits can be found in strings of Christmas-tree lights. In the old-fashioned tree lights, one burned-out bulb caused the entire string to go out (series wiring). With modern strings of lights, a bulb can burn out and the remainder will stay lit (parallel wiring).

Sheet 2: The symbolism of 0 and 1 introduced in sheet 1 is now further extended by introducing the "•" and "+" symbols to represent series and parallel cases, respectively. Also, a third switch is introduced, thus expanding the number of combinations that must be checked in completing the charts. The discussion should address the question of why three switches yield eight combinations, and later generalizations can point to the conclusion that n switches will entail $2n$ combinations.

Note that in completing the chart for exercise 9, column 4 requires that one consider the effect on the bulb due to the A–B branch alone, as though switch C was not in the circuit at all. Column 6 of the chart then regards the effect of the A–B branch as though it was a single switch in state 0 or 1 (from column 4) and asks for the effect of this single element in parallel with switch C. A similar relationship exists between columns 5 and 7 of the chart.

The concept of an equivalent circuit should be clarified with the class. Two circuits are considered equivalent if the effect on the bulb is the same for all possible off-on combinations of the switches. Thus, for example, the circuits in exercises 7b and 8 all are equivalent, since they produce the same charts. This equivalence becomes the basis for establishing an associative property for the circuits.

Sheet 3: The completion of exercise 12 leads to the establishment of the distributive properties. By completing charts for each of the four circuits, students will see that the first two circuits are equivalent, as are the last two. Note that this result produces *two* distributive laws:

$$A + (B \cdot C) = (A + B) \cdot (A + C)$$
$$A \cdot (B + C) = (A \cdot B) + (A \cdot C)$$

The significance of this double distributive property should be discussed and contrasted with the more familiar distributive property for real numbers. The circuits depicted in exercise 13 illustrate the conclusions that A • (A + B) = A and A + (A • B) = A. These new properties have no counterparts in the algebra of real numbers. They say that the effect on the circuits in both cases depends only on the state of switch A, regardless of the state of switch B. The idea of the relevance or irrelevance of various switches in a circuit allows us to simplify circuit diagrams by removing unnecessary switches. This concept is explored further in the next activity sheet.

Sheet 4: The fourth sheet introduces the idea of complements: two switches with the relationship that whenever one is open, the other must be closed, and vice versa. A pair of complementary switches in series always produces an open circuit, just as the combination of a statement and its negation (A and –A) always produces a contradiction. Likewise, complementary switches in parallel always produce a closed circuit and, thus, are analogous to the statement "A or not A," which is always true. For exercise 15, pupils should be encouraged to begin to work symbolically and to apply principles already derived, as was done in the example. Note that the simplification of the circuit in this example makes use of a distributive property and the property of complements. The "1" in the expression $1 \cdot (B + A)$ can be interpreted as meaning a closed path in series with the B–A parallel path. You may wish to have students draw the simplified circuit diagrams. This experience will serve as a nice bridge to exercises 16 and 17 on sheet 5.

Sheet 5: This final sheet focuses on a synthesis of the results of the investigations in the preceding sheets and a comparative analysis of the algebra of circuits and the algebra of real numbers. It is recommended that pupils first work alone or in pairs so that each is responsible for producing personal responses to the questions. Total class discussion should follow to assure that essential concepts are correctly and adequately presented.

Extensions: This activity is rich in potential follow-up projects in both science and mathematics. Some students could be encouraged to research actual situations in which series (parallel) circuits are used and discuss reasons for the choice of circuit design. Other students may enjoy actually wiring circuits represented in this activity.

Two important relationships that were not investigated here are knows as De Morgan's laws. They are stated as follows:

$$(A \cdot B)' = A' + B' \qquad (A + B)' = A' \cdot B'$$

You might suggest that some students create circuit diagrams and charts to establish each of these laws and prepare a report for the class.

Several possible extensions of these activities can be made to other areas of mathematics that are particularly appropriate for secondary school students. A unit in symbolic logic, for example, can be organized using the circuit example as a concrete model for true and false propositions and their composition. Other topics that can flow from this activity include set theory (with intersection and union analogous to series and parallel circuits, respectively); the structure of algebraic groups; binary arithmetic; Boolean algebra; the applications of direct and inverse variation in electrical circuits (known as Ohm's law); and the application of binary systems in computers.

ANSWERS

Sheet 1: 1) Unlit, unlit, unlit; 2) Lit, lit, lit, unlit; 3)

Series Circuit			Parallel Circuit		
A	B	Bulb	A	B	Bulb
1	1	1	1	1	1
1	0	0	1	0	1
0	1	0	0	1	1
0	0	0	0	0	0

4) When both switches are closed; 5) When at least one switch is closed.

Sheet 2: 6) a. See charts for exercise 3; b. Breaks the circuit, bulb is unlit; c. Completes the circuit, bulb is lit. 7) a.

A	B	C	(A · B · C)
1	1	1	1
1	1	0	0
1	0	1	0
1	0	0	0
0	1	1	0
0	1	0	0
0	0	1	0
0	0	0	0

b.

A	B	C	(A + B + C)
1	1	1	1
1	1	0	1
1	0	1	1
1	0	0	1
0	1	1	1
0	1	0	1
0	0	1	1
0	0	0	0

c. Yes. In the case of the series circuit the bulb is unlit whenever a switch is open; in the case of the parallel circuit the bulb is lit whenever at least one switch is closed; 8) A + (B + C).

9)

A	B	C	(A+B)	(B+C)	(A+B)+C	A+(B+C)
1	1	1	1	1	1	1
1	1	0	1	1	1	1
1	0	1	1	1	1	1
1	0	0	1	0	1	1
0	1	1	1	1	1	1
0	1	0	1	1	1	1
0	0	1	0	1	1	1
0	0	0	0	0	0	0

10) Yes; parallel circuit in exercise 7.

Sheet 3: 11) a. $(A \cdot B) \cdot C = A \cdot (B \cdot C)$

b.

A	B	C	(A·B)	(B·C)	(A·B)·C	A·(B·C)
1	1	1	1	1	1	1
1	1	0	1	0	0	0
1	0	1	0	0	0	0
1	0	0	0	0	0	0
0	1	1	0	1	0	0
0	1	0	0	0	0	0
0	0	1	0	0	0	0
0	0	0	0	0	0	0

12) a.

A	B	C	A+(B·C)	(A+B)·(A+C)	A·(B+C)	(A·B)+(A·C)
1	1	1	1	1	1	1
1	1	0	1	1	1	1
1	0	1	1	1	1	1
1	0	0	1	1	0	0
0	1	1	1	1	0	0
0	1	0	0	0	0	0
0	0	1	0	0	0	0
0	0	0	0	0	0	0

b. $A+(B \cdot C)=(A+B) \cdot (A+C)$; $A \cdot (B+C)=(A \cdot B)+(A \cdot C)$;

13) a.

A	B	A·(A+B)	A+(A+B)
1	1	1	1
1	0	1	1
0	1	0	0
0	0	0	0

b. $A \cdot (A+B) = A$; $A+(A \cdot B) = A$; or $A \cdot (A+B) = A+(A \cdot B)$;

Sheet 4: 14) a.

A	A′	A·A′	A + A′
1	0	0	1
0	1	0	1

b. $A \cdot A' = 0$; $A+A' = 1$

15) a. $(A' \cdot B) + (A \cdot B) + (A \cdot B') = (A' + A) \cdot B + (A \cdot B') = 1 \cdot B + (A \cdot B) = B + (A \cdot B')$, or simply $B + A$; b. $(A' + B') \cdot (A' + B) \cdot (A+B) = [A' + (B' \cdot B)] \cdot (A+B) = [A' + 0] \cdot (A+B) = A' \cdot (A+B) = A' \cdot A + A' \cdot B = 0 + A' \cdot B = A' \cdot B$; *c* $(A \cdot B) + (A \cdot B') + (A' \cdot B) + (A' + B') = A \cdot (B+B') + A' \cdot (B+B') = A \cdot 1 + A' \cdot 1 = A + A' = 1$.

Sheet 5: 16)

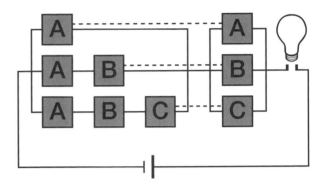

17)

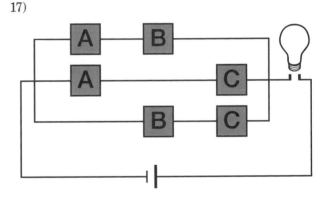

18) Yes;

A	B	A·B	B·A	A+B	B+A
1	1	1	1	1	1
1	0	0	0	1	1
0	1	0	0	1	1
0	0	0	0	0	0

From the circuit chart it can be seen that $A \cdot B = B \cdot A$ and $A+B = B+A$.

19) a. $a(b+c) = ab + ac$; b. Yes, $A \cdot (B+C) = (A \cdot B) + (A \cdot C)$; c. No; d. Yes; $A+(B \cdot C) = (A+B) \cdot (A+C)$;

20) a. A, A; b No

21) a. Circuit is always open; b. Circuit is always closed; c. No.

REFERENCES

House, Peggy A. *Interactions of Science and Mathematics*. Columbus, Ohio: ERIC Clearinghouse for Science, Mathematics, and Environmental Education, 1980. (Available from the School Science and Mathematics Association)

National Council of Teachers of Mathematics (NCTM). *An Agenda for Action: Recommendations for School Mathematics of the 1980s.* Reston, Va.: NCTM, 1980.

Introduction to Electrical Circuits

Sheet 1

The figures below illustrate the two possible states of a switch in a simple electrical circuit.

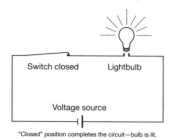

"Closed" position completes the circuit—bulb is lit.

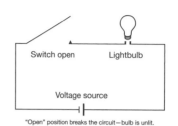

"Open" position breaks the circuit—bulb is unlit.

1. The following figure represents a series circuit, since the current must flow through both switches A and B in sequence in order to light the bulb. Complete the chart to the right of the figure to indicate the condition of the bulb as A and B are opened or closed independently.

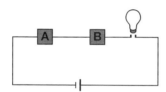

Switch A	Switch B	State of Bulb
Closed	Closed	Lit
Closed	Open	
Open	Closed	
Open	Open	

2. The figure below represents a parallel circuit with two switches, A and B. Complete the chart to show the state of the bulb in this parallel circuit as A and B are opened or closed.

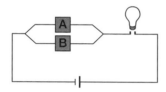

Switch A	Switch B	State of Bulb
Closed	Closed	
Closed	Open	
Open	Closed	
Open	Open	

3. Let us use the symbols 0 to represent an open switch (since no current can flow) and 1 to represent a closed switch (current flowing). Similarly, let 0 indicate that the bulb is not lit, and let 1 indicate that it is lit.

Series Circuit			Parallel Circuit		
A	B	Bulb	A	B	Bulb
1	1	1	___	___	___
1	0	___	___	___	___
___	___	___	___	___	___
___	___	___	___	___	___

Using this representation, translate the information in the completed charts into symbolic form and record the results in the corresponding chart above.

4. Under what conditions will current flow completely (and hence light the bulb) in a series circuit?_____

_____

5. Under what conditions will current flow to the bulb in a parallel circuit?_____

_____

Representing Current Flow

6. Let us use the symbol " • " to represent current flowing in a series circuit and " + " to represent current flowing in a parallel circuit.

    a. Use this new convention together with the information you recorded in exercise 3 to complete the following charts.

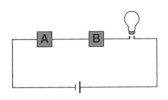

Series Circuit

$1 \cdot 1 = 1$

$1 \cdot 0 = $ ___

$0 \cdot 1 = $ ___

$0 \cdot 0 = $ ___

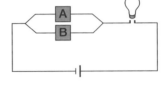

Parallel Circuit

$1 + 1 = 1$

$1 + 0 = $ ___

$0 + 1 = $ ___

$0 + 0 = $ ___

    b. What is the effect of an open switch in a simple series circuit?

    c. What is the effect of a closed switch in a simple parallel circuit?

7. For each circuit, complete the corresponding chart to determine the state of the bulb.

    a.

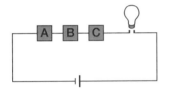

A	B	C	(A • B • C)
1	1	1	___
1	1	0	___
1	0	1	___
1	0	0	___
0	1	1	___
0	1	0	___
0	0	1	___
0	0	0	___

    b.

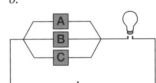

A	B	C	(A + B + C)
1	1	1	___
1	1	0	___
1	0	1	___
1	0	0	___
0	1	1	___
0	1	0	___
0	0	1	___
0	0	0	___

    c. Do your answers for exercises 6b and 6c hold when the electrical circuit has three switches? _____ Explain.

8. If we represent the parallel circuit on the left below as (A + B) + C, how should we represent the parallel circuit on the right?

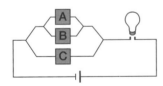

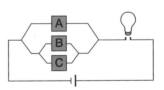

9. On a separate sheet of paper, complete a chart as in exercise 7b for the two circuits above. Use the following column headings.

A	B	C	(A+B)	(B+C)	(A+B)+C	A+(B+C)

10. Two circuits are *equivalent* provided they have the same effect on the light bulb for every line in the circuit chart. Are the two parallel circuits in exercise 8 equivalent? _____ Are they equivalent to any other circuit on this sheet? _____

Complex Circuits

11. In exercise 9 you verified the *associative* property for parallel circuits: (A + B) + C = A + (B + C).

    a. Formulate in symbols a corresponding state for series circuits. _____

    b. Make a chart similar to the one you made for exercise 9 to verify the property formulated in 11a.

12. More complex circuits are depicted below. The dotted lines in the second and fourth figures indicate that the two switches are coupled together so that they open and close simultaneously. These switches are given the same name because whenever one is closed the other is also closed.

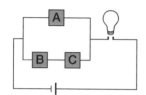 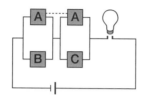 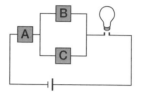 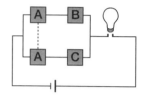

    a. Complete the portion of the chart corresponding to each figure. In the last two figures, supply the corresponding column headings as well.

A	B	C	A + (B · C)	(A + B) · (A + C)		
1	1	1				
1	1	0				
1	0	1				
1	0	0				
0	1	1				
0	1	0				
0	0	1				
0	0	0				

    b. Analyze the chart above and state two additional properties of circuits. _____
    _____

13. a. Describe each circuit below symbolically. Use these representations as the corresponding column headings for the chart to the right and then complete this chart.

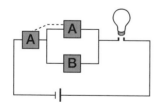 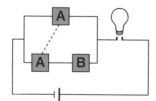

A	B		
1	1		
1	0		
0	1		
0	0		

    b. Write two generalizations suggested by an analysis of this chart. _____
    _____

Additional Complex Circuits

14. Another type of circuit switch is a "double-throw switch." In this situation, two switches are connected so that whenever one is closed, the other is open, and vice versa. We will represent such switches in pairs A and A'. The following will always be true of these pairs:

If A = 1, then A' = 0.
If A = 0, then A' = 1.

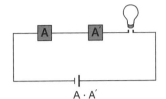
$$A \cdot A'$$

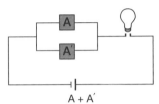
$$A + A'$$

    a. On a separate sheet of paper, make a chart for each of these circuits.
    b. In general, A · A' = _____ and A + A' = _____ .

15. Describe each of the following circuits symbolically. Use the principles you have derived in this activity to simplify the expressions and thus the design of the circuits.

Example

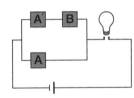

$$\begin{aligned}(A' \cdot B) + A &= (A'+A) \cdot (B + A) \\ &= 1 \cdot (B + A) \\ &= B + A\end{aligned}$$

    a.

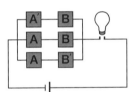

$$(A' \cdot B) + (A \cdot B) + \underline{\hspace{1cm}} = \underline{\hspace{1cm}}$$

    b.

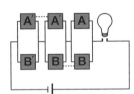

$$(A' + B') + (A \cdot B) \cdot \underline{\hspace{1.5cm}} \cdot \underline{\hspace{1cm}} = \underline{\hspace{1cm}}$$

    c.

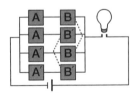

Pulling It Together

16. Draw the circuit diagram represented by

$$[A + (A \cdot B) + (A \cdot B \cdot C)] \cdot [A + B + C]$$

17. A panel of three judges is to review applicants for a talent show, and each judge will vote electronically either to accept or reject the potential contestant. Draw a diagram of a circuit that will give a signal of acceptance whenever the majority of judges votes in favor of the contestant.

The circuits you have investigated and the symbolic representations given them form a special mathematical structure called a Boolean algebra after its developer, the British mathematician George Boole (1815–1864). The following questions will help you put together some of the similarities and some of the differences between Boolean algebra and the algebra of real numbers.

18. In exercises 9 and 11 you verified the associative properties for " + " and " · " in circuit algebra. This property also holds for real-number addition and multiplication. Another important property is the *commutative* property. Is this a property found in the algebra of circuits? _____ On a separate sheet of paper, explain your answer in terms of circuit diagrams and their charts.

19. *a.* State the *distributive property for multiplication over addition* of real numbers. _____

    b. Does a similar property exist for circuit algebra? _____

    c. Does there exist a distributive property for addition over multiplication of real numbers? _____

    d. Does the property in part c exist in circuit algebra? _____

20. *a.* What conclusions can you draw about the following two cases in circuit algebra?

        $A \cdot (A + B) =$ _____ $A + (A \cdot B) =$ _____

    b. Do there exist corresponding properties in the algebra of real numbers? _____
_____

21. Pairs of switches such as A and A′ are called complementary.

    a. What is the effect of complementary switches in a series circuit (A · A′)? _____

    b. What is the effect in a parallel circuit (A + A′)? _____

    c. Do corresponding properties exist in the algebra of real numbers? _____

Making $-x$ Meaningful

David R. Johnson

How do your algebra students read the symbol $-x$? Common responses are "negative x," "minus x," "the opposite of x," or "the additive inverse of x." The most common response is "negative x." But are these responses meaningful? Definitely not! In fact, the first two responses are very misleading, if not incorrect. In many classrooms, teachers are quite careful to name a real number less than zero (or to the left of zero on the real-number line) a *negative number*. Students quite easily grasp the meaning of the phrase *negative number*. But suddenly we bring out the expression $-x$ and read it "negative x"! Trouble begins. Students immediately assume that this symbol stands for a number less than zero simply because its verbal name contains the word *negative.*

If you don't believe that "negative x" is a misleading name, try writing the symbol $-x$ on the board and asking your students how it relates to zero on the number line. You will probably hear a resounding cry, "to the left of zero!"

Too many students do not understand that if x is a real number, $-x$ could be positive, negative, or zero, and that more information is needed before a decision can be made. Students are so inured to the use of the word *negative* and the definition of a negative number that they do not appreciate the meaning of the expression $-x$. Students do appreciate, however, that additive inverses or opposites do not have to be negative. We would probably be better off if we taught our students to read the expression $-x$ as "the opposite of x." in fact, the main point of this article is that discontinuing the use of the term *negative* would be beneficial.

If we do not read $-x$ as "the opposite of x," the problem is further complicated when we teach the definition of absolute value. Though the concept of absolute value can be defined in many homespun ways, some are confusing and often incorrect, for example: "The absolute value can be found by dropping off the sign." That idea is deadly. If b is less than zero, then $|b|$, in this instance, equals $-b$. No sign was chopped off. In fact, one was added. When it comes to the definition of such terms as *absolute value*, a mathematically sound definition is necessary. See definition 1.

Definition 1. For all real numbers a, (1) if $a = 0$, then $|a| = 0$, so a remains unchanged; (2) if $a > 0$, $|a| = a$, so a remains unchanged; (3) if $a < 0$, $|a| = -a$, so the result is the opposite of a.

This definition is difficult for students to understand. First, they must know the size of the real number in relationship to zero. Secondly, the student must understand that $-x$ is simply a symbol for the "inverse of x" and obeys the property of trichotomy. That is, $-x$ could be positive, negative, or zero. If, for example, a student is asked to define the "absolute value of a" where a is less than zero, it follows that the $|a|$ equals $-a$. But for students who believe that a negative sign must be dropped to obtain the absolute value of a number, or for those who do not appreciate that in this instance "$-a$" is really a positive number, correctly applying the definition of absolute value to this expression will be difficult.

Students will do well on the examples that follow if they have a good understanding of the definition of absolute value and if they understand that the symbol $-x$ represents the inverse of x.

Simplify the following:

1. $|-b^3|$, if $b < 0$.
    (*Answer:* $-b^3$, because $(-b) > 0$)

2. $|-3 - x|$, if $x > 0$
    (*Answer:* $-(-3 - x)$, because $(-3 - x) < 0$)

3. $|-b| + |-b|$, if $b < 0$
    (*Answer:* $-2b$, because $(-b) > 0$)

4. $|-3b|$, if $b < 0$
    (*Answer:* $-3b$, because $3b < 0$)

Expressions with variables should be introduced when the concepts of absolute value are taught in first-year algebra. Using only constants in your examples may lead to students' using poor techniques to simplify absolute-value expressions. That is, students may be able to get the correct answer but never realize that they do not understand the definition.

Practice in determining the size of the expression prior to teaching an algebraic definition of absolute value will help make the definition more meaningful. See the activity sheets for a series of questions that will help students understanding of $-x$.

We should read the expression $-x$ as the "opposite of x" and insist that students do the same. The correct reading of this expression should begin in the early grades because an incorrect reading of the symbol is not easily changed. A student's use of "negative x" complicates and confuses the concepts that an algebra teacher must teach. It's time to tell it like it is!

Understanding the Meaning of –x

Consider the following questions:

1. What is the value of the expression $(x - 6)$ if $x < 0$?

    a. Always less that 0._____

    b. Always greater than 0._____

    c. Zero._____

    d. Sometimes less than and sometimes greater than zero.

On a number line place "$(x - 6)$" to the left of zero:

That is, if $x < 0$, then $(x - 6) < 0$. $(x - 6)$ is a negative number for any value of x that is negative.

$$x - 6 \qquad\qquad 0$$

2. Determine the sizes of the following expressions given the information about the values of the variables. Place a check in the appropriate column.

Expression	Value of Variable	Less than Zero	Equal to Zero	Greater than Zero
1. x^3	$x < 0$			
2. $x^2 + 6$	$x > 0$			
3. $x^2 + 6$	$x < 0$			
4. $x^2 + 6$	$x = 0$			
5. $-3x$	$x < 0$			
6. $5y$	$y < 0$			
7. $-5x + y$	$x < 0,\ y > 0$			
8. $(x - 14)^2$	$x < 0$			
9. $(x - 14)^2$	$x > 0$			
10. $(x - 14)^2$	$x = 0$			
11. $-f$	$f < 0$			
12. $-f^2$	$f > 0$			
13. $-f + (-g)$	$f < 0,\ g < 0$			
14. $f^3 + (-2g)$	$f > 0,\ g < 0$			
15. $-3x^2$	$x > 0$			
16. $-2(x - 1)^2$	$x < 0$			

Understanding the Meaning of $-x$ Sheet 2

3. Given the following information, place the nonzero real numbers $-f, g,$ and h on the proper side of zero and in the proper order on a real-number line. Assume the following: $f < 0, f < g, g < 0, -h > -f, -h > 0, h < f.$

$$0$$

Using this information, decide on which side of zero the following numbers are located:

Expression	Left	Right
f^3		
$f - g$		
$-g^2$		
h^2		
$h + g$		
$(-g) + (-f)$		

Solving Linear Equations Physically

Barbara Kinach

TEACHER'S GUIDE

Introduction: The "Activities" section of the *Mathematics Teacher* first appeared in 1972 as a means of providing classroom teachers ready-to-use discovery lessons and laboratory experiences in worksheet form. The importance and efficacy of these alternative instruction methodologies were reaffirmed in NCTM's *An Agenda for Action* (1980, 12). The Agenda specifically recommended that teachers employ diverse instructional strategies, materials, and resources, including the following:

- The provision of situations that provide discovery and inquiry as well as basic drill.

- The use of manipulatives, where suited, to illustrate or develop a concept or skill.

The following materials provide opportunities for students to investigate solving linear equations using both a physical and a pictorial model and in the process discover a method that will permit them to solve such equations at the symbolic level.

Grade levels: 7–10.

Materials: Scissors, cardboard, a lighter-weight tagboard, and a set of activity sheets for each student. Transparency cutouts of the strips, squares, and balance scale on sheet 4 would be useful for purposes of demonstration and for discussion of students' solutions.

Objectives: Students will (1) represent first-degree polynomials physically (with strips and squares) and pictorially; (2) solve linear equations both physically through the use of a balance scale and pictorially; and (3) discover a method for solving linear equations using algebraic manipulation.

DIRECTIONS

On the day before this lesson, distribute a copy of sheet 4 to each student. As indicated on sheet 1, instruct them to glue the row of strips on a piece of cardboard, glue the rows of squares on a piece of lighter-weight tagboard, and then cut out the pieces. Provide students with an envelope in which to keep their equation-solving pieces.

On the day of the lesson, distribute the worksheets one at a time. All students should be able to complete the first two sheets during a single forty-five-minute class period if the strips and squares have been precut. Depending on the time available, sheet 3 can be completed during the next class period.

Sheet 1

Sheet 1 emphasizes the distinction between a variable and a constant. It is important that the students realize that the strips were purposely constructed so that their weights were a variable quantity in relation to the weights of the squares. In addition, this first worksheet uses the strip-square diagrams to clarify the use of grouping symbols in algebraic expressions. Students thus distinguish the difference between $2S + 3$ and $2(S + 3)$ pictorially.

Sheet 2

Sheet 2 establishes the analogy between a balancing scale and solving a linear equation of the form $ax + b = cx + d$ where a, b, c, and $d \geq 0$. In this activity pupils first attempt to maintain the scale's balance while replacing each strip with the same number of squares. This experience reinforces the conditional nature of an equation. Specifically, it demonstrates that the asserted equality between two expressions need not hold for all replacements of the variable S. The second portion of this sheet provides an algorithm for physically determining the value for S that will maintain the scale's balance. Students may solve the equations in exercise 8 by making pictorial representations of their physical manipulations or by using standard symbolic methods. At this stage, either method would be accepted. You might also have students verify their solutions for exercises 7 and 8 by actually substituting the values obtained for S into the equations and then performing the arithmetic indicated.

Sheet 3

Finally, sheet 3 introduces a pictorial method for solving linear equations as a bridge between the physical model and the ultimate goal of algebraic manipulation. It would be instructive to demonstrate how the equation $4S + (-3) = 3S + (-4)$ in the example can also be solved pictorially by adding three white squares to both sides of the configuration in step 2 and using the fact that a white square and a gray square cancel each other. The solution of exercise 9c will require students to add three gray squares to each side of the configuration. Encourage pupils to verify their solutions for exercises 9 and 10 by substituting values obtained for S into the equations and then performing the arithmetic indicated.

After all students have completed sheet 3, carefully establish the algebraic methods for solving linear equations in terms of the corresponding pictorial manipulations. In the example at the top of sheet 3, this correspondence can be illustrated as follows.

Step 1: Represent the equation. See (1).

Step 2: Subtract (remove) 3 gray squares. See (2).

Step 3: Subtract (remove) 3 strips. See (3).

$$
\begin{array}{llll}
(1) & 4S + (-3) & = & 3S + (-4) \\
(2) & \underline{\quad -(-3)} & & \underline{\quad -(-3)} \\
 & 4S & = & 3S + (-1) \\
(3) & \underline{-3S} & & \underline{-3S} \\
 & S & = & -1
\end{array}
$$

Students should note the pictorial distinction made between a negative number and the operation of subtraction. A negative number is represented by a gray square. The operation of subtraction is indicated by crossing out with an "X" the strip or square to be removed. This distinction of notation should remind pupils that the symbol (–) has different meanings.

Supplementary Activities: Introduce notation for subtracting a constant from a variable. For example, to indicate the subtraction $3S - 4$, place (or draw) four white squares on top of the strips,

Note the distinction with the representation of $3S + (-4)$,

Challenge students to solve each of the equations first pictorially and the using the corresponding algebraic manipulations.

    a. $4S - 3 = S + 3$

    b. $5(S - 1) = 2S + 7$

    c. $-3 + 2S = 3(S - 2)$

Later in the year you may wish to use a modification of this strip-square model as described by Hirsch (1982) to factor polynomials physically.

SOLUTIONS

Sheet 1:

2) b.

    c. The rectangular region forms consists of two rows of identical shapes. Since the weight of one row is $S + 3$, the weight of the region is $2(S + 3)$.

3) a.   b.

    c.   d.

    e.   f.

4) a. $3S + 6$;

    b. $4(S + 1)$;

    c. $3S + (-9)$

Sheet 2:

5) b. No

    f. No;

6) d. Three

7) a. $S = 1$

    b. $S = 7$

    c. $S = 5$

    d. $S = 2$

    e. $S = 3$

8) a. $S = 4$

    b. $S = 7$

    c. $S = 2.$

Sheet 3:

9) a.

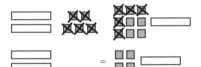

The solution is $S = -4$.

    b.

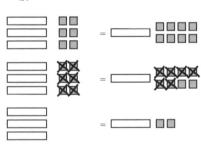

The solution is $S = -1$.

e.

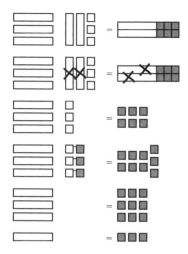

The solution is $S = -3$.

10) a. $S = -4$

   b. $S = 14$.

REFERENCES

Hirsch, Christian R. "Finding Factors Physically." *Mathematics Teacher* 75 (May 1982): 399–93, 419.

National Council of Teachers of Mathematics (NCTM). *An Agenda for Action: Recommendations for School Mathematics for the 1980s.* Reston, Va.: NCTM, 1980.

Solving Linear Equations Physically

Sheet 1

1. Sheet 4 of this activity consists of a series of strips, white and gray squares, and a diagram of a balance scale.

   a. Cut the sheet along the two dashed lines.

   b. Glue the series of strips onto a piece of cardboard, and then carefully cut out the strips.

   c. Glue the series of squares onto a piece of lightweight tagboard, such as a file folder, and then carefully cut out the white and gray squares.

For this activity, assume that the weight of—

   a. A white square is a positive one (+1) unit,

   b. A gray square is a negative one (−1) unit,

   c. A strip is a variable quantity S.

2. The expression $2S + 6$ can be represented by two strips and six white squares.

   a. Place these eight pieces on your desk.

   b. Rearrange the pieces to form a rectangle.

   c. Explain why the weight of these eight pieces can also be expressed as $2(S + 3)$. _____

3. Represent each algebraic expression with strips and squares. Draw diagrams of your solutions in the spaces provided.

   a. $2S + 4$ b. $2S + (−4)$ c. $2(S + 4)$

   d. $3S + (−1)$ e. $3[S + (−1)]$ f. $3(2S + 1)$

4. Write an algebraic expression for each diagram.

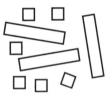   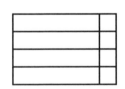   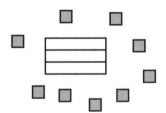

a. _____   b. _____   c. _____

Solving Linear Equations Physically

5. To represent the linear equation $3S + 2 = 2S + 5$ physically, place the strip-square representation of $3S + 2$ on the left pan in your diagram of a balance scale and the corresponding representation of $2S + 5$ on the right pan.

    a. Replace each strip with one white square (that is, assume $S = 1$).

    b. Would your scale remain balanced? _____

    c. Restore the original pieces to the scale.

    d. Now replace each strip with two gray squares (that is, assume $S = -2$).

    e. Simplify by using the fact that a gray square and a white square cancel each other out, since $(-1) + (+1) = 0$.

    f. Describe the contents of the left pan: _____ The right pan: _____
        Does the scale balance? _____

6. To solve the linear equation $3S + 2 = 2S + 5$ physically, we must determine the number of squares that can be used to replace each strip and keep the scale balanced. This can be accomplished by removing *equally weighted* pieces from each side of the scale until you have only one strip remaining on the scale.

    a. Represent the equation $3S + 2 = 2S + 5$ on your balance scale.

    b. Remove two squares from each side.

    c. Now remove two strips from each side.

    d. The weight of one strip equals the weight of _____ white squares. Thus, the solution of the equation is $S = 3$.

7. Use the method in exercise 6 to solve each of the following linear equations. Write your solutions in the spaces provided.

    a. $2S + 4 = S + 5$ $S =$ _____

    b. $4S = 3S + 7$ _____

    c. $5S + 3 = 4S + 8$ _____

    d. $3S = 6$ _____

    e. $4S + 1 = 2S + 7$ _____

8. Try solving each of the following equations without using the materials from sheet 4. Physically check your solutions by using the method in exercise 6.

    a. $S + 5 = 9$ $S =$ _____

    b. $2S + 6 = S + 13$ _____

    c. $5S + 2 + 3S + 6$ _____

Solving Linear Equations Physically

Sheet 3

Since we do not usually think of putting a weight of -1 on a scale, it is helpful also to look at pictorial methods for solving linear equations. For example, to solve the equation $4S + (-3) = 4S + (-4)$ pictorially, we again use the process of moving *equally valued* pieces from (or adding *equally valued* pieces to) each side of the configuration.

Step 1

Step 2

Step 3

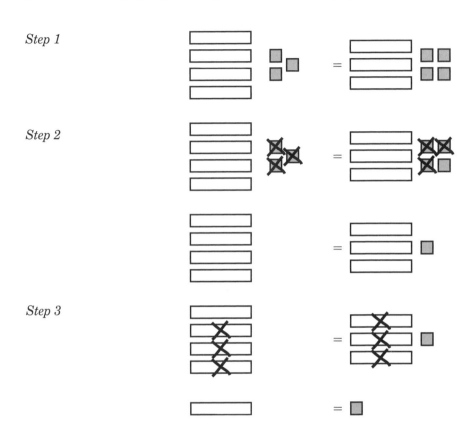

The solution is $S = -1$.

Note that to remove (or subtract) a square or strip, we cross it out with an "X." In step 2 we could have added three white squares to each side instead of removing three gray ones.

9. Use the method described above to solve each of the following linear equations.

   a. $2S + (-5) = -9 + S$

   b. $3S + (-6) = S + (-8)$

   c. $5S + 3 = 2[S + (-3)]$

10. Try solving each of the following equations without using pictorial representations. Check your solutions by using the method shown at the top of this sheet.

    a. $4S + (-2) = 2S + (-10)$ $S=$ _____

    b. $5S + (-6) = 4S + 8$ _____

Solving Linear Equations Physically

Sheet 4

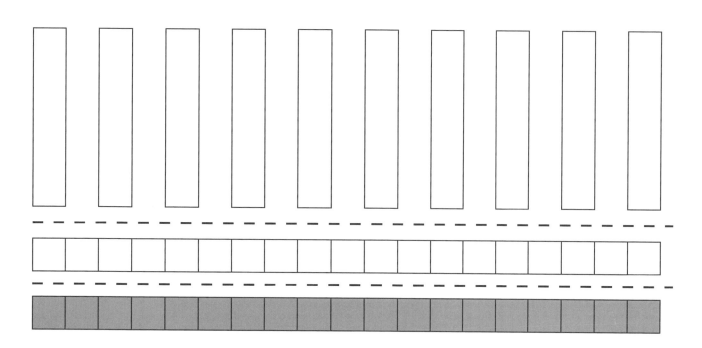

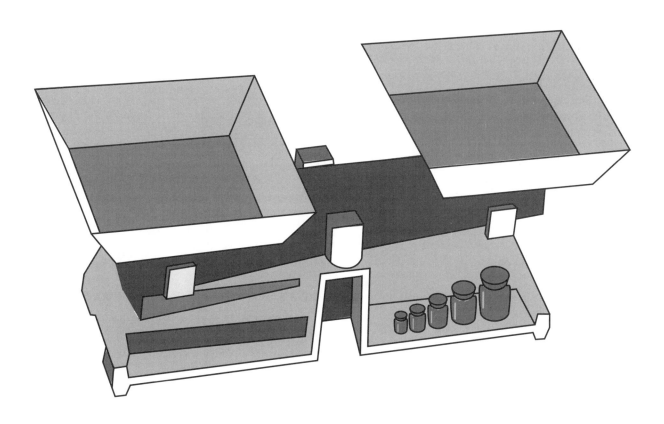

Relations among Powers of 2, Combinations, and Symbolic Algebra

Yukio Kobayashi

The aim of this article is to derive the formula

$$\sum_{k=0}^{n-1} 2^k = \sum_{k=1}^{n} {}_nC_k$$

in addition to the well-known formula

$$2^n = \sum_{k=0}^{n} {}_nC_k,$$

where ${}_nC_k$ is the number of k-element subsets of an n-element set, through an approach that uses symbolic algebra. Let us consider the following problem:

Assume there are six separate and distinct sites fixed in a plane. Each site is marked by ♣ or ♥. In how many ways can the sites be so labeled?

COUNTING ALL THE LABELS

Each site has two states, ♣ or ♥. Hence, the total number of arrangements of the six symbols is $2 \times 2 \times 2 \times 2 \times 2 \times 2 = 2^6$, and there are 2^6 possible states. We can also count the number of sites by classifying them by the number of ♣s they contain.

Arrangements with six ♣s: ${}_6C_6 = 1$
(six sites, choose all six to be ♣)

Arrangements with five ♣s: ${}_6C_5 = 6$
(six sites, choose five to be ♣)

Arrangements with four ♣s: ${}_6C_4 = 15$
(six sites, choose four to be ♣)

Arrangements with three ♣s: ${}_6C_3 = 20$
(six sites, choose three to be ♣)

Arrangements with two ♣s: ${}_6C_2 = 15$
(six sites, choose two to be ♣)

Arrangements with one ♣s: ${}_6C_1 = 6$
(six sites, choose one to be ♣)

Arrangements with no ♣s: ${}_6C_0 = 1$
(six sites, choose none to be ♣)

Total $\sum_{k=0}^{6} {}_6C_k = 64$

And in general,

(1) $\sum_{k=0}^{n} {}_nC_k = 2^n$

We may use the following simple notation for a single state of the system of six sites:

(2)   ♣ ♥ ♥ ♥ ♣ ♥

We can also number the sites in sequence from left to right:

     ○   ○   ○   ○   ○   ○
     1   2   3   4   5   6

This leads to a convenient notation. State (2) can be denoted by

(3) $♣_1 ♥_2 ♥_3 ♥_4 ♣_5 ♥_6$.

Every distinct state of the system is a term in a symbolic product of six factors:

(4) $(♣_1 + ♥_1)(♣_2 + ♥_2)(♣_3 + ♥_3)(♣_4 + ♥_4)(♣_5 + ♥_5)(♣_6 + ♥_6)$.

The (noncommuting) multiplication rule is defined by

(5) $(♣_1 + ♥_1)(♣_2 + ♥_2) = ♣_1 ♣_2 + ♣_1 ♥_2 + ♥_1 ♣_2 + ♥_1 ♥_2$.

If product (4) is expanded using this rule, all the terms are distinct, and there will be 2^6 terms. More generally, if we have n factors, there will be 2^n terms.

ANOTHER EXPRESSION OF POWERS OF 2

2^6 can be written as

$$
\begin{aligned}
2^6 &= 2 \times 2^5 = 2^5 + 2^5 \\
&= 2^5 + 2 \times 2^4 = 2^5 + 2^4 + 2^4 \\
&= 2^5 + 2^4 + 2 \times 2^3 = 2^5 + 2^4 + 2^3 + 2^3 \\
&\vdots \\
\end{aligned}
$$

(6) $= 2^5 + 2^4 + 2^3 + 2^2 + 2^1 + 2^0 + 1.$

Our symbolic product is convenient for visualizing equation (6).

$(♣_1 + ♥_1)(♣_2 + ♥_2)(♣_3 + ♥_3)(♣_4 + ♥_4)(♣_5 + ♥_5)(♣_6 + ♥_6)$ 2^6 terms

$= ♣_1(♣_2 + ♥_2)(♣_3 + ♥_3)(♣_4 + ♥_4)(♣_5 + ♥_5)(♣_6 + ♥_6)$ 2^5 terms,
                                   all starting with $♣_1$

$+ ♥_1(♣_2 + ♥_2)(♣_3 + ♥_3)(♣_4 + ♥_4)(♣_5 + ♥_5)(♣_6 + ♥_6)$ 2^5 terms,
                                   all starting with $♥_1$

Leave the first summand alone (with 2^5 terms, all starting with $♣_1$), and expand the second summand. Then keep doing this, each time keeping the first summand and expanding the second one:

$$= \clubsuit_1(\clubsuit_2 + \heartsuit_2)(\clubsuit_3 + \heartsuit_3)(\clubsuit_4 + \heartsuit_4)(\clubsuit_5 + \heartsuit_5)(\clubsuit_6 + \heartsuit_6) \; 2^5$$

$$+ \; \heartsuit_1\clubsuit_2(\clubsuit_3 + \heartsuit_3)(\clubsuit_4 + \heartsuit_4)(\clubsuit_5 + \heartsuit_5)(\clubsuit_6 + \heartsuit_6) \qquad 2^4 \text{ terms,}$$

all starting with $\heartsuit_1\clubsuit_2$

$$+ \; \heartsuit_1\heartsuit_2(\clubsuit_3 + \heartsuit_3)(\clubsuit_4 + \heartsuit_4)(\clubsuit_5 + \heartsuit_5)(\clubsuit_6 + \heartsuit_6) \qquad 2^4 \text{ terms,}$$

all starting with $\heartsuit_1\heartsuit_2$

$$\vdots$$

$$= \clubsuit_1(\clubsuit_2 + \heartsuit_2)(\clubsuit_3 + \heartsuit_3)(\clubsuit_4 + \heartsuit_4)(\clubsuit_5 + \heartsuit_5)(\clubsuit_6 + \heartsuit_6) \; 2^5$$

$$+ \; \heartsuit_1\clubsuit_2(\clubsuit_3 + \heartsuit_3)(\clubsuit_4 + \heartsuit_4)(\clubsuit_5 + \heartsuit_5)(\clubsuit_6 + \heartsuit_6) \qquad 2^4$$

$$+ \; \heartsuit_1\heartsuit_2\clubsuit_3(\clubsuit_4 + \heartsuit_4)(\clubsuit_5 + \heartsuit_5)(\clubsuit_6 + \heartsuit_6) \qquad 2^3$$

$$+ \; \heartsuit_1\heartsuit_2\heartsuit_3\clubsuit_4(\clubsuit_5 + \heartsuit_5)(\clubsuit_6 + \heartsuit_6) \qquad 2^2$$

$$+ \; \heartsuit_1\heartsuit_2\heartsuit_3\heartsuit_4\clubsuit_5(\clubsuit_6 + \heartsuit_6) \qquad 2^1$$

$$+ \; \heartsuit_1\heartsuit_2\heartsuit_3\heartsuit_4\heartsuit_5\clubsuit_6 \qquad 2^0$$

$$(7) + \; \heartsuit_1\heartsuit_2\heartsuit_3\heartsuit_4\heartsuit_5\heartsuit_6 \qquad 1$$

Combining results (1) and (7), we have

$$2^6 = \sum_{k=0}^{6} {}_6C_k = 1 + \sum_{k=0}^{5} 2^k$$

and more generally

$$(8) \qquad 2^n = \sum_{k=0}^{n} {}_nC_k = 1 + \sum_{k=0}^{n-1} 2^k$$

Additionally, by excluding $\heartsuit_1\heartsuit_2\heartsuit_3\heartsuit_4\heartsuit_5\heartsuit_6$ from our symbolic product, we can find another formula:

$$(9) \qquad \sum_{k=0}^{n-1} 2^k = \sum_{k=1}^{n} {}_nC_k$$

This formula may be also derived from the sum of a geometric series,

$$\sum_{k=0}^{n-1} 2^k = 2^n - 1.$$

A comparison of

$$2^n = 1 + \sum_{k=0}^{n-1} 2^k$$

with

$$2^n = \sum_{k=0}^{n} {}_nC_k$$

also yields equation (9). It is instructive to consider various approaches in solving a problem. The symbolic expansion is a nice example of how one can use algebra as a tool to think about counting problems.

SOLUTIONS

1)

Number of $\heartsuit$s	Combinatorial Symbol	Number of States
3	${}_3C_3$	1
2	${}_3C_2$	3
1	${}_3C_1$	3
0	${}_3C_0$	1
Total	$\sum_{k=0}^{3} {}_3C_k$	8

2) a. $\clubsuit_1\clubsuit_2\clubsuit_3$
   b. $\clubsuit_1\clubsuit_2\heartsuit_3$
   c. $\clubsuit_1\heartsuit_2\clubsuit_3$
   d. $\clubsuit_1\heartsuit_2\heartsuit_3$
   e. $\heartsuit_1\clubsuit_2\clubsuit_3$
   f. $\heartsuit_1\clubsuit_2\heartsuit_3$
   g. $\heartsuit_1\heartsuit_2\clubsuit_3$
   h. $\heartsuit_1\heartsuit_2\heartsuit_3$

3)
$$\begin{aligned} 2^3 &= 2 \times 2^2 \\ &= 2^2 + 2^2 \\ &= 2^2 + 2 \times 2^1 \\ &= 2^2 + 2^1 + 2^1 \\ &= 2^2 + 2^1 + 2 \times 2^0 \\ &= 2^2 + 2^1 + 2^0 + 1 \end{aligned}$$

4)
$$2^3 = {}_3C_3 + {}_3C_2 + {}_3C_1 = 2^2 + 2^1 + 2^0$$

REFERENCE

Kittel, Charles, and Herbert Kroemer. *Thermal Physics.* San Francisco: W. H. Freeman, 1980.

Understanding Powers of 2

Consider the following questions:

1. Each of three separate and distinct sites fixed in a plane has two states, ♣ or ♥. Classify 23 possible states by the number of ♥s they contain.

Number of ♥s	Combinatorial Symbol	Number of States
3		
2		
1		
0		
Total	$\displaystyle\sum_{k=?}^{?} {}_?C_k$	

2. Expand the symbolic product of these three factors: $(\clubsuit_1 + \heartsuit_1)(\clubsuit_2 + \heartsuit_2)(\clubsuit_3 + \heartsuit_3)$

= a. _____ + b. _____ + c. _____ + d. _____

+ e. _____ + f. _____ + g. _____ + h. _____

3. Fill in the blanks.

$$2^3 \;=\; 2 \times 2^{\Box}$$
$$=\; 2^{\Box} + 2^{\Box}$$
$$=\; 2^{\Box} + 2 \times 2^{\Box}$$
$$=\; 2^{\Box} + 2^{\Box} + 2^{\Box}$$
$$=\; 2^{\Box} + 2^{\Box} + 2 \times 2^{\Box}$$
$$=\; 2^{\Box} + 2^{\Box} + 2^{\Box} + 1$$

4. Fill in the blanks.

$$2^3 \;=\; {}_\Box C_\Box + {}_\Box C_\Box + {}_\Box C_\Box = 2^{\Box} + 2^{\Box} + 2^{\Box}$$

Math Magic

Miriam A. Leiva

TEACHER'S GUIDE

Grade level: 7–10.

Materials: One set of worksheets for each student. Calculators would be helpful for sheet 3.

Objectives: To generate interest in mathematics while providing arithmetical experiences that lead to the introduction of the concept of a variable.

Directions: Distribute copies of the activity sheets, one at a time, to each student. Sheet 1 may be used independently of the other sheets to provide self-checking computational practice by having pupils complete the tricks using additional numbers, perhaps including fractions. You may wish to indicate that using many numbers to verify a trick do not, however, assure us that it will *always* work. Sheet 2 introduces us to the notion of a variable as a means to discover why the tricks work as they do. Within this context, a variable is simply a symbol used to denote any one of a given set of numbers. Sheet 3 provides opportunities for pupils to exercise a little "magic" of their own. They should be encouraged to verify their tricks using numbers and then to show that they will always work by using a variable. The final trick on this sheet should convince students that an understanding of the use of variables is not sufficient, however, for removing all the magic from mathematics. Numbers, other than powers of 2, frequently produce long sequences of results. For example, the choice of the number 9 produces a sequence of eighteen results before a 1 is obtained. Pupils who have had an introduction to computer programming should be encouraged to write a program that will carry out the instructions. This will expedite the testing of more and larger numbers. An analysis of the output will yield some interesting patterns. A discussion of some of these patterns may be found in Nievergelt, Farrar, and Reingold (1974).

SOLUTIONS

1) The results will always be the number chosen.

2) The result of this trick will always be 2.

3) The result will always be 6.

4) n, $3n$, $3n + 8$, $4n + 8$, $n + 2$, 2, n, $2n + 1$, $2n + 12$, $n + 6$, 6

5) Subtract 18, divide by 3.

6) Answers will vary.

7) It is an unproved conjecture that the instructions will always produce a 1.

REFERENCE

Nievergelt, Jürg, Joel Craig Farrar, and Edward M. Reingold. *Computer Approaches to Mathematical Problems*. Englewood Cliffs, N.J.: Prentice-Hall, 1974.

Math Magic

Sheet 1

Recently I found myself in the company of Matt-E-Magic, a skillful magician. He involved me in some interesting math magic tricks, some of which I want to share with you.

1.

Instructions	Number Choices
Choose any number	5 12 36 81
Multiply by 2	10
Add 5	15
Multiply by 5	75
Subtract 25	50
Divide by 10	5

   a.   Follow Matt-E-Magic's instructions for the numbers 12, 36, and 81 above. Write each step as in the example.

   b.   Suppose the magician's instructions were carried out for a decimal such as 0.62. What do you think the final result would be? Carry out the instructions and test your prediction.

   c.   Do you think the result of this number trick will always be the number chosen?

2.

Instructions	Number Choices
Choose any number	9 22 −5 0.3
Multiply by 3	27
Add 8	35
Add your original number choice	44
Divide by 4	11
Subtract your original number choice	2

   a.   Follow Matt-E-Magic's instructions for the other numbers, 22, −5, and 0.3.

   b.   Will the results of this magical trick always be 2? Check it one more time with a number of your choice.

Math Magic

3.

Instructions	Number Choices
Choose any number	14 30 −7 0.4
Add the number one larger than your original number choice	29
Add 11	40
Divide by 2	20
Subtract your original number choice	6

a. Follow Matt-E-Magic's instructions for the numbers 30, −7, and 0.4.

b. Do you think that the result of this magical trick will always be 6? Check it one more time using a number of your choice.

What is the secret behind these math tricks?

To find out, let us reconsider the first trick on sheet 1. Suppose we refer to the number chose as "number," or more simply just as n.

Choose any number	Number	n
Multiply by 2	2(number)	$2n$
Add 5	2(number) + 5	$2n + 5$
Multiply by 5	10(number) + 25	$10n + 25$
Subtract 25	10(number)	$10n$
Divide by 10	number	n

Aha! Now it is easy to see why the magic works. Since "number" or n can represent any particular choice of number, the result must always be the original number.

4. Discover the magic behind tricks 2 and 3 by using the word "number" or the letter n for the number chosen in each case. Write each step on a separate line as was done above.

Math Magic

Now it is your turn to work some magic.

5. Each of the math tricks below is missing one instruction. Fill in the instruction so that, in each case, the result will always be the original number chosen.

Instructions	Number Choices
Choose any number	Choose any number
Multiply by 4	Multiply by 2
Add 6	Add 9
Multiply by 3	Add your original number choice
_____	_____
Divide by 12	Subtract 3

6. Make up a math magic trick in which the final result will always be 7, no matter what number is chosen.

7. Your experiences in this activity have dispelled much of the magic in Matt-E-Magic's program. However, he concluded his performance with the following trick, which I have yet to unravel.

Instructions

1. Choose any whole number other than zero.

2. *a.* If the number is even, divide by 2 and then carry out instruction 3.

   b. If the number is odd, multiply by 3, add 1, and then carry out instruction 3.

3. *a.* If your result is 1, stop.

   b. If your result is not 1, repeat instruction 2 using your result as the number.

Example: If your number choice is 5, the sequence of results is $5 \rightarrow 16 \rightarrow 8 \rightarrow 4 \rightarrow 2 \rightarrow 1$ (stop).

No one in the audience could find a number for which the instructions did not eventually produce a 1. Can you?

Teaching Aid for Developing $(a + b)(a - b)$

Carl Uth

Editor's Note: The concept-developing technique included in this contribution is as clever as any that the postman has ever dropped into this department editor's mailbox. So that the reader may experience the "feel" of the development as presented it is suggested that he supply himself with a sheet of 8 1/2 × 11-inch typing paper, and observe the directions in the order listed. Doing this will help the reader understand what the experience may be like for his students. The plan outlined below should prove to be a profitable classroom activity for ninth-grade algebra classes.

The purpose of the procedure outlined in this article is to help students develop an understanding of the concept embodied in the following relation:

$$(a + b)(a - b) = a^2 - b^2$$

The approach is geometric and depends on the area concept. The only materials needed for this development are a rectangular sheet of typing paper, a pencil a straightedge, and a scissors. The plan of the development is contained in the following sequence of steps:

1. Place the sheet of typing paper on a working table or desk so that the reader will be oriented with respect to its position in the same way that it appears to him in **figure 1**.

2. Make the assumption that the shorter dimension is a units and that the longer one is $a + b$ units. To determine b, fold the upper left-hand vertex of the rectangular sheet down to meet the bottom edge (**fig. 2**). Draw a line along edge EE′ as indicated.

3. Turn the upper left-hand vertex of the sheet back to its original position (**fig. 3**). Note that the original rectangle has now been divided into two figures; the figure on the left is a square which is a units on a side, and the figure on the right is a rectangle whose dimensions are b and a. Thus the magnitude of b is determinate. (To facilitate communication the writer has adopted the convention of referring to rectangles with unequal adjacent sides as "rectangles," and of reserving the word "square" for reference to rectangles with equal adjacent sides.)

4. Fold the upper right-hand vertex down as shown in **figure 4**, and with the aid of a straightedge, draw line DD′.

5. Label the various edges and lines with their dimensions as indicated in **figure 5**, and, as a simple expedient for clarifying references, "key" the four figures formed with the letters W, X, Y, and Z. Check to insure that the dimensions recorded are consistent with the assumptions made in step 2.

6. With the aid of a scissors (or by creasing and tearing) separate the three rectangles W, Y, and Z, and square X into four separate pieces as shown in **figure 6**.

7. Rectangle W may be discarded; it is not needed for this development. From the remaining three pieces X, Y, and Z, select two which may be used to form a rectangle which will have dimensions $(a + b)$ and $(a - b)$. This requirement will be satisfied if rectangles Y and Z are placed adjacent to each other in such a way that their common side is $(a - b)$. See **figure 7**.

8. An inspection of **figure 7** will reveal that the area of Y is $a(a - b)$ or $a^2 - ab$, and that the area of Z is $b(a - b)$ or $ab - b^2$.

9. Thus $(a + b)(a - b) = a^2 - b^2$.

The device described in the foregoing steps may also be used to provide a geometric interpretation of the fact that the factors of $a^2 - b^2$ are $(a + b)(a - b)$. Arrange rectangles Y and Z and square X as illustrated in **figure 8**. The area of the three pieces is a^2 certainly. By removing (subtracting) b^2 (keyed X), and again arranging rectangles Y and Z as in **figure 7**, the desired result is immediately apparent.

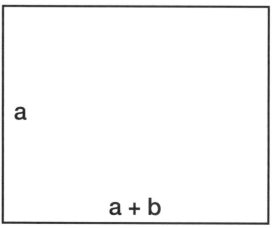

Fig. 1

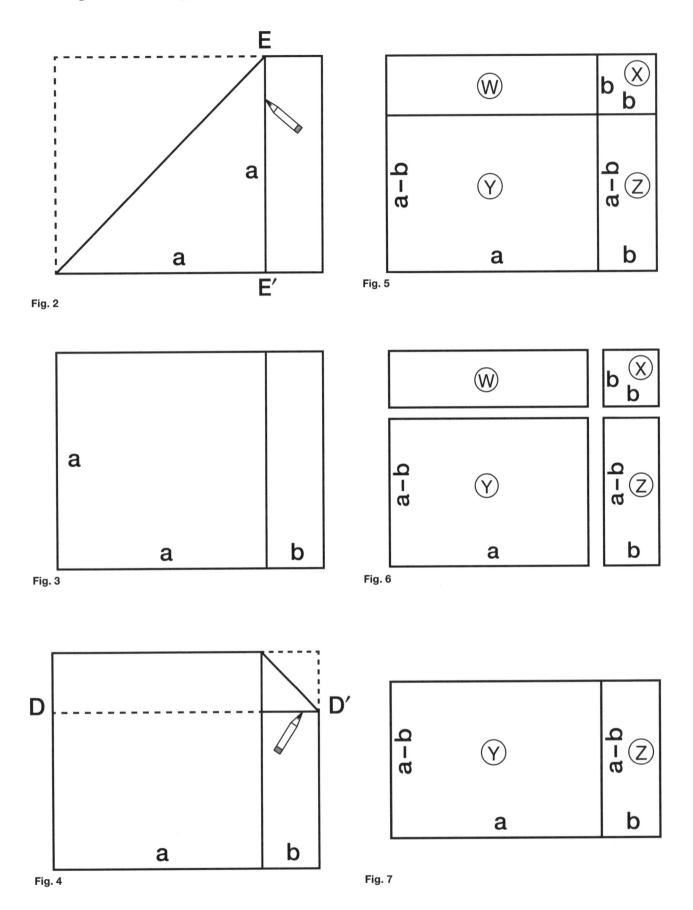

Fig. 2

Fig. 3

Fig. 4

Fig. 5

Fig. 6

Fig. 7

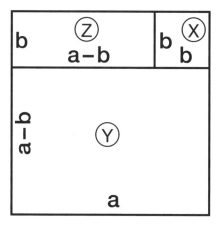

Fig. 8

Developing (*a* + *b*)(*a* − *b*)

1. Begin with a piece of 8 1/2 × 11-inch paper placed horizontally on the table in front of you as it is oriented in **figure 1**. Assume that the shorter dimension is *a* units and the longer one is *a* + *b* units.

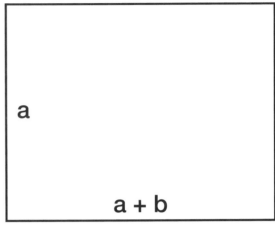

Fig. 1

2. To determine *b*, fold the upper left-hand vertex of the rectangle down to meet the bottom edge (**fig. 2**). Draw a line along edge EE′ as indicated.

3. Turn the upper left-hand vertex of the sheet back to its original position (**fig. 3**). Note that the original rectangle has now been divided into two figures; the figure on the left is a square that is *a* units on a side, and the figure on the right is a rectangle whose dimensions are *b* and *a*. Thus the magnitude of *b* is determined.

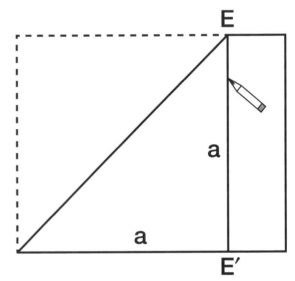

Fig. 2

4. Fold the upper right-hand vertex down as shown in **figure 4**, and with the aid of a straightedge, draw line DD′.

5. Label the various edges and lines with their dimensions as indicated in **figure 5**, and the three rectangles and one square with the letters W, X, Y, and Z. Check to be sure that the dimensions as labeled are consistent with the assumptions in step 2.

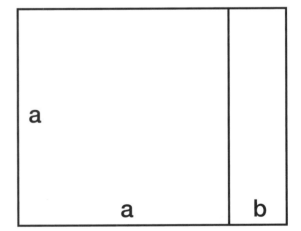

Fig. 3

Developing (a + b)(a − b)

6. Use scissors to separate the three rectangles W, Y, and Z, and the square X into four separate pieces as shown in **figure 6**.

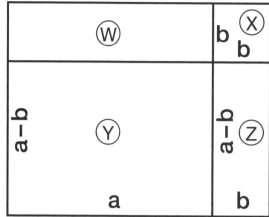

Fig. 5

7. At this point you may discard rectangle W; you will not need it anymore in this activity. From the remaining three pieces X, Y, and Z, select two that may be used to form a rectangle that will have dimensions ($a + b$) and ($a − b$).

Draw the resulting rectangle and label the two rectangles that you used.

This requirement will be satisfied if rectangles Y and Z are placed adjacent to each other in such a way that their common side is ($a − b$). See **figure 7**.

8. What are the areas of each rectangle?_____

9. If the area of the rectangle consisting of Y and Z can be represented as ($a − b$)($a + b$), then complete the following:

   ($a − b$)($a + b$) = _____

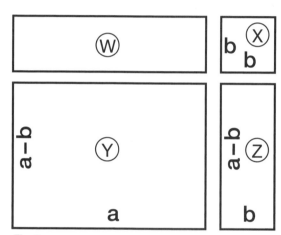

Fig. 6

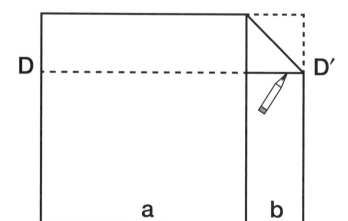

Fig. 4

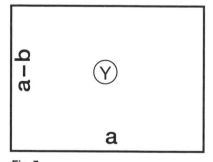

Fig. 7

Developing $(a + b)(a - b)$

10. We can also see this relationship in another way.

    a. Arrange rectangles Y and Z and square X as illustrated in **figure 8**.

    b. What is the area of the three pieces? _____

    c. Remove square X and you are left with an area of $a^2 - b^2$.

    d. Again rearrange the rectangles Y and Z as in **figure 7**.

    e. What do you observe about the areas? _____

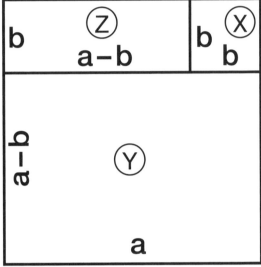

Fig. 8

Expressions, Equations, and Inequalities

Randall P. Vandyk

TEACHER'S GUIDE

Introduction: The *Curriculum and Evaluation Standards for School Mathematics* (NCTM 1989) call for an increased emphasis on informal explorations at the middle school level designed to develop the algebraic concepts of variation, expression, equation, and inequality. These concepts and their relationships can be developed through an effective use of tables (Heid and Kunkle 1988), which is the focus of this activity.

Grade levels: 8–10

Materials: One set of activity sheets for each student; calculators to aid in the computations (optional)

Objectives: To improve students' understanding of the concepts of variable, range and zeros of an expression, the roots of an equation, and the solution set of an inequality; to enable students to calculate the range of an expression given a specific domain, compare the ranges of two expressions, and formulate generalizations concerning the expressions.

Procedure: These activities assume that the students are familiar with the concept of real number, have a basic understanding of absolute value, and know the meaning of the symbols of inequality.

Distribute sheet 1 first. The teacher and the students should do sheet 1 together, discussing the answers thoroughly. After sheet 1 is completed, review the meanings of the terms *variable*, *expression*, *domain*, *zeros*, *roots*, and *solution set*. When you are satisfied with the students' understanding of the concepts, distribute sheets 2 and 3.

You should stress that the domain chosen for the tables is a set of values that is a subset of the real numbers. Explain that it would not be feasible to make a table listing all the real numbers in the domain. Therefore, we try to find a set of values that will be in the neighborhood of the critical values in the analysis. After reviewing sheets 2 and 3, have students construct tables to solve such equations as $x^2 = 2x + 3$, $|x + 2| = |x - 4|$, and so on, and the related inequalities to reinforce the concepts developed in this activity.

This activity could very easily be supplemented with the plotting of the points generated by the tables. The graphing would reinforce the concepts of preimages, images, and (preimage, image) pairs. It would also visually reinforce the concepts of zeros of an expression, roots of an equation, and the solution sets of inequalities.

You might also consider using a computer to generate the tables with a spreadsheet or other computer software or with a computer program.

SOLUTIONS:

Sheet 1:

1) $x = 4$

2) Yes, $x = 2$

3) $\{-3, -2, -1, 0, 1\}$

4) Yes, because each value of $3x - 6$ is obtained by multiplying 3 by a smaller number, resulting in a smaller product

5) $x = 5$

6) Yes, because each succeeding term is obtained by subtracting 4 from a larger number, resulting in a larger difference

7) $x = 1$

8) $\{2, 3, 4, 5\}$

9) $x > 1$.

Sheet 2:

x	x^2	$3x + 4$
−5	25	−11
−4	16	−8
−3	9	−5
−2	4	−2
−1	1	1
0	0	4
1	1	7
2	4	10
3	9	13
4	16	16
5	25	19

1) None

2) None: when you square any number (other than 0), the result is positive (0 is neither positive nor negative).

3) Yes

4) No
   a. $\{0, 1, 2, 3, 4, 5\}$

b. $\{-5, -4, -3, -2, -1, 0\}$

c. $x > = 0$

5) $\{-1, 4\}$

6) $\{0, 1, 2, 3\}$

7) $x > -1$ and $x < 4$.

Sheet 3:

x	$2x-1$	$|2x-1|$
3	-7	7
-2	-5	5
-1	-3	3
0	-1	1
1	1	1
2	3	3
3	5	5
4	7	7

1) No

2) Yes, $x = 0.5$

3) No, $|2x - 1|$ decreases in the first half of the table and increases in the second half.

4) $x = 0.5$, zero

5) $x \leq 0.5$

6) $\{-2, 3\}$

7) $\{-1, 0, 1, 2\}$

8) $x > -2$ and $x < 3$

9) Never, the absolute value of an expression is always greater than or equal to zero.

REFERENCES

Heid, M. Kathleen, and Dan Kunkle. "Computer-Generated Tables: Tools for Concept Development in Elementary Algebra." *The Ideas of Algebra, K–12,* 1988 Yearbook of the National Council of Teachers of Mathematics (NCTM), edited by Arthur F. Coxford, pp. 170–77. Reston, Va.: NCTM, 1988.

National Council of Teachers of Mathematics (NCTM), Commission on Standards for School Mathematics. *Curriculum and Evaluation Standards for School Mathematics.* Reston, Va.: NCTM, 1989.

Expressions, Equations, and Inequalities   Sheet 1

Consider the expressions $x - 4$ and $3x - 6$. In the table below, we have evaluated the expressions for a given set of values of the variable called the domain.

x	$x - 4$	$3x - 6$
-3	-7	-15
-2	-6	-12
-1	-5	-9
0	-4	-6
1	-3	-3
2	-2	0
3	-1	3
4	0	6
5	1	9

1. What domain elements result in a value of zero in the expression $x - 4$? _____
   We call this value a zero of the expression.

2. Does $3x - 6$ have a zero in the given domain? _____

   If so, what is it? _____

3. What values of the variable x result in values for $3x - 6$ that are negative? _____

4. As the values in the domain decrease beyond -3, will $3x - 6$ always decrease? _____

   Why or why not? _____

5. What values in this domain result in positive values for $x - 4$? _____

6. As the domain values increase beyond 5, will the values of the expression $x - 4$ always increase? _____

   Why or why not? _____

7. What value in this domain results in equal values for $x - 4$ and $3x - 6$? _____
   We call this number a root or solution of the equation $x - 4 = 3x - 6$.

8. For what values of this domain are the values of $x - 4$ less than the values of $3x - 6$? _____
   These values form the solution set of $x - 4 < 3x - 6$ over the given domain.

9. Assuming that the domain now includes all real numbers, state a rule in terms of x that would describe the solution set of $x - 4 < 3x - 6$. _____

   Test your rule by finding the values of the two expressions for a value of x between 1 and 2.

Expressions, Equations, and Inequalities   Sheet 2

The values of an expression that result from evaluating that expression over a given domain form the *range*. Given the following domain, complete the table by calculating the values of the ranges for the two expressions x^2 and $3x + 4$.

x	x^2	$3x + 4$
-5		
-4		
-3		
-2		
-1		
0		
1		
2		
3		
4		
5		

1. Which elements of this domain are zeros for the expression $3x + 4$? _____

2. What values of this domain result in negative range values for x^2?_____

   Why?_____

3. Are the range values of $3x + 4$ always increasing as x increases over the given domain? _____

4. Are the range values of the expression x^2 always increasing as x increases over the given domain? _____

   a. For which values in the given domain are the range values of x^2 increasing as the domain values increase? _____

   b. For which values in the given domain are the range values of x^2 decreasing as the domain values increase? _____

   c. Assume the domain consists of all the real numbers. For what real values of x are the range values for x^2 always increasing? _____

5. What are the roots of $x^2 = 3x + 4$? _____

6. For what elements of the given domain are the range values for x^2 less than the range values for $3x + 4$?_____

   _____

7. Assume the domain consists of all real numbers. Write inequalities that describe the solution set of $x^2 < 3x + 4$.

   _____

   Test your answers for real values of x in the neighborhood of -1 and 4.

Expressions, Equations, and Inequalities   Sheet 3

Recall that the absolute value of any real number is always positive (except for 0, where $|0| = 0$), that is, $|4| = 4$ and $|-4| = 4$. Given the following domain, complete the table by calculating the range values of the expressions $2x - 1$ and $|2x - 1|$. (Note that the importance of $2x - 1$ is primarily in its use in calculating the values of $|2x - 1|$.

x	$2x-1$	$|2x-1|$
-3		
-2		
-1		
0		
1		
2		
3		
4		

1. Are any of the elements of the given domain zeros for $|2x - 1|$? _____

2. If we expand our domain to include all real numbers, do you think a zero exists for $|2x - 1|$?_____ If so, approximate it and check.

3. Are the range values of $|2x - 1|$ always increasing as the values of the domain increase?_____
    Why or why not? _____

4. Assume the domain consists of all real numbers. Where do you think the range values will switch from decreasing to increasing? _____
    (*Note:* We have seen this value before; it was the _____ of the expression.)

5. Define a rule for x so that the range elements will always be decreasing. _____

6. What are the roots of $|2x - 1| = 5$? _____

7. For the original domain, what elements in the domain are solutions of $|2x - 1| < 5$? _____

8. If the domain contains all real numbers, write inequalities that describe the solution set of $|2x - 1| < 5$. _____

9. For what real values of x is $|2x - 1| < 1$?_____
    Explain._____

CHAPTER 3

Functions

Introduction

"Functions are one of the most important mathematical tools for helping students make sense of the world around them, as well as preparing them for further study in mathematics (Yerushalmy and Shternberg 2001). Functions appear in most branches of mathematics and provide a consistent way of making connections between and among topics. Students' continuing development of the concept of function must be rooted in reasoning, and likewise functions are an important tool for reasoning" (NCTM 2009, p. 41).

Even though functions are one of the key tools students use in many branches of mathematics, not many activities in *Mathematics Teacher* focused on functions until the end of 1980. The 1990s saw a surge in functions activities. One reason for this increase may be the inclusion of functions as a separate content standard in NCTM's 1989 *Curriculum and Evaluation Standards*. The activities in this book are presented in a variety of contexts that may help students make sense of the mathematics they are learning. For example, in Edwards and Chelst (1999) students act as directors of manufacturing and try to maximize their companies'

TABLE 3.1			
Functions Activities			
Author and title	Mathematical topic(s)	Context(s)	Materials
Andersen (1973), "Griefless Graphing for the Novice"	Graphing linear functions	Mathematics	Student activity sheets
Davidenko (1997), "Building the Concept of Function from Students' Everyday Activities"	Introduction to functions	Graphs and tables from newspapers	Newspaper clippings, student activity sheets
Day (1993), "Solution Revolution"	Graphing functions	Mathematics	Graphing calculator, spreadsheet software, student activity sheets
Edwards and Chelst (1999), "Promote Systems of Linear Inequalities with Real-World Problems"	Systems of linear inequalities	Product-mix problems	Lego plastic construction toys, student activity sheets
Hershkowitz, Arcavi, and Eisenberg (1987), "Geometrical Adventures in Functionland"	Quadratic functions	Mathematics	Student activity sheets
Moore-Russo and Golzy (2005), "Helping Students Connect Functions and their Representations"	Sum and product of functions	Mathematics	Graphing calculator, student activity sheets
Moyer (2006), "Non-Geometry Mathematics and The Geometer's Sketchpad"	Composition of functions, quadratic functions, trigonometric functions, piecewise functions, derivatives	Technology	The Geometer's Sketchpad, student activity sheets
Peterson (2006), "Linear and Quadratic Change: A Problem from Japan"	Functions (general)	Mathematics	Student activity sheets
Van Dyke (2003), "Using Graphs to Introduce Functions"	Introduction to functions	Distance from an object as a function of time	Student activity sheets

profits. Another reason behind the recent proliferation of activities may be the ubiquity of technology in classrooms, especially the graphing calculator. Moyer (2006), Moore-Russo and Golzy (2005), and Day (1993) use technology to show aspects of functions that are difficult to show without technology. Activities vary regarding their topic, context, and level of difficulty. **Table 3.1** summarizes the characteristics of the activities.

Focus in High School Mathematics (NCTM 2009) suggests three key elements of reasoning and sense making within functions:

1. *Multiple representations of functions.* Representing functions in various ways, including tabular, graphic, symbolic (explicit and recursive), visual, and verbal; making decisions about which representations are most helpful in problem-solving circumstances; and moving flexibly among those representations

2. *Modeling by using families of functions.* Working to develop a reasonable mathematical model for a particular contextual situation by applying knowledge of the characteristic behaviors of different families of functions

3. *Analyzing the effects of parameters.* Using a general representation of a function in a given family (e.g., the vertex form of a quadratic) to analyze the effects of varying coefficients or other parameters, converting between different forms of functions (e.g., the standard form of a quadratic and its factored form) according to the requirements of the problem-solving situation (e.g., finding the vertex of a quadratic or finding its zeros)

Most activities in this chapter encourage multiple *representations of functions*. For example, Van Dyke (2003) uses the "theme of distance from an object as a function of time … to help students deepen their understanding of functions, graphs, and the underlying equivalence of the numeric, tabular, and algebraic representations of functions" (p. 126). Similarly, Peterson (2006) adopts an interesting problem from Japan and discusses how students see the same problem from many different vantage points and use a variety of representations (e.g., tables, graphs, equations, geometric figures) to express a solution. This activity also gives a helpful example of how a teacher can use an open-ended question to elicit students' own solution strategies that help them make sense of the problem.

On the way to solving the equation $2^x = x^{10}$, Day (1993) uses technology to investigate algebraic, numeric, and graphical viewpoints. He emphasizes the importance of understanding the behaviors (i.e., complete, hidden, and end behaviors) of the functions to apply graphical strategies. Moore-Russo and Golzy (2005) approach the sum and product of functions by using graphical

representations followed by algebraic representations. They emphasize the order of representations because "students can do the algebraic approach with little thought about what the symbols actually represent … [so] by solving the problem graphically before algebraically, they seem to make a better connection between the equation and the graph" (Moore-Russo and Golzy 2005, p. 158). Finally, Functionland with Hershkowitz, Arcavi, and Eisenberg (1987) presents a geometric problem that students can represent with either an algebraic concept of function or the geometric concept of quadrilaterals.

Four activities address the *families of functions for modeling element*. Edwards and Chelst (1999) describe how students can use systems of linear inequalities to determine "the production levels of different products to maximize a company's profit" (p. 119). They ask students to model two problems and find optimal solutions. Davidenko's (1997) activity expects students to collect tables and graphs from newspapers that can be described as functions. This introductory activity targets "the information already known and understood by students" (p. 149). For her, students are already using concepts of functions in their lives and "teachers can play an important role in helping students become aware of their own thought processes" (p. 144). After students become familiar with the structure of functions, they are ready to model real-life situations presented in the activity.

Andersen, in his 1973 article, *analyzes the effects of parameters* and illustrates how to introduce graphs of functions. His activity presents important graphing concepts, such as translation, rotation, intersection, and vertical and horizontal shifting, as well as "encourages a student to ask 'What would happen if …?'" by manipulating the parameters of the functions and observing their effects. In contrast, Moyer (2006) introduces a technological approach to investigate the effects of parameters. He demonstrates the use of The Geometer's Sketchpad beyond geometry class and presents activities that animate functions to help students visualize the effects of changing coefficients and parameters.

These activities require students to use the four mathematical reasoning habits described in *Focus in High School Mathematics: Reasoning and Sense Making* (NCTM 2009, pp. 9–10): (1) analyzing a problem, (2) implementing a strategy, (3) seeking and using connections, and (4) reflecting on a solution. For instance, in Davidenko's (1997) activity, students work in groups to "*analyze* the information in terms of domain, range(s), and functions" (p. 145). Edwards and Chelst (1999) suggest a concrete exploration of a problem by building with Lego blocks first and then conducting an abstract exploration where they define variables, model the problem's objective, and identify the constraints of the problem. In Day (1993), students *implement* graphical and numeric *strategies* to approximate the solution. Moore-Russo and

Golzy (2005) *seek connections* between algebraic solutions and graphical solutions. Finally, Peterson (2006) motivates teachers to encourage students to "*justify* how each component of their equation related to the figure and numerical patterns" and facilitate a *discussion of students' solutions* that are "focused on similarities and differences between the solutions" (p. 211).

REFERENCE

National Council of Teachers of Mathematics (NCTM). *Focus in High School Mathematics: Reasoning and Sense Making*. Reston, Va.: NCTM, 2009.

Griefless Graphing for the Novice

Harold Andersen

Typically, a student in the first year of high school algebra finds graphing a hassle of "cleaning up" equations. While the skill of manipulating algebraic quantities is important, it detracts from the important aspects of graphing. For example, in order to graph $\{(x, y) \mid 3x - 5y = 30\}$ a student would have to do a lot of housekeeping to find enough points to determine the graph. There is a good chance that the concept a teacher is trying to have the student discover would be lost in the maze of manipulative machinations necessary to get an equation in workable form. I suggest that we leave out the manipulative details and concentrate of the set of points under consideration.

Students prepared with basic graphing skills of locating and naming ordered pairs in a coordinate plane are ready for the following development. Simply it is this: have students graph sets of ordered pairs that are of a given form. For example, graph the ordered pairs (k, k) where k is a real number. Students have no difficulty finding enough points to get the idea that these points lie on a line bisecting the first and third quadrants.

If the aim of a lesson is to lead students into the idea of slope, the following exercise is a good way to motivate that discovery.

Graph in the same grid (**fig. 1**) all ordered pairs of the following forms:

1. (k, k)
2. $(k, 2k)$
3. $(k, \frac{1}{2}k)$
4. $(k, -k)$
5. $(k, -\frac{5}{3}k)$

Group discussion of the resulting graphs can generate some important ideas. Some questions for the group to consider follow:

What common characteristics do the five graphs possess?

What effect does the coefficient of k in the second coordinate have on a given set of points when compared with (k, k)? In general, how does (k, ak) compare to (k, k)?

Have a student pick five points from the graph of $(k, \frac{1}{2}k)$ and form the slope ratio for pairs of these points. Say the points are the following: $(2, 1)$, $(6, 3)$, $(-4, -2)$, $(3, \frac{3}{2})$, and $(-\frac{1}{2}, -\frac{1}{4})$.

Ask the student how the y-values change when he goes from $(2, 1)$ to $(6, 3)$, then how the x-values change from $(2, 1)$ to $(6, 3)$. Do this for any distinct pair of points. Discussion should lead the group to the generalized concept of slope ($\frac{1}{2}$ in this case).

Translation and rotation can also be investigated the same way. Consider the following exercises. Graph ordered pairs of the forms shown, each on a different set of axes.

Set A (**fig. 2**)

1. (k, k)
2. $(k, k + 2)$
3. $(k, k - 3)$

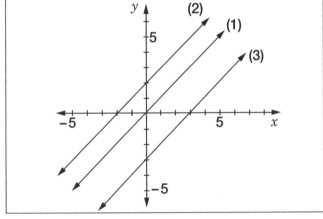

Fig. 2

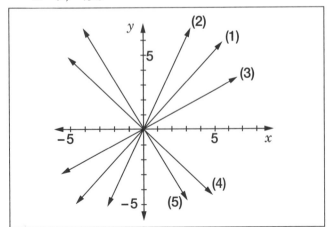

Fig. 1

Set B (**fig. 3**)

1. $(k, -\frac{1}{2}k)$
2. $(k, 2k + 2)$
3. $(k, -3k - 3)$

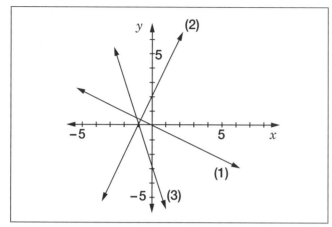

Fig. 3

Set C (**fig. 4**)

1. $(k, 2k - 3)$
2. $(k, 2(k - 3))$
3. $(k, 2)$
4. $(-2, k)$

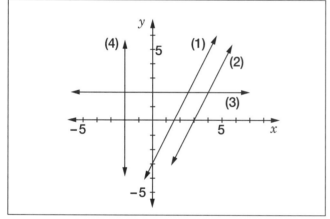

Fig. 4

Clearly the message in set A is that all of the lines are parallel and that each of them can be thought of as a vertical shift or translation of the $(k, |k|)$ set of points. set B illustrates rotation; each line may be thought of as the correspondingly numbered line in set A rotated on its point of intersection with the *y*-axis. From set C and similar exercises devised by the teacher the students can see the difference between adding a constant before and after multiplying, the kind of line obtained when one of the coordinates is a constant, and so forth. The exercises also provide an excellent means of introducing the ideas of intercepts.

Once a class has become familiar with this method, it is easy to consider other functions that have interesting graphs. For example, absolute value is always a bit of a mystery. Having students graph functions containing

absolute value can be helpful. Consider the following possibilities.

Set D (**fig. 5**)

1. $(k, |k|)$
2. $(k, 2|k|)$
3. $(k, -\frac{1}{3}|k|)$

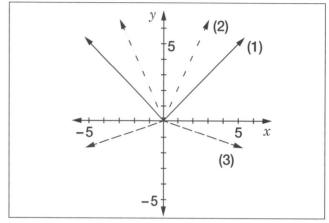

Fig. 5

Set E (**fig. 6**)

1. $(k, |k| + 2)$
2. $(k, |k| - 3)$
3. $(k, 2|k| - 5)$

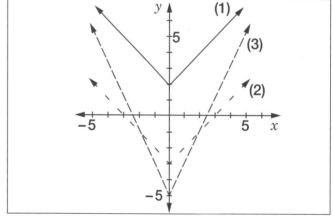

Fig. 6

In this section it would be wise to reemphasize rotation and translation. Set D, as illustrated in **figure 5**, demonstrates the effect of a constant multiplier $|x|$. Students should note that the graph is "open" or "closed" as compared with the graph of $(k, |k|)$ depending on whether the constant is greater than one or between zero and one, respectively. The effect of the negative sign is to produce a reflection or a "flip" about the *x*-axis. sets E and F are designed to be discussed together; notice that the example 1 in set E and example 1 in set F are very nearly the same. However, when a student examines the corresponding graphs, he or she should discover the first graph is a vertical shift of 2 when compared to the graph of $(k, |k|)$, whereas the second graph is horizontal shift of -2 when compared to the graph of $(k, |k|)$.

Example 3 in set F is a combination horizontal and vertical shift.

Set F (**fig. 7**):

1. $(k, |k + 2|)$
2. $(k, |k - 3|)$
3. $(k, 2|k - 5| - 3)$

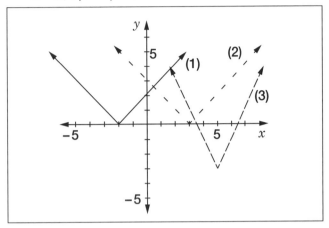

Fig. 7

If the class experiences little difficulty with the linear sets of ordered pairs, they should experience some non-linear pairs. Students should see that not all graphs of ordered pairs are graphs of straight lines. The idea of translation can be reinforced, and the student can see how the coefficient a in $y = ax^2$ affect the graph relative to the graph $y = x^2$. Consider the following set of problems. It is necessary to point out to the student that he or she will need about ten or more points to get a general idea of the sketch of a curve.

Set G (**fig. 8**):

1. (k, k^2)
2. $(k, 2k^2)$
3. $(k, -k^2)$ Note: $-k^2 \neq (-k)^2$
4. $(k, \frac{1}{4}k^2)$

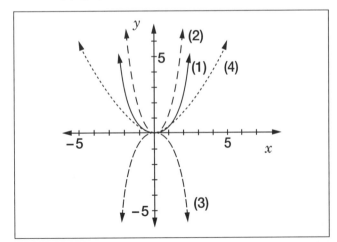

Fig. 8

Set H (**fig. 9**):

1. (k, k^2)
2. $(k, k^2 + 3)$
3. $(k, k^2 - 1)$
4. $(k, 2k^2 - 5)$

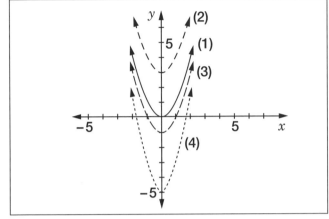

Fig. 9

Set I (**fig. 10**):

1. (k, k^2)
2. $(k, (k - 1)^2)$
3. $(k, (k + 3)^2)$
4. $(k, 2(k - 5)^2)$

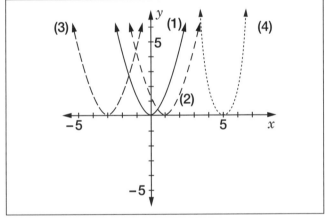

Fig. 10

The benefits to this approach are many. Principally, it allows the student to get on with graphing without the interference of manipulative requirements. Later the student will do the necessary work willingly because he or she has an overview of graphing techniques to make his work lighter. The method encourages the student to ask "What would happen if ...?" Finally, it lays the intuitive groundwork for the function concept. Next time you begin graphing in algebra, try this approach and see if you don't find it one of the highlights of the year.

Griefless Graphing for the Novice

Sheet 1

1. Graph on the same grid (below) all ordered pairs of the following forms:

   a. (k, k)

   b. $(k, 2k)$

   c. $(k, \frac{1}{2}k)$

   d. $(k, -k)$

   e. $(k, -\frac{5}{3}k)$

2. What are the common characteristics of the five graphs?

3. What effect does the coefficient of k in $(k, 2k)$ have on a given set of points when compared with (k, k)? In general, how does (k, ak) compare with (k, k)?

4. Pick five points from the graph of $(k, \frac{1}{2}k)$ and form the slope for pairs of these points.

5. Pick two of the points from question 4 (*point 1* and *point 2*). How do the *y*-values change when you go from *point 1* to *point 2*? How do the *x*-values change from *point 1* to *point 2*? Pick a different set of two points and again observe the changes in the *x*- and *y*-values. What do you observe?

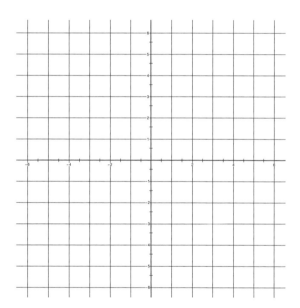

Graph 1

Griefless Graphing for the Novice

Sheet 2

1. Graph ordered pairs of the forms shown:

   a. (k, k)

   b. $(k, k + 2)$

   c. $(k, k - 3)$

2. What do you observe about the relationship between the lines formed by ordered pairs in graph 2?

3. Graph ordered pairs of the forms shown:

   a. $(k, -\frac{1}{2}k)$

   b. $(k, 2k + 2)$

   c. $(k, -3k - 3)$

4. What do you observe in graph 3?

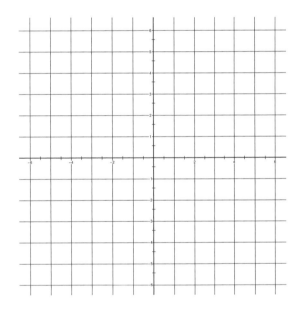

Graph 2

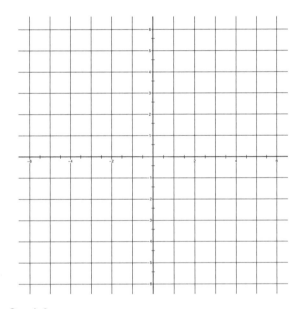

Graph 3

Griefless Graphing for the Novice

5. Graph ordered pairs of the forms shown:

   a. $(k, 2k - 3)$

   b. $(k, 2(k - 3))$

   c. $(k, 2)$

   d. $(-2, k)$

6. What do you observe about the relationship between the lines formed by ordered pairs in graph 4?

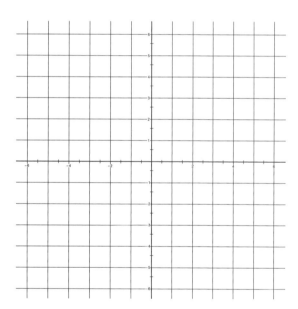

Graph 4

Griefless Graphing for the Novice

1. Graph ordered pairs of the forms shown:

   a. $(k, |k|)$

   b. $(k, 2|k|)$

   c. $(k, -\frac{1}{3}|k|)$

2. What is the effect of the constant multiplier of $|k|$?

3. What do you observe about the relationship between the lines formed by ordered pairs in graph 5?

4. Graph ordered pairs of the forms shown:

   a. $(k, |k| + 2)$

   b. $(k, |k| - 3)$

   c. $(k, 2|k| - 5)$

5. What do you observe in graph 6?

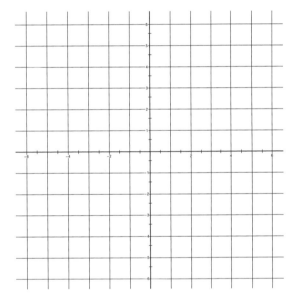

Graph 5

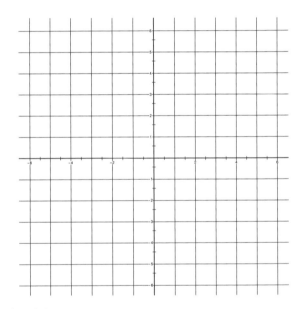

Graph 6

Griefless Graphing for the Novice

Sheet 5

6. Graph ordered pairs of the forms shown:

   a. $(k, |k + 2|)$

   b. $(k, |k - 3|)$

   c. $(k, 2|k - 5| - 3)$

7. What do you observe about the relationship between the lines formed by ordered pairs in graph 7?

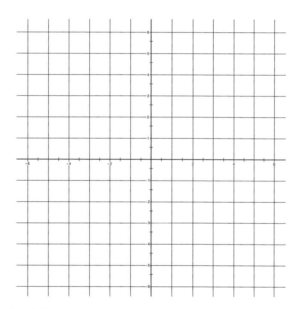

Graph 7

Griefless Graphing for the Novice   Sheet 6

1. Graph ordered pairs of the forms shown:

    a. (k, k^2)

    b. $(k, 2k^2)$

    c. $(k, -k^2)$ [Note: $-k^2 \neq (-k)^2$]

    d. $(k, \frac{1}{4}k^2)$

2. What do you observe about the relationship between the lines formed by ordered pairs in graph 8?

3. Graph ordered pairs of the forms shown:

    a. (k, k^2)

    b. $(k, k^2 + 3)$

    c. $(k, k^2 - 1)$

    d. $(k, 2k^2 - 5)$

4. What do you observe in graph 9?

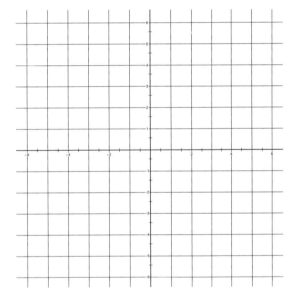

Graph 8

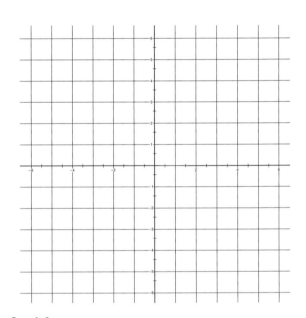

Graph 9

Griefless Graphing for the Novice

5. Graph ordered pairs of the forms shown:

   a. (k, k^2)

   b. $(k, (k-1)^2)$

   c. $(k, (k+3)^2)$

   d. $(k, 2(k-5)^2)$

6. What do you observe about the relationship between the lines formed by ordered pairs in graph 10?

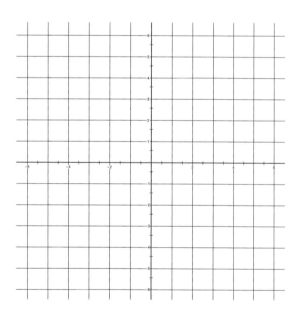

Graph 10

Building the Concept of Function from Students' Everyday Activities

Susana Davidenko

In everyday life, people analyze information using algebraic thinking, often being unaware of doing so. Teachers can play an important role in helping students become aware of their own thought processes. Students should have the opportunity to bring their experiences into the mathematics class, reflect on their own thinking, and deepen their understanding of real problems.

The concepts of variable and function are the building blocks of algebra. Why do students have difficulties with these ideas when, indeed, they use them all the time? At high school or college levels, functions are usually introduced by stating the definition and emphasizing that *each* element of the domain has a *unique* image. Then several examples and counterexamples are given.

I believe, though, that students' first introduction to functions should emphasize the versatility of functions to describe the relationship among two or more sets of objects. Mathematics teachers should encourage students to develop their capability to discover mathematical relations embedded in any piece of information. Most functions used and perceived in out-of-school experiences are not clearly defined with domain, range, and a rule that defines how each element of the domain is mapped onto its image. This information is implicit in these functions.

The goals of this article are (1) to propose activities that promote students' awareness of the structure of the information involved in their everyday activities, (2) to build the concept of function from students' mental representations of phenomena, and (3) to recommend the use of spreadsheets to provide external representations of these functions.

FUNCTIONS ARE EVERYWHERE

Variables and functions are constantly used, even by young children.

Example 1

Suppose that we ask a kindergartner to tell us the name of her classmates as she points to them on the class picture. She will interpret "to name" as a function of f to be "evaluated" on each student in the picture. Here we have a function f defined by—

- Domain: the set of students in the picture,
- Range: the set of names of students,
- Definition: f(picture) = name, and
- Example: f(boy seated on the first row, wearing a white shirt) = Michael.

If we ask the student, "Why didn't you mention Patty?" the child might say, "Remember? Patty is not in my class this year!" She is telling us that her friend is not in the domain of the function.

Example 2

When a second grader tells his parents what the prices at the school store are, he will say something like, "The small notebooks are 79 cents, the large notebooks are $1.49, pencils are 20 cents, and erasers are 5 cents each." That is, the boy will spontaneously interpret "to price" as a function to be "evaluated" on each product. For this function p we have—

- *Domain:* the set of products available at the school store,
- *Range:* the set of prices,
- *Definition:* p(product) = price, and
- *Example:* p(eraser) = 5 cents.

If we ask the student, "Is there anything that costs $100.00?" he might tell us, "No, nothing at the school store costs more than $20.00." Thus, he is suggesting an upper bound for the range of the function.

Example 3

We all have had a student who says, "I'm not good in math." This student may be good at organizing parties. He can make a list of friends to invite to a party and write their telephone numbers, the snack that each of them will bring, and the cost of the snack. This student is defining several functions on the same domain: a group of friends. The image of the functions can be either qualitative, for instance, the snack, or quantitative, for example, the price.

We should take advantage of students' ability to organize their own data. They can use their mathematical models to understand their own algebraic thinking.

In the previous examples, the students did not describe the information as $f(x) = \ldots$ or $y = f(x)$. When students depict an event, they have it structured in their minds somehow. Talking about some calculations, students often say, "I did it in my head." They mean that they mentally performed a computation using their own rules of associativity, distributivity, or both. Similarly, in the examples of functions, the students had personal representations of the information. We need to develop activities that build on students' mental representations and then help students to develop higher levels of algebraic thinking.

USING NEWSPAPERS TO DEVELOP ALGEBRAIC THINKING

In a college-algebra course, I first introduced the concept of functions by using examples of daily-life experiences. We discussed how to organize the answer to this question: What is the price of gasoline? After a class discussion, we decided to make tables first by writing the price of gasoline by types and then by listing the prices of a particular type of gasoline but varying the gas station. I suggested that students fix the type of gasoline and the gas station and look for changes in the prices depending on days of the week or months of the year. The students realized that even in a well-known situation, we deal simultaneously with many variables without being aware of what we are doing. Through this situation, I informally introduced the concepts of variables, functions, domain, and range. As follow-up activities, I gave students several graphs and tables from newspapers. They worked with partners for a couple of classes to analyze the information in terms of domain, range(s), and functions. Then we held class discussions.

As a homework assignment, the students had to collect newspaper clippings on situations that they could describe as functions. In all cases they had to make a written description of the function; a table for the function, if a table was not part of the clipping; and, if possible, a graph. I carried out this activity soon after I presented the definition of functions. I did not spend much time on the appropriate mathematical language needed for the activity, to avoid interfering with the students' spontaneous interpretation of the data. By looking at some students' answers, we can see that they had some misconceptions. Other mistakes show the lack of appropriate vocabulary, but they are not misconceptions. Some students' responses follow.

Example 1

In the example in **figure 1**, "age" and "annuity rate" are the names of two variables involved in the information. The domain is the set of positive integers greater than or equal to 65; the range is a set of real numbers from 7.3 to 12.0 inclusive. This student did not use appropriate language to define domain and range. However, he does not show a misconception. He has clearly distinguished the dependent and independent variables. Note that the student was aware of the units of each variable. Labeling the axis with the units was important for communicating the information. This student added two questions to be answered using the information, as we had done in class.

Example 2

Figure 2 contains an interesting example because it simultaneously shows four functions. The graph makes a visual impact, compared with the table. We can see that the student did not keep the proportion along the

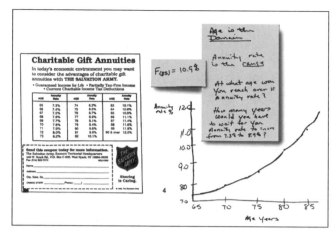

Fig. 1. Student's graph based on the information given by the table.

entire y-axis in her attempt to fit all the values in the graph. Note that although the range of the first three functions represents the number of votes, the range of the fourth function represents the number of points; that is, the y-axis is used to represent values of two different units. The total number of points can be obtained as the sum of the votes times the points per vote for first, second, and third place. Asking students to write a formula for these values in the last column could generate interesting results.

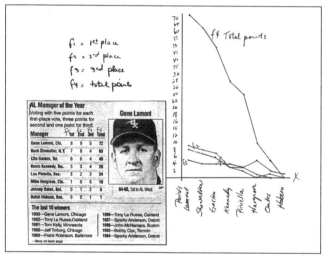

Fig. 2. Graph of four functions based on the information given by one table

ASSESSING ALGEBRAIC THINKING

As an assessment activity—still before a more formal introduction of the algebraic vocabulary—I gave the students the map in **figure 3**, which was taken from a national newspaper. The students worked individually in class to answer the questions of the quiz shown in **figure 4**. Some students' responses follow.

Example 1

Figure 5 shows a very well organized response. The student shows a lack of a precise language of functions when she writes, for example, domain—Boston, range—61/44°F. Still, the concepts are clear when she defines

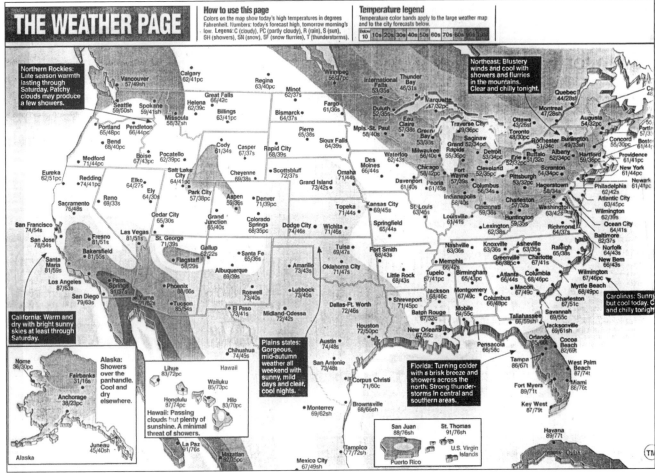

Fig. 3. Map with the information of low and high temperatures

---

Quiz on Functions

1. Summarize the information given in the map by defining two functions on the same domain. Call them f_1 and f_2. Give, in your own words, a definition of each function.

2. Find the domain and range(s) of the functions.

3. Evaluate f_1 (Miami) and f_2 (Miami).

4. Graph both functions on the same graph. Take only seven points from the domain to graph.

5. Estimate f_1 (Syracuse) and f_2 (Syracuse). Explain.

6. Make four questions that can be answered using the graph you made. Answer the questions.

Fig. 4. Quiz on functions based on the information given by the map

domain as different cities and range as different temperatures in degrees Fahrenheit. The student's explanation of how Syracuse's temperature was estimated is interesting and reveals how she made sense of the information. A good response demonstrates the student's use of common sense when doing linear interpolation.

Example 2

What I found most interesting in the response in **figure 6** were the student's questions. By answering the first question, the student uses the information from the map to make some conjectures about variables not explicit in the graph, for example, in inferring the location of a city given its temperature. Question 3 suggests how the information from the map could be useful for real purposes. In question 4, the student found that a bar graph was a better way to communicate the information.

Example 3

This student misinterpreted part 4 of the quiz (**fig. 7**). However, he did make a good attempt at a graph. Note that the labeling of the horizontal axis was not done to scale and that the information was distorted. The student graphed the information of high-low temperatures as points in the plane. The domain was not part of the graph itself, although the student showed the domain by writing the names of the cities below their maximum temperatures. Other students made similar graphs. I discussed with the class the difference between this graph and the one I had expected. We talked about what novel information could be obtained from this graph. I made some comments about statistical analysis by comparing two functions on the same domain.

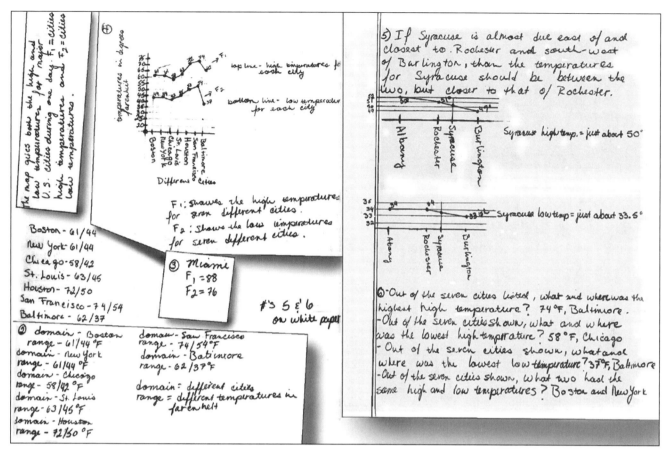

Fig. 5. A thorough answer to the quiz

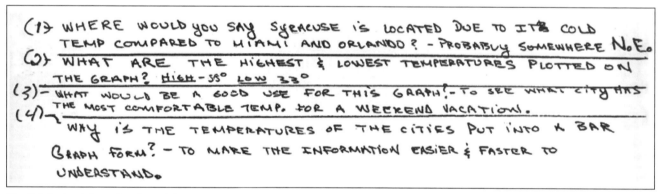

Fig. 6. Inferring information from data

After this first exploratory stage of working with functions, I tried to enrich the language and vocabulary related to functions. We talked about *domain*, *range*, *image*, *preimage*, and *mapping*. I also introduced algebraic functions given by formulas, and I began to deal with linear functions.

USING SPREADSHEETS TO EXTEND STUDENTS' MODELING SKILLS

The structure of the information of the examples previously analyzed—picture/name, product/price, . . . , city/temperatures—can be made explicit with the structure of a spreadsheet. Using spreadsheets to model real-life situations is a natural extension of these types of

activities, which build on daily uses of the concepts of variables and functions. Let us see with an example how spreadsheets can help students not only model the problem but deepen their understanding and gain new information.

The mathematics behind spreadsheets

Suppose that we give our students a special-sale flyer from a nearby store. Just by looking at the information, the students will at once interpret two functions:

$$r(x) = \text{"the regular price of"} \text{ product } x,$$

and

$$s(x) = \text{"the sale price of"} \text{ product } x.$$

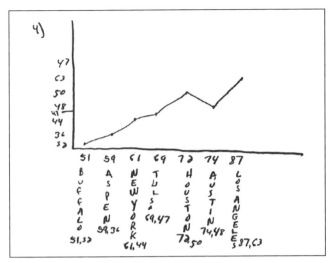

Fig. 7. A common misinterpretation of part 4 of the quiz

	A	**B**	**C**
	Product	Reg. Price	Sale Price
1	14" Planter	4.50	3.25
2	15q. Pail	3.45	2.00
3	Patio Chair	15.99	10.99
4	Square grill	10.99	4.99

Fig. 8. Spreadsheet with given information

The domain of these functions is the set of the articles on sale.

Suppose that we assign the students to design a spreadsheet on which to enter the information given in the flyer. Let us examine what kind of algebraic thinking would be done by the students while doing this assignment.

Designing the spreadsheet

Most students will design a spreadsheet with three columns (**fig. 8**): column A: Product, column B: Regular Price, and column C: Sale Price. Some students might create only columns for the prices, without reference to the products. Let us analyze the first case, where three columns are defined. In this case, the students use column A to enter the data of the *independent variable*, the name of the product. Columns B and C represent *dependent variables*. They clearly depend on column A, although it is likely that the sale price also depends on the regular price. The students might say, for example, that the regular price depends on the product; that is, the regular price is a *function* of the product. Each row of the spreadsheet represents the information of one

subject or unit under analysis. That is, each row displays the values of the functions at one particular element of the domain.

If we ask the students how much can be saved on each sale item, they will calculate the difference between regular and sale prices. To write this information on the spreadsheet, the students need to create a new column where the difference of the regular price and the sale price is being calculated. While doing this task, the students are actually formalizing the *function difference*, $d(x)$, as $d(x) = r(x) - s(x)$. This process is natural for students, since they probably calculate that difference of prices while shopping. It is important to use students' mental representations as a base on which to build concepts like sum and difference of functions.

Another mathematical concept that is easy to understand using spreadsheets is the composition of functions. For example, we might suggest that students create a column to display the sale price plus the corresponding tax (**fig. 9**). In this way, the customer can estimate the final cost more accurately. A new column could be defined as

$$\text{column E} = t(\text{column C}) = \text{column C} \cdot 1.07,$$

which represents New York state tax, for example. Thus, column E depends on column C, which, in turn, depends on column A.

Creating graphs

When students make graphs using data from a spreadsheet, they select one column for the horizontal axis x and one numerical column for the vertical axis y. The students are actually defining the domain and the range of a function.

When using a numerical range to define a function, teachers should emphasize that the numerical value of a function is not necessarily given by a formula, as is the case if the students make the graph for the product and its regular price.

When the students plot the regular and sale prices on the same graph, they will immediately see the difference between prices. The numerical value of the difference is represented by the vertical distance between the two graphs at each point (**fig. 10**).

If the students simultaneously graph the sale price and the sale price plus tax, they will realize how the new function depends on the sale price. It is not only an

	A	**B**	**C**	**D**	**E**
	Product	Reg. Price	Sale Price	Difference	S. Price + Tax
1	14" Planter	4.50	3.25	1.25	3.48
2	15q. Pail	3.45	2.00	1.45	2.14
3	Patio Chair	15.99	10.99	5.00	11.76
4	Square grill	10.99	4.99	6.00	5.34

Fig. 9. Spreadsheet with novel information

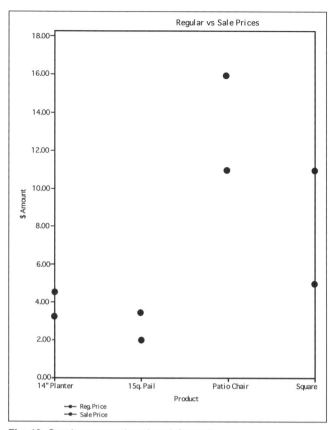

Fig. 10. Graph representing given information

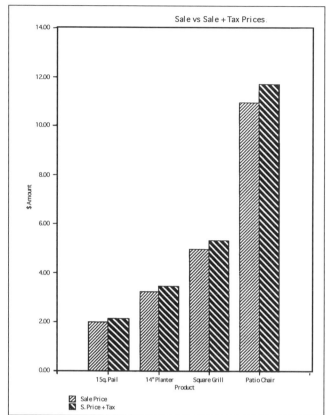

Fig. 11. Bar graph representing given and novel information

addition, it is an addition of a proportional quantity. If the columns of the spreadsheet are sorted by the regular price, then the students can have a visual representation of the proportionality of the "plus tax" (**fig. 11**).

The use of spreadsheets is ideal for students to interpret the correspondence between tables and graphs of functions. The students can display different kinds of graphs for the same set of data and decide which graph best represents the data and communicates their results.

CONCLUSION

Algebraic thinking is constantly used in everyday activities. Students receive information and organize it in their minds. Teachers should build on students' intuitive thinking to make them aware of the structure of the information. Using material from newspapers or magazines, or any data in which the students are interested, teachers can explain the concepts of variables and functions. The structure of a function can then be used to describe the information already known and understood by students. This process is more natural than teaching functions as relationships between two sets and then "filling the structure with content."

Using spreadsheets affords the opportunity to progress toward a deeper understanding of the concept of functions. The students can create spreadsheets to depict their mental representations of the information and begin to describe the information in terms of domain, range, independent and dependent variables, and transformations. Performing a higher level of algebraic thinking and using the language of functions allow students to explore the data beyond the descriptive level. The use of spreadsheets is ideal to promote connections between mathematics and students' everyday experiences, to develop mathematical language and reasoning skills, and to create a cooperative environment in the classroom.

Directions

Sheet 1

1. Read the information you have received (e.g., text, tables, graph) about the given topic.

   (Alternatively, decide with your group the topic you want to explore and the information you want to obtain. Once you have obtained the information, continue with the following steps.)

2. Discuss with your group and answer the following questions:

   a. Could you describe the information in *terms of two or more variables?*

   b. What does each variable represent—

       i. An object, a person, an item?

       ii. An attribute, a property, a measurement?

   c. What is the unit of each variable?

   d. Is there a variable you could consider the *independent variable?* If so, name it; let's say you name it *Indep.*

   e. Could you interpret any of the other variables as a *dependent variable?* In other words, could you describe any other variable as a *function of the independent variable?* In that case, name the function and write its definition. For example:

   FTN (*Indep*) = "Definition of FTN in terms of Indep"

   f. Describe the *domain* and *range* of each variable you have defined.

3. Discuss with your group: Is there any *new* information we would like to obtain or *derive* from the information you already have? If so, could you define this new information as another function of the independent variable?

Directions

4. *Design* and then *fill* a spreadsheet as follows:

   a. Title of the first column: name of the independent variable

   b. Entries that go in the first column: the values of the independent variable that you obtained from the given information or found on your own

   c. Titles of the other columns: names of the functions you defined in the previous step

   d. Entries of the column for a given function: the value or information you obtain by *evaluating the function on corresponding items in the independent variable column.*

Once your spreadsheet is complete, create one or more *graphs* based on it. You can graph more than one function on each graph if you think that visualizing those functions together could yield insights about the topic.

Discuss with your group what can be inferred from the graphs: Do any patterns exist? What do they mean? Could they be further explored? How? Does the information in the spreadsheet and from the graph allow you to make predictions, make conclusions, and make decisions?

Write a brief group report: explain what new information you have learned by completing the previous process.

Example 1

Issue/Question	Independent Variable	Original Dependent Variables	Derived Variables and New Questions
Temperatures around the country	A set of **cities** included in a USA weather map	**FMax (city)** = Maximum temperature forecasted for day N	**ARanTemp (city)** = Range of temperature on day N = (**AMax (city)** − **AMin (city)**)
		FMin (city) = Minimum temperature forecasted for day N	**AccMax (city)** = Difference between the forecasted and the actual max temperature = **FMax (city)** − **AMax (city)**
		AMax (city) = Actual maximun temperature for day N (obtained on day $N + 1$)	**AccMin (city)** = Difference between the forecasted and the actual max temperature = **FMin (city)** − **AMin (city)**
		AMin (city) = Actual minimum temperature for day N (obtained on day $N + 1$)	**Possible questions based on the graphs:** 1. What are the ranges of the temperatures? 2. How accurate were the forecasts? 3. Where did the temperatures vary the most? and the least?

Example 2

Issue/Question	Independent Variable	Original Dependent Variables	Derived Variables and New Questions
Last week's temperatures in our city	**Days** of the week, Monday to Sunday	**AMx (Monday)** = last Monday's actual max temperature	**AccMax (Monday)** = Difference between the forecasted and the actual max temperature = **AMx (Monday)** – **AMax (city)**
		AMi (Monday) = last Monday's actual min temperature . . .	**AccMin (city)** = Difference between the forecasted and the actual max temperature = **FMin (city)** – **AMin (city)**
		AMx (Sunday) = last Sunday's actual max temperature	**AccMin (city)** = Difference between the forecasted and the actual max temperature = **FMin (city)** – **AMin (city)**
		AMi (Sunday) = last Sunday's actual min temperature	**Possible questions based on the graphs:** 1. How did the temperature change throughout the week? (observe graph) 2. How did the range of temperatures vary throughout out the week? 3. What were the average **AMx** and **AMi** during the week?

Example 3

Issue/Question	Independent Variable	Original Dependent Variables	Derived Variables and New Questions
How did the basketball team do in the last game?	**Names** of the players (M or W) of the high school varsity basketball team	**PPG (Player n)** = Points by Player n during last game	After creating the spreadsheet, if using a program such as Excel, sort the data by each variable—one at a time— to find the best three players for each category.
		RPG (Player n) = Rebounds by Player n during last game	1. Is there a player who is the first in more than one category?
		APG (Player n) = Assists by Player n during last game	2. Find the average for the team for each category. Compare with the statistics of an NBA team.
		SPG (Player n) = Steals by Player n during last game	
		FG (Player n) = Free throw points by Player n during last game	3. How would you say the team performs in each category from observing the graphs? Are there big disparities among players?
		3P (Player n) = Three-pointers made by Player n during last game	
		MPG (Player n) = minutes played by Player n during last game	

Solution Revolution

Roger P. Day

In *Consortium*, the newsletter for the Consortium for Mathematics and Its Applications, Froelich (1988) explored graphical strategies for solving the equation $2^x = x^{10}$. He described his use of a function plotter as he sought a graphical solution for that equation. The NCTM's *Curriculum and Evaluations Standards for School Mathematics* (1989) has proposed the increased use of function plotters and other technology to investigate and solve problems. Equations such as $2^x = x^{10}$ can be used to illustrate the potential of that technology and to consider its implications for the teaching and learning of mathematics.

The meaningful application of technology strongly suggests the development of a new collection of basic skills for algebra. In this article, I review Froelich's graphical strategies and offer additional alternatives to algebraic methods. These alternative strategies will help us to identify the new basic skills our students need to master as technology-rich environments emerge within and outside our classrooms.

ALTERNATIVE STRATEGIES FOR SOLUTION

On realizing that $2^x = x^{10}$ does not lend itself to traditional algebraic methods for isolating the unknown, we may wonder what solution options are available. Although no strategy is available for deriving an exact solution, several graphical and numerical approximation techniques exist. Before going on, readers are encouraged to explore strategies for solving $2^x = x^{10}$.

Graphical strategies

When applying graphical approximation techniques, a typical first step is to use a function plotter to graph the expressions on each side of the equals sign and to examine those graphs for points of intersection. The graphs of $y = 2^x$ and $y = x^{10}$ are shown in **figure 1a**. It appears that two solutions exist, one near $x = -1$ and one near $x = 1$. Using the zoom-in and trace features of the function plotter, we can better estimate the coordinates of the intersection points and use the x-coordinates as approximate solutions for $2^x = x^{10}$. **Figures 1b** and **1c** illustrate the zoom-in sequence near $x = -1$.

If we check this solution using another strategy or make a seemingly futile algebraic attempt to use logarithms, we can transform the original equation to get $x(\ln 2) = 10(\ln x)$. Graphing the expressions on each side of the equals sign yields the graphs shown in **figure 2**. Notice that the solution near $x = -1$ does not appear on this graph, although a new solution may exist, near $x = 60$.

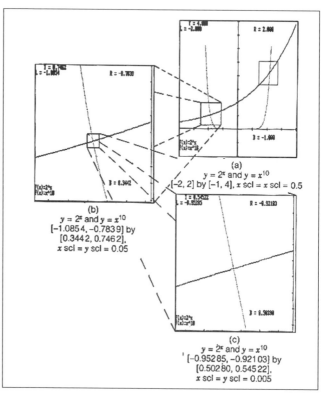

Fig. 1. Graphical zoom-in applied to $2^x = x^{10}$

At least three distinct solutions to te equation $2^x = x^{10}$ seem to exist. Are there others?

Let us analyze what we have discovered. In our second solution attempt, the apparent solution near $x = -1$ is not shown because $y = \ln x$ is not defined for $x \le 0$. Therefore, the absence of any second- or third-quadrant intersection points in **figure 2** is justified. Looking at **figure 2**, suppose that a solution does exist near $x = 60$. Substituting, we have $2^{60} \approx 1.15 \times 10^{18}$. Using this estimate, we reset the viewing rectangle of the function plotter with an x-range of $[-5, 65]$ and a y-range $[0, 1.15 \times 1018]$. We view the graphs as in **figure 3**, showing $y = 2^x$ and $y = x^{10}$. Our third solution, near $x = 60$, is revealed.

How is that third solution justified? We see that x^{10} is indeed greater than 2^x for x immediately greater than p^1, where $x = p^1$ is the solution of $2^x = x^{10}$ near $x = 1$. How do the two functions behave as x increases? The function $y = 2^x$ increases more rapidly than $y = x^{10}$. At p^2, where $x = p^2$ is the solution near $x = 60$, $y = 2^x$ and $y = x^{10}$ are equal. For $x > p^2$, $y = 2^x$ continues to increase more rapidly than $y = x^{10}$, and therefore $y = 2^x$ remains greater than $y = x^{10}$ as x grows infinitely large.

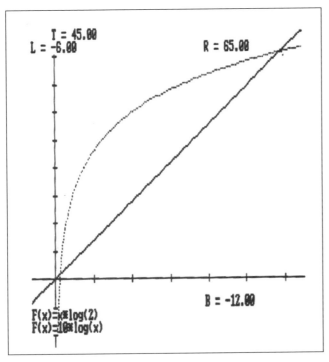

Fig. 2. Graphical zoom-in applied to $y = x \ln 2$ and $y = 10 \ln x$
[–6, 65] by [–12, 45], x scl = 10, y scl = 5

Using calculus we can justify that to compare more formally the rates of increase of the two functions in support of our argument.

Can we be sure that no more solutions exist to the original equation? A view within accurate scaling is drawn in **figure 4**. Both functions are continuous. The function $y = 2^x$ is ever increasing and always positive, and $y = x^{10}$ like a typical even-powered monomial function

with positive coefficient; it has a minimum at the origin and increases positively as $|x|$ grows large. **Figure 4** thus represents the complete behavior of each function, and we conclude that the two functions share exactly three common points. Zooming in on each intersection point yields solution estimates of -0.937, 1.078, and 58.770, each within one one-thousandth unit of an exact solution.

We have used two different graphs to determine solutions. Froelich (1988), however, sought a way to picture these three solutions on one viewing rectangle of a function plotter. Can readers determine a strategy for doing so? They are encouraged to try it before proceeding.

Rewriting the equation as $2^x - x^{10} = 0$, we plot the left expression using a viewing rectangle that covers the range of x-values for which solutions are known to exist. We see no x-intercept near $x = 60$ (see **fig. 5**). Why not? Even for values of x close to the desired solution, $|2^x - x^{10}|$ is extremely large. The finite number of pixel columns in the function plotter corresponds to x-values. Here these values range over a large interval and prevent us from getting as close as we would like to a value of x that makes $2^x = x^{10}$ close to zero.

Using this analysis, we need to minimize the effect of the large magnitude of the function values near $x = 60$. When 2^x does equal x^{10}, the ratio of the two expressions is 1. Thus we can examine the graphs of the left and right sides of the equation $2^x/x^{10} = 1$, shown in **figure 6**, and search for intersection points of the graphs. We can then see all three solutions on the same viewing rectangle of our function plotter.

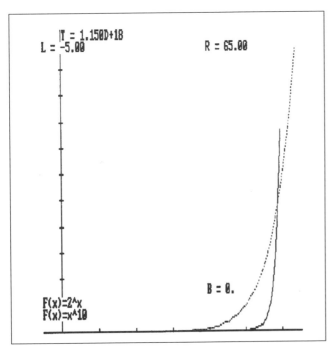

Fig. 3. $y = 2^x$ and $y = x^{10}$
[–5, 65] by [0, 1.15×10^{18}]
x scl = 10, y scl = 1.05×10^{17}

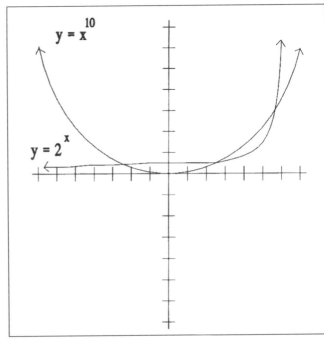

Fig. 4. An exaggerated look at the graphs of $y = 2^x$ and $y = x^{10}$
(scales inaccurate)

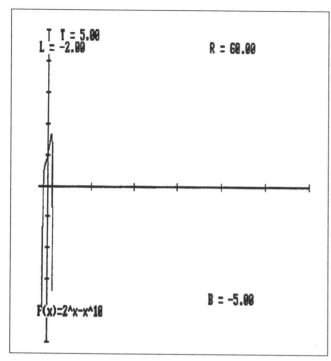

Fig. 5. $y = 2^x - x^{10}$
[–2, 60] by [–5, 5], x scl = 10, y scl = 1

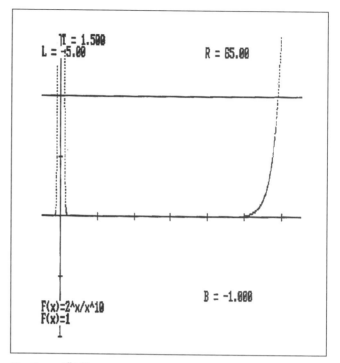

Fig. 6. $y = 2^x/x^{10}$ and $y = 1$
[–5, 65] by [–1, 1.5], x scl = 10, y scl = 0.5

Similarly, graphing the left expression of $2^x/x^{10} - 1 = 0$ resulting in all three solutions being visible in one viewing rectangle (**fig. 7**).

Numeric strategies

Numeric methods offer additional alternatives for solution approximations. Coupled with graphical strategies and the use of computer technology, numeric methods supply a reliable means for quickly generating approximations to a high degree of accuracy.

When viewing the graphs of $y = 2^x$ and $y = x^{10}$ in **figure 1**, we saw that a second-quadrant solution appeared near $x = -1$. We can create a spreadsheet table to examine the values of these two functions. (Software such as Mathematics Exploration Toolkit or Derive can also be used to generate tables, as can the calculator.) In **table 1a**, our initial x-value is -1 and we increment x by 0.1. Column A shows the x-values, and columns B and C display the associated values of $y = 2^x$ and $y = x^{10}$, respectively.

We search for the solution by considering where $y = 2^x$ changes from being less than $y = x^{10}$ to being greater than $y = x^{10}$, as seen in **figure 1**. Note that the change occurs between $x = -1.0$ and $x = -0.9$. In the spreadsheet we reset our initial value of x to the left endpoint of this interval—in this example, not changing it from the earlier value—and the increment to 0.01. This procedure divides into ten parts the interval within which the solution exists. **Table 1b** shows this step. We can continue this numerical zoom-in process until we reach our desired degree of accuracy, given the accuracy limitations of the spreadsheet program. **Tables 1c** and **1d** exhibit two more steps in this process.

We can refine this process by using another form of the equation $2^x - x^{10} = 0$. The left expression was graphed in **figure 5**. Next we seek the x-axis intercepts, values for x that make $2^x - x^{10} = 0$. **Table 2** shows the spreadsheet process applied to $2^x - x^{10}$.

This numeric zoom-in process, a variation on the guess-and-check strategy, can be compared with other numeric methods. As students, we may not have learned the

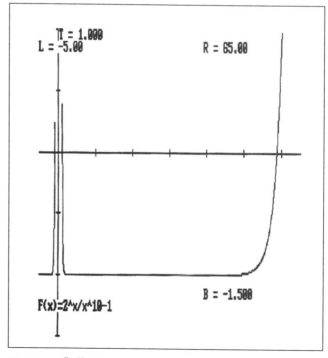

Fig. 7. $y = 2^x/x^{10} - 1$
[–5, 65] by [–1, 1.5], x scl = 10, y scl = 0.5

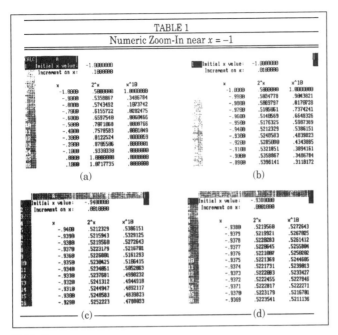

Table 1

TABLE 1
Numeric Zoom-In near x = -1

(a) Initial x value: -1.0000000 Increment on x: .1000000

x	2^x	x^10
-1.0000	.5000000	1.0000000
-.9000	.5358867	.3486784
-.8000	.5743492	.1073742
-.7000	.6155722	.0282475
-.6000	.6597540	.0060466
-.5000	.7071068	.0009766
-.4000	.7578583	.0001049
-.3000	.8122524	.0000059
-.2000	.8705506	.0000001
-.1000	.9330330	.0000000
.0000	1.0000000	.0000000
.1000	1.0717735	.0000000

(b) Initial x value: -1.0000000 Increment on x: .0100000

x	2^x	x^10
-1.000	.5000000	1.0000000
-.990	.5034778	.9043821
-.980	.5069797	.8170728
-.970	.5105061	.7374241
-.960	.5140569	.6648326
-.950	.5176325	.5987369
-.940	.5212329	.5386151
-.930	.5248583	.4839023
-.920	.5285090	.4343085
-.910	.5321851	.3894161
-.900	.5358867	.3486784
-.890	.5396141	.3118172

(c) Initial x value: -.9400000 Increment on x: .0010000

x	2^x	x^10
-.9400	.5212329	.5386151
-.9390	.5215943	.5329125
-.9380	.5219568	.5272643
-.9370	.5223179	.5216781
-.9360	.5226801	.5161293
-.9350	.5230425	.5106415
-.9340	.5234051	.5052083
-.9330	.5237681	.4998232
-.9320	.5241312	.4944918
-.9310	.5244947	.4892117
-.9300	.5248583	.4839023
-.9290	.5252223	.4780803

(d) Initial x value: -.9380000 Increment on x: .0001000

x	2^x	x^10
-.9380	.5219568	.5272643
-.9379	.5219921	.5267025
-.9378	.5220283	.5261412
-.9377	.5220645	.5255804
-.9376	.5221007	.5250202
-.9375	.5221369	.5244605
-.9374	.5221731	.5239013
-.9373	.5222093	.5233427
-.9372	.5222455	.5227846
-.9371	.5222817	.5222271
-.9370	.5223179	.5216781
-.9369	.5223541	.5211136

Table 2

TABLE 2
Numeric Zoom-In near x = 1

(a) Initial x value: 1.0000000 Increment on x: .1000000

x	2^x - x^10
1.0000	1.0000000
1.1000	-.4501955
1.2000	-3.8943397
1.3000	-11.3235604
1.4000	-26.2864497
1.5000	-54.8366119
1.6000	-106.9197296
1.7000	-198.3568885
1.8000	-353.5645204
1.9000	-609.3744938
2.0000	>>>>>>>>>>
2.1000	>>>>>>>>>>

(b) Initial x value: 1.0000000 Increment on x: .0100000

x	2^x - x^10
1.0000	1.0000000
1.0100	.9892090
1.0200	.8889245
1.0300	.6981079
1.0400	.5759834
1.0500	.4416352
1.0600	.2948038
1.0700	.1322828
1.0800	-.0448889
1.0900	-.2386233
1.1000	-.4501955
1.1100	-.6809645

(c) Initial x value: 1.0700000 Increment on x: .0010000

x	2^x - x^10
1.0700	.1322828
1.0710	.1152756
1.0720	.0981145
1.0730	.0807973
1.0740	.0633231
1.0750	.0456985
1.0760	.0278984
1.0770	.0099457
1.0780	-.0081690
1.0790	-.0264468
1.0800	-.0448889
1.0810	-.0634966

(d) Initial x value: 1.0770000 Increment on x: .0001000

x	2^x - x^10
1.0770	.0099457
1.0771	.0081415
1.0772	.0063357
1.0773	.0045283
1.0774	.0027193
1.0775	.0009087
1.0776	-.0009036
1.0777	-.0027175
1.0778	-.0045331
1.0779	-.0063582
1.0780	-.0081690
1.0781	-.0099894

bisection method or how to carry out *fixed-point iteration* until we enrolled in a calculus course. Computing technology changes that situation. These numeric strategies are available to, and comprehensible by, algebra students who can use technology to help approximate solutions.

We return to the equation $2^x - x^{10} = 0$ to illustrate the *bisection method*. This method requires that we determine an x-interval within which a continuous function changes sign. **Table 2a** shows that [1, 2] is such an interval. We then calculate the midpoint of that interval, x = 1.5, and determine the value of the function at that midpoint. If that value is 0, we have found a solution. Here, the solution is approximately −54.84. The midpoint creates two x-axis subintervals—[1, 1.5] and [1.5, 2]. We choose the interval over which a sign change occurs, in this example [1, 1.5]. We repeat the procedure using this interval. We calculate the midpoint, evaluate the function at that midpoint, and choose one of two new x-axis subintervals over which we again repeat the same procedure. We continue until either the function value at the midpoint is 0 or the interval length meets our required degree of accuracy. **Table 3** shows a spreadsheet application of the bisection method, reaching an accuracy level of 0.001.

Fixed-point iteration is another powerful numeric strategy. We illustrate and discuss this process as we search for the solution of $2^x - x^{10}$ near x = 60. First we transform the equation into the form x = g(x):

$$2^x = x^{10}$$
$$\ln(2^x) = \ln(x^{10})$$
$$x\ln 2 = 10\ln x$$
$$x = \frac{10\ln x}{\ln 2}$$

Next we choose an initial value for x. We intend to determine an x-value that remained unchanged by the function g, a so-called fixed point. Where this value is found, we have x = g(x) and therefore a solution to the original equation. If we let x = 60, g(60) ≈ 59.0689 and we see that 60 ≠ g(60). We have not found a fixed point.

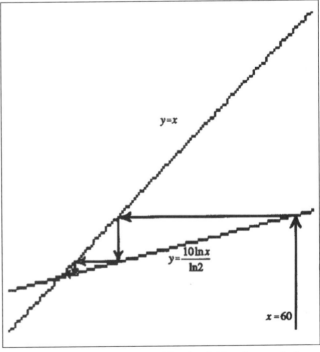

Fig. 8. Graphical representation of fixed-point iteration applied to $x = 10\ln x/\ln 2$ near x = 60

Then we begin iteration. We input $g(60) \approx 59.0689$ into $g(x)$ and again check whether the output is equal to the input. We have $g(g(60)) \approx g(59.0689) \approx 58.8433$. We still have no fixed point, but the difference between the input and the output has decreased. **Table 4** shows several more iterations, very quickly converging to an apparent solution near 58.7701059. This computation can be done using a spreadsheet program, by writing a short computer program, or by using the answer key on your calculator (TI-81 or Casio). The Appendix at the end of this discussion includes suggestions for using the TI-81 for numeric approximation techniques.

Figure 8 illustrates this process graphically. Here are graphs of $y = x$ and $y = g(x) = 10\ln x/2$, where the directed horizontal and vertical segments represent the iteration steps. Each vertical segment represents the calculation of a value of $g(x)$ for a particular x-value; each horizontal segment represents the difference between $g(x)$ and x. As the horizontal segments decrease in length, we converge on the desired fixed point.

Will fixed-point iteration always converge on a solution? Readers are invited to explore that question. They are encouraged to try to find a different solution to $2^x - x^{10}$ using the fixed-point method.

TECHNOLOGY IMPLICATIONS: NEW BASIC SKILLS FOR ALGEBRA

The application of technology to graphical and numeric investigation makes accessible powerful strategies and techniques that students will employ as they explore problems. Zooming in on an intersection point of two graphs, searching for sign changes in a table of values, and converging on a solution by using fixed-point iteration extend the traditional symbolic solution methods. In examining these graphical and numeric strategies, it becomes clear that for those strategies to be meaningful, students must comprehend mathematical concepts and skills. Without that understanding, our students may simply be turning a new crank on the same old incomprehensible solution machine. The technology that opens the door to such powerful methods thus forces us to emphasize the mathematics that justified the methods. We must help our students develop the following basic skills if we intend them to apply graphical and numeric methods with meaning.

A toolkit of functions

Critical to the meaningful use of graphical methods is a basic understanding of elementary functions, functions we call toolkit functions (North Carolina School of Science and Mathematics 1986, p. 25). In our illustrations, we needed to recognize specific toolkit functions as well as transformations of those function; $y = x$, $y = x^a$, $y = a^x$, and $y = \ln x$. Students should be able to sketch the graphs of basic constant, linear, polynomial, logarithmic, nth root, reciprocal, absolute-value, greatest-integer, exponential, and trigonometric functions. They

TABLE 3				
Solution Search Using the *Bisection Method*				

1.0000 left endpoint (start)		1.0000	value of 2^x - x^10 @ L Endpt	
2.0000 right endpoint (start)		-1020.0000	value of 2^x - x^10 @ R Endpt	

Interval No.	Left Endpoint	Right Endpoint	Interval Midpoint	Value of 2^x - x^10 at Midpoint
0	1.0000000	2.0000000	1.5000000	-54.8366
1	1.0000000	1.5000000	1.2500000	-6.9348
2	1.0000000	1.2500000	1.1250000	-1.8663
3	1.0000000	1.1250000	1.0625000	.2550
4	1.0625000	1.1250000	1.0937500	-.3158
5	1.0625000	1.0937500	1.0781250	-.0104
6	1.0625000	1.0781250	1.0703125	.1270
7	1.0703125	1.0781250	1.0742188	.0595
8	1.0742188	1.0781250	1.0761719	.0248
9	1.0761719	1.0781250	1.0771484	.0073
10	1.0771484	1.0781250	1.0776367	-.0016
11	1.0771484	1.0776367	1.0773926	.0029
12	1.0773926	1.0776367	1.0775147	.0006
13	1.0775146	1.0776367	1.0775757	-.0005
14	1.0775146	1.0775757	1.0775452	.0001

Table 3

TABLE 4	
Solution Search Using *Fixed-Point Iteration*	

begin iteration at x =	60

x	g(x)
60.0000000	59.0689060
59.0689060	58.8432699
58.8432699	58.7880551
58.7880551	58.7745115
58.7745115	58.7711874
58.7711874	58.7703714
58.7703714	58.7701711
58.7701711	58.7701219
58.7701219	58.7701099
58.7701099	58.7701069
58.7701069	58.7701062
58.7701062	58.7701060
58.7701060	58.7701060
58.7701060	58.7701059
58.7701059	58.7701059
58.7701059	58.7701059
58.7701059	58.7701059

Table 4

should know each function's domain and range and be able to identify graphs that correspond to these toolkit functions. Manipulating these functions with transformations that include horizontal and vertical shifts and stretches is also a basic expectation.

Knowledge of the behavior of these toolkit functions is essential for a meaningful application of graphical strategies and is one of the new basic skills in evolving algebra curriculum. Without knowing the end behavior of the graphs of $y = 2^x$ and $y = x^{10}$, for instance, determining the existence of all solutions of $2^x = x^{10}$ is much more difficult. Demana and Waits (1990) stress the need to know the *complete behavior* of functions when using electronic function plotters to demonstrate how the function behaves at extremely large absolute values of x. Those authors also emphasize examining the behavior of a function as x approaches some point on the border of the function's domain, such as recognizing that $y = \ln x$ grows indefinitely negative as x approaches zero from the positive side. *Hidden behavior* is another consideration, a condition wherein the behavior of a function may not be readily apparent from one particular view of its graph (Demana and Waits 1988). Fundamental knowledge and understanding of the behavior of toolkit functions are essential components in a meaningful application of graphical strategies.

Connecting Algebraic, Graphical, and Numeric Representations

Another basic skill students must develop as they exploit technology is an understanding of the connection among the symbolic, the graphical, and the numeric representations of a mathematical relation. Students must come to associate a graph and a table of values with a symbolic statement, as illustrated in **figure 9**. Students must recognize the equivalence of those three representations, be able to use each representation to its advantage, and use technology to manipulate correctly those representations as desired in exploring problems.

Accuracy

When using traditional algebraic strategies to solve equations, we virtually always generated an exact solution. Checking accuracy meant only that our solution correctly satisfies the original equation or that we could reverse the algebraic steps we had applied. As technology unleashes powerful graphical and numeric approximation strategies, such as those illustrated here, the issue of accuracy becomes much more significant.

How will we use our approximate solution? The degree of accuracy required when practicing solution techniques in the classroom may differ from that required by a structural engineer determining the maximum weight to be borne by a bridge. What degree of accuracy is required to make us confident in our solution? How can we be sure that we indeed have a legitimate solution? What methods generate the greatest degree of accuracy most efficiently? Facing new questions of accuracy will quickly become fundamental as we employ technology to help solve problems.

Problem solving

The meaningful application of concepts and skills required to use graphical and numeric methods comprises a unique collection of problem-solving strategies. Dion (1990) believes that more analysis and decision making are required when we apply graphical solution strategies than when we apply algebraic manipulations, and the same can be said about applying numeric strategies. Techniques and insights such as those illustrated here will become part of an expanded collection of heuristics for the technology-rich environments emerging in our classrooms.

For example, we would seldom transform $2^x - x^{10}$ to yield $2^x/x^{10} = 1$ when applying traditional algebraic methods. That procedure moves us no closer to isolating the variable, which, after all, is what we stress in using algebraic manipulations. For Froelich (1988), however, this manipulation was a critical step in reducing the extreme absolute value magnitude of $2^x - x^{10}$ for a large x. He used the ratio to help overcome that obstacle as

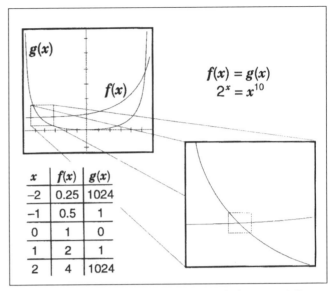

$$f(x) = g(x)$$
$$2^x = x^{10}$$

x	$f(x)$	$g(x)$
−2	0.25	1024
−1	0.5	1
0	1	0
1	2	1
2	4	1024

Fig. 9. Connecting symbolic-graphical-numerical representations

he sought a solution employing graphical strategies.

Likewise, rewriting $2^x - x^{10}$ as $x = 10 \ln x/\ln 2$ would not be encouraged in a purely algebraic solution, yet it is required for the applications of fixed-point iteration.

When graphically solving equations of the form $f(x) = g(x)$, we typically apply difference and ratio functions. On the TI-81 graphing calculator, a student may enter $f(x)$ as Y1 and $g(x)$ as Y2, then Y3 as Y1 − Y2 and Y4 as (Y1/Y2) −1. Looking for zeros of either Y3 or Y4 helps students use graphical strategies with less concern for the appropriate viewing rectangle required for a particular function.

Choosing among approximation techniques is itself an important factor. Fixed-point iteration is remarkably efficient yet cannot be applied to all equations. The bisection method is straightforward in application but may be meaningless for some students without an accompanying graph to help them see the axis intercept. Applying algebraic techniques will remain important, either as the sole method of solution or as a tool to transform an equation before applying graphical or numeric strategies. Appropriate decisions regarding what technique to apply become more frequent when technology offers alternative strategies.

SUMMARY AND RECOMMENDATIONS

With the power of technology at our fingertips, graphical and numeric strategies offer great potential as alternative methods for solving equations. As classroom teachers, we need to master these skills and techniques ourselves and help our students become adept at applying these methods. As we help our students master the procedures, we must also consider the implications for the *meaningful* application of these strategies and thereby afford our students opportunities to develop and value these new basic skills of algebra.

For the meaningful interpretation of graphical representations, students who use function plotters must have a basic knowledge of toolkit functions. They need to recognize these functions, transform them, and describe their behavior. Students must know that tables, graphs, and equations can be used to represent the same phenomenon; they should comfortably move among those representations; and they need to master the application of technology to manipulate those representations. Questions of a solution's accuracy become more critical as students generate approximate, rather than exact, solutions. They must determine the desired level of accuracy, partly on the basis of how a particular solution will be used. Finally, with our help, students will develop and use novel strategies to aid the graphical and numeric methods that help exploit the power and overcome the limitations of technology.

REFERENCES

Demana, Franklin, and Bert K. Waits. "Pitfalls in Graphical Computation, or Why a Single Graph Isn't Enough." *College Mathematics Journal* 19 (March 1988): 177–83.

———. *Precalculus Mathematics: A Graphing Approach.* Reading, Mass.: Addison-Wesley Publishing Co., 1990.

Dion, Gloria. "The Graphics Calculator: A Tool for Critical Thinking." *Mathematics Teacher* 83 (October 1990): 564–71.

Froelich, Gary. "The IBM Toolkit." *Consortium* 27 (Fall 1988): 8.

———. "Iteration." *Consortium* 37 (Spring 1991): 4–5.

National Council of Teachers of Mathematics (NCTM). *Curriculum and Evaluation Standards for School Mathematics.* Reston, Va.: NCTM, 1989.

North Carolina School of Mathematics and Science (NCSSM). *Functions.* Durham, N.C.: NCSSM, 1986.

APPENDIX

1. Here is a TI-81 program that performs the bisection method:

PrgmB:BISCTION	If Y/X $\geq$ 0
Disp "LEFT PT"	Goto 2
Input L	M $\rightarrow$ R
Disp "RIGHT PT"	Goto 1
Input R	Lbl 2
Lbl 1	M $\rightarrow$ L
(L + R)/2 $\rightarrow$ M	Goto 1
2^L – L^10 $\rightarrow$ X	Lbl 3
2^M – M^10 $\rightarrow$ Y	Disp "ROOT IS"
2^R – R^0 $\rightarrow$ Z	DISP M
If abs Y $\leq$ .0001	End
Goto 3	

In lines 7 through 9, the user must supply the function in question. The accuracy level of .0001 can be changed in line 10.

2. To iterate $x = 10 \ln x/\ln 2$ using the TI-81 program, use the following procedure:

a. Enter the initial value of x on the screen and press ENTER.

```
60
                    60




```

b. Enter the function to be iterated, with ANS (the answer key) substituted for x, and press ENTER.

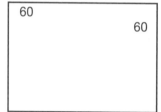

```
60
                    60
101ln Ans/ln 2
            59.06890596
```

c. Touch the ENTER key repeatedly until you've reached the desired accuracy.

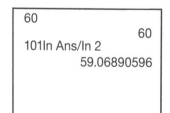

```
60
                    60
101ln Ans/ln 2
            59.06890596
            58.84326986
            58.78805513
            58.77451145
```

3. Froelich (1991) supplies this TI-81 program for a geometric demonstration of the iteration process.

```
Prgm2:ITERATE

ClrDraw

Input A

DispGraph

Line (Xmin, Xmin, Xmax, Xmax)

A → X

Line (X, 0, X, Y1)

Pause

Lbl A

Pause

Y1 → X

Line (X, X, X, Y1)

Pause

Goto A

End
```

The function to iterate is typed as Y1 using the Y = key. The range is set using the RANGE screen. The program prompts the user for the initial value of x and pauses after each segment is drawn. The user must press ENTER to see the next segment. Touching the calculator's ON key will cause [the] program to terminate (Froelich 1991, p. 5).

4. The graphs illustrated throughout this article were created using Master Grapher on am IBM-compatible computer. Those graphs were captured using Pizazz Plus and imported as PCX graphics into WordPerfect 5.1. Some graphs were enhanced using Accudraw for WordPerfect. The spreadsheet illustrations were programmed using SuperCalc4 and imported into the word processor as described in the forgoing. All figures were originally printed on a LaserJet IIp laser printer.

Solution Revolution

1. What strategies might we use to solve $2^x = x^{10}$?

2. Can technology help in solving the equation? How might a graphing calculator be useful? How might a spreadsheet be useful?

3. Can you describe the two functions in the equations $y = 2^x$ and $y = x^{10}$? What do the graphs of these functions look like? Can this information help in finding solutions?

4. How can we decide whether solution(s) exist? How do we know whether we've found them all?

Solution Revolution

5. If solution(s) exist, how can we determine how many exist?

6. Can we find an approximate solution? If so, how? If not, why not?

7. Can we find an exact solution? If so, how? If not, why not?

8. Can we check whether our solutions (approximate or exact) are correct or nearly correct? How would we determine this? How close do we need to be? Why?

Promote Systems of Linear Inequalities with Real-World Problems

Thomas G. Edwards and Kenneth R. Chelst

A key underlying principle of NCTM's *Curriculum and Evaluation Standards* (1989) is the notion that instruction ought to arise out of problem situations. Moreover, since mathematics is a foundation for an ever-widening array of disciplines, "the curriculum for all students must provide opportunities to develop an understanding of mathematical models, structures, and simulations applicable to many disciplines" (NCTM 1989, 7).

Mathematical models, structures, and simulations are precisely the tools of operations research. The field of operations research is a rich source of real-world problem situations to which students can easily relate and within which mathematical concepts may be developed or skills practiced. For example, graphing systems of linear equations and inequalities, often without a meaningful practical context, has long been a staple of beginning high school algebra courses. At the same time, the graphs of such systems are typically used in operations-research textbooks to develop the concepts of linear programming, which are essential to understanding the solution of complex optimization problems (see, e.g., Winston [1994]).

We believe that bringing scaled-down real-world problem situations similar to those tackled by operations researchers into high school mathematics better motivates students to learn mathematics. Problems set in everyday situations have great potential for attracting and holding the attention of high school students because they deal with situations with which students have experience, for example, the clothes they wear, the places they work, or the lines in which they wait.

A BRIEF HISTORY OF OPERATIONS RESEARCH

Operations research has its roots in the years just before World War II, when the British prepared for the anticipated air war. In 1937, a new device, later called *radar*, was field-tested. The following summer, experiments began that explored how the information provided by radar could be used to direct the deployment and use of fighter planes. Until that time, the word experiment had conjured up the picture of a scientist carrying out a controlled *experiment* in a laboratory. In contrast, the multidisciplinary team of scientists working on this radar-fighter-plane project studied the actual operating conditions of these new devices and designed experiments in the field of operations. Thus was born the new term *operations research*. The team's goal was to understand the operations of the complete system of equipment, people, and such environmental conditions as weather or darkness and then improve on it. The work was an important factor in winning the Battle of Britain, and operations research eventually spread to all the military services.

In the United States, the first team working on anti-submarine tactics paralleled this approach. That group developed a series of mathematical models that they called *search theory*, which was used to develop optimal patterns of air search. Like their British counterparts, they got close to the action by riding in airplanes on patrol, just as the modern operations researcher might ride in a police car or spend time in an automotive-assembly plant. Every branch of the military currently has its own operations-research group that includes both military and civilian personnel. These groups play a key role in both long-term strategy and weapons development, as well as in directing the logistics of such actions as Operation Desert Storm. In addition, the National Security Agency has its own Center for Operations Research.

Operations research moved into the industrial domain in the early 1950s, and its growth paralleled the growth of the computer as a business-planning and management tool. As the field evolved, the core moved away from interdisciplinary teams to focus on developing mathematical models that can be used to model, improve, and even optimize real-world systems. These mathematical models include both deterministic models, such as mathematical programming, routing, or network flows; and probabilistic models, such as queuing, simulation, and decision trees.

MATHEMATICAL PROGRAMMING

The father of linear programming is George Dantzig, who developed its basic concepts between 1947 and 1949. During World War II, Dantzig worked on developing various plans, or proposed schedules, of training, logistical supply, and deployment, which the military calls *programs*. After the war, he was challenged to find an efficient way to develop these programs. Dantzig recognized that the planning problem could be formulated as a system of linear inequalities.

His next challenge involved the concept of a goal. When managers thought of goals at that time, they generally meant rules of thumb for carrying out a goal. For example, a naval officer might say, "Our goal is to win the war, and we can do that by building more battle-ships." Dantzig was the first to express the criterion for selecting a good or best plan as an explicit mathematical function, which we now call the *objective function*. All this work would have had limited practical value without an efficient method, or *algorithm*, for finding the best, or *optimal*, solution to a set of linear inequalities that maximizes an objective function, such as profit, or minimizes an objective function, such as cost. Dantzig developed the *simplex algorithm*, which efficiently solves this problem. Interestingly, in 1939 the Soviet mathematician and economist L. V. Kantorovich formulated and solved a linear-programming problem dealing with production planning. However, his work was essentially unknown even in the Soviet Union for twenty years and had no impact on the post–World War II development of linear-programming.

As mainframe computers grew more powerful, the first major users of the simplex algorithm to solve practical problems were the petroleum and chemical industries. One use was in minimizing the cost of blending gasoline to meet performance and content criteria. The field of linear programming grew rapidly and led to the development of nonlinear programming, in which inequalities or the objective function are nonlinear functions. In another extension of linear programming, called *integer programming*, some variables may take on only integral values. These techniques are known collectively as *mathematical programming*.

In addition to the blending example, other applications of mathematical programming include scheduling workers to minimize labor costs, using a pattern or template that minimizes waste in cutting stock, and determining the production levels of different products to maximize a company's profit. In this article, we develop two problem situations of the last type, known as *product-mix problems*, through a series of student activities. Both examples were originally adapted for classroom use by an operations researcher. Although these examples are scaled-down versions of actual problems from industry, where such problems typically involve thousands of variables, they retain a real-world flavor.

MODELING A PRODUCT-MIX PROBLEM: THE LEGO® FACTORY

Pendegraft (1997) suggests a way to use Lego plastic construction toys to model a product-mix problem situation for college students studying linear programming. A similar approach is easily accessible to high school algebra students.

The problem

Suppose that a factory manufactures only tables and chairs and that the profit on one chair is $15 and on one table is $20. Each chair requires one large piece of stock and two small pieces of stock, that is, one large Lego or other construction block and two small ones.

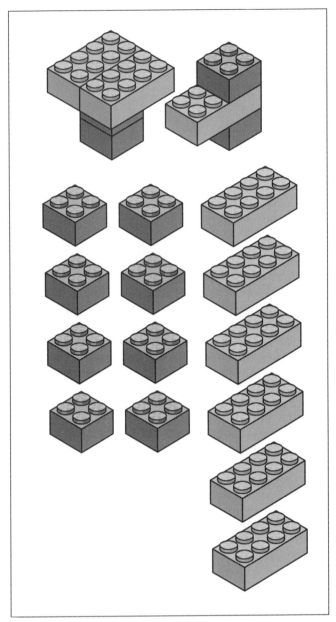

Fig. 1. A Lego table, chair, and available resources.

Each table requires two large and two small pieces of stock. **Figure 1** shows a table and a chair. Finally, suppose that you have only six large and eight small pieces of stock. How many chairs and how many tables should you build to maximize profit?

A concrete exploration

The student activity works best if students actually have construction blocks such as Lego blocks with which to build "tables" and "chairs." However, if they are unavailable, other materials could be substituted, for example, two different sizes or colors of plastic tiles or cubes. It is important to allow students an opportunity to see what they can build using the available materials and to determine the profit for each possibility. Later, during a more abstract exploration of this problem, the abstract concepts can easily be linked to this more concrete exploration.

Students might compete in pairs or in small groups to obtain the optimal solution. Students will quickly see that only four logical possibilities for the optimal solution exist—(1) four chairs, (2) three chairs and one table, (3) two chairs and two tables, and (4) three tables—and they should have no trouble determining the profit for each possibility. It will then be evident that building two chairs and two tables maximizes profit. **Figure 2** shows this solution.

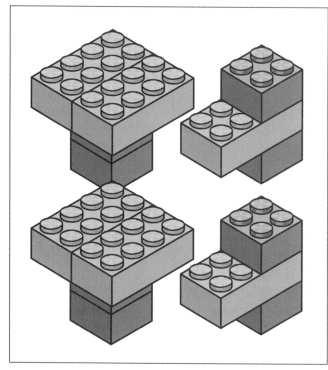

Fig. 2. Two of each is the optimal solution

An abstract exploration

After students have completed their concrete exploration of the Lego problem, that exploration can be linked to a more abstract one that uses the terminology of mathematics and operations research. The steps in the abstract exploration involve—

- Defining a set of *decision variables* that completely describe the decision to be made;

- Modeling the problem's objective by using the decision variables to define an *objective function*;

- Identifying any *constraints*, or restrictions, on the decision variables, such as the limited resources available;

- Graphing the system of constraints to locate a *feasible region*; and

- Determining which solution within the feasible region is the optimal solution.

On the basis of their concrete exploration of the Lego example, students should be able to identify two decision variables—

$$C = \text{the number of chairs built}$$

and

$$T = \text{the number of tables built.}$$

Students should next model the objective of the problem by defining the profit, P, as the objective function $P = 15C + 20T$. They should also identify two constraints on what they were able to build, because they were given only six large and eight small Lego blocks. Translating these constraints into inequalities using the decision variables may take some probing. Many students may try to write such constraints as—

$$1 \text{ table} = 2 \text{ large pieces} + 2 \text{ small pieces}$$

or

$$1 \text{ chair} = 1 \text{ large piece} + 2 \text{ small pieces}$$

because that is how each item is constructed. However, the constraints concern the consumption of limited resources, so students will need to focus separately on the number of large and small Lego pieces and how each of these resources is consumed in constructing a table or a chair. Since a chair requires one large piece and a table requires two large pieces, students should find that $1C + 2T \leq 6$, since a chair and a table each require two small pieces, $2C + 2T \leq 8$.

After students identify the system of constraints, they should graph the system to locate the feasible region. Because students often do not view graphs in the same ways that teachers do (Dunham and Osborne 1991), you may want to include some activities to help students decide which half-plane to include in the graph of each inequality. For example, students might test representative points to see whether they satisfy the inequality. **Figure 3** shows a graph of the feasible

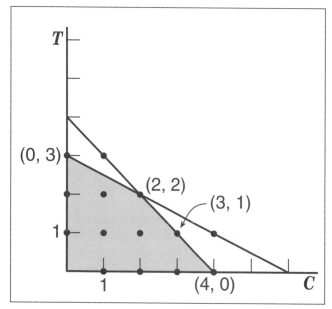

Fig. 3. The feasible region

region with the four logical possibilities for the optimal solution labeled with their coordinates. Students should be asked to interpret each of those points as they relate to the problem situation, so that they begin to link the graph with their previous exploration.

You may next want to discuss with the students the discrete nature of the feasible region for this example. Since the decision variables must take on integral values, the feasible region actually consists only of the lattice points in the shaded region. In real-world applications, these problems are usually formulated in terms of hourly or weekly production rates, and continuous variables are acceptable. The feasible region is then the entire shaded region.

You should also determine whether students understand why the possibility of building two chairs and two tables renders building one chair and two tables nonoptimal. When students comprehend this concept, the teacher can ask whether they notice anything about the location in the feasible region of the four points that were logical possibilities for the optimal solution. The location of these points on the boundary of the feasible region will play an important role in the next student exploration.

SOLVING A PRODUCT-MIX PROBLEM: THE HIGH STEP SHOE CORPORATION

The example that follows presents students with a somewhat more abstract exploration of a product-mix problem situation. This example develops the principle that a unique optimal solution to a linear-programming problem always occurs at a *corner point* on the boundary of the feasible region.

The problem

Mr. M. Jordan, director of manufacturing for the High Step Sports Shoe Corporation, wants to maximize the company's profits. High Step makes two brands of sport shoes, Airheads and Groundeds. Each pair of Airheads returns a $10 profit, and each pair of Groundeds returns a profit of $8.50.

The steps in manufacturing the shoes include cutting the materials on a machine and having workers assemble the pieces into shoes. Six machines cut the materials, 850 workers assemble the shoes, and the factory operates forty hours per week.

Each cutting machine can actually perform only fifty minutes of work in an hour because of the time required to calibrate and maintain the machines. Each pair of Airheads requires three minutes of cutting time, whereas each pair of Groundeds requires two minutes of cutting time. A worker takes an average of seven hours to assemble a pair of Airheads and an average of eight hours to assemble a pair of Groundeds.

Mr. Jordan's goal is to maximize High Step's profits, subject to all those constraints.

The student exploration

After students have read and discussed the problem situation, they should identify the decision variables and define an objective function. In this problem, the decision variables are—

A = the number of pairs of Airheads manufactured each week

and

G = the number of pairs of Groundeds manufactured each week.

The objective function is then profit, $P = 10A + 8.5G$.

Next, students must explore the constraining conditions. First, one constraint is related to the total machine time available per week. Since each machine can perform only fifty minutes of work each hour, the six machines working together can provide only

$$6 \times 50 \text{ min/h} \times 40 \text{ h/wk.} = 12{,}000 \text{ min/wk.}$$

of cutting time. We know that each pair of Airheads requires three minutes of cutting time and each pair of Groundeds requires two minutes of cutting time, so

$$3A + 2G \leq 12{,}000.$$

A second constraint is related to the total worker time available each week for assembly. The 850 assembly workers provide

$$850 \times 40 = 34{,}000 \text{ h/wk.}$$

of assembly time. Again, we know that seven hours are required to assemble each pair of Airheads and eight hours to assemble each pair of Groundeds. Thus,

$$7A + 8G \leq 34{,}000.$$

Some students may not see why it is desirable to express the machine and assembly times available on a weekly basis. The class could discuss the arbitrary nature of this aspect of the problem formulation. These rates, as well as the decision variables, could just have well been expressed on a daily or hourly basis. It is important only to express these rates consistently.

Some students may also be bothered because the machine constraint is expressed in minutes per week, whereas the assembly constraint is expressed in hours per week. You may want students to reformulate the assembly constraint using minutes per week as the basic unit. Converting 7, 8, and 34 000 hours to minutes yields

$$420A + 480G \leq 2{,}040{,}000.$$

But then factoring 60 out of each term in the last inequality and dividing both sides by 60 leaves the original assembly constraint intact:

$$7A + 8G \leq 34{,}000$$

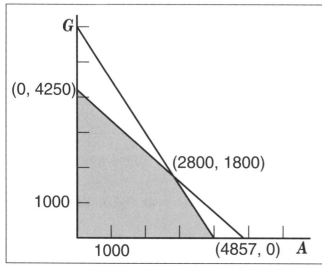

Fig. 4. High Step Shoes feasible region

Two constraints remain that are not explicitly stated in the description of the problem. Since A and G both represent a number of pairs of shoes manufactured in a week, each variable is nonnegative. Thus, we have $A \geq 0$ and $G \geq 0$. You may want to ask students whether either or both of these variables could be zero.

The system of constraints defines the feasible region, which is shown in **figure 4**. The magnitude of appropriate values for A and G makes the scale of the graph an important consideration. Because the exploration of optimality that follows depends on an accurate graph, you will want to be certain that each student or group of students has such a graph before proceeding to the optimal solution. It is also helpful for students to find the coordinates of the vertices of the feasible region at this time, as those points will be important in developing the optimal solution.

The best solution for this problem gives the High Step Sports Shoe Company a maximum profit. To determine this optimal solution, a number of strategies could be used. However, if the problem has a unique optimal solution, that solution must lie at one of the corner points of the feasible region. This principle is known as the corner principle.

One way to help students understand the corner principle is to identify a number of points in the feasible region then use the objective function to compute the profits for each point selected. Small groups of students could be assigned the task of finding in the feasible region the point that generates the maximum profit. After students have computed the profit, P_i, for a number of points in the feasible region, they should graph an equation $10A + 8.5G = P_i$ corresponding to each point. Students will then need only to observe that $10A + 8.5G = P_i$ defines a family of parallel lines and that the further right and higher the line lies in the coordinate system, the greater the profit is. Thus, the line that is highest and farthest right while still intersecting the

feasible region always passes through a corner point, and this point corresponds to the optimal solution.

Figure 5 shows the feasible region of the High Step Shoe problem and a family of parallel lines, including the line passing through (2800, 1800), the optimal solution.

When the family of parallel lines defined by the objective function is actually parallel to the boundary of the feasible region, the optimal solution may no longer be unique. Any point lying on that portion of the boundary is optimal. In an example where the decision variables must take on integral values, as in the High Step example, more than one point along that boundary could have integral coefficients. The optimal solution then is not unique. If the nature of the problem does not restrict the decision variables to integral values, then any point along that portion of the boundary parallel to the family of lines defined by the objective function is optimal.

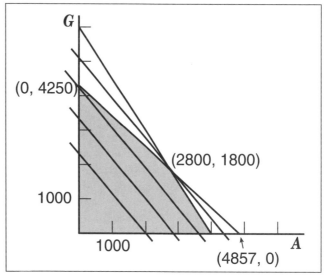

Fig. 5. A family of parallel lines

A PRODUCT-MIX PROBLEM FROM THE REAL WORLD

Brown, Graves, and Honczarenko (1987) describe the use of a linear-programming model to solve a real problem for Nabisco. Production in the Biscuit Division of Nabisco involves baking and such secondary operations as sorting, packaging, and labeling.

Scheduling and operating bakeries are difficult tasks. Each oven can produce some, but not all, of the products. The efficiency of the ovens varies. Several ovens can simultaneously share the facilities for secondary operations at one site. In addition, production must be planned to keep manufacturing and transportation costs at a minimum.

Some questions addressed by the mathematical model at Nabisco include the following:

- Where should each product be produced?
- How much of each product should be assigned to each oven?
- From where should each product be shipped to each customer?

A realistic problem at Nabisco could involve 150 products, 218 facilities, 10 plants, and 127 customer zones. A problem of this size involves more than 40,000 decision variables and almost 20,000 constraints.

Obviously, a problem of this magnitude cannot be solved using a paper-and-pencil graphical approach. However, the same sort of problem-formulation skills that are developed in such examples as the Lego factory or High Step Shoes are used to construct mathematical models of such problems as Nabisco's. After the problem has been mathematically formulated, it can be solved routinely using a computer with appropriate software. Indeed, when the Nabisco problem was solved in 1983 on an IBM 3033 computer, less than sixty seconds of CPU time was required. Today, the same problem is solvable on a microcomputer.

CONCLUSION

Mathematical programming is just one in an array of mathematically based techniques that are used in the field of operations research and are accessible to high school students. Other techniques include routing, queuing, logistics, and simulation. Because operations researchers solve problems in the real world, operations-research-based problems have rich connections to the world in which students live and work. Drawing on such problem situations is one way in which teachers can let applications of mathematics drive instruction. We believe that doing so will better motivate students to learn the mathematics they encounter in the classroom.

BIBLIOGRAPHY

Dantzig, George B. "Reminiscences about the Origins of Linear Programming." *Operations Research Letters* 1 (November 1982): 43–48.

> The reader may enjoy Dantzig's own account of the birth of linear programming.

Miser, Hugh J. "Operations Research." *In International Encyclopedia of Business and Management*, edited by Malcolm Warner, 3743–57. Boston: International Thomson Business Press, 1996.

> Miser provides a brief history of operations research.

Smith, Karan B. "Guided Discovery, Visualization, and Technology Applied to the New Curriculum for Secondary Mathematics." *Journal of Computers in Mathematics and Science Teaching* 15 (summer 1996): 383–99.

> Teachers with access to a computer laboratory and a mathematical software package, such as Derive, may be interested in Smith's use of technology to facilitate student exploration of concepts of linear programming.

REFERENCES

Brown, Gerald G., Glenn W. Graves, and Maria Honczarenko. "Design and Operation of a Multi-commodity Production/Distribution System Using Primal Goal Decomposition." *Management Science* 33 (November 1987): 1469–80.

Dunham, Penelope H., and Alan Osborne. "Learning How to See: Students' Graphing Difficulties." *Focus on Learning Problems in Mathematics* 13 (fall 1991): 35–49.

National Council of Teachers of Mathematics (NCTM). *Curriculum and Evaluation Standards for School Mathematics*. Reston, Va.: NCTM, 1989.

Pendegraft, Norman. "Lego of My Simplex." *OR/MS Today* 24 (February 1997): 128.

Winston, Wayne L. *Operations Research: Applications and Algorithms*. 3rd ed. Belmont, Calif.: Duxbury Press, 1994.

The Lego Factory

We can use Lego pieces to model production in a furniture factory. The company produces only tables and chairs. A table is made of two large and two small pieces, whereas a chair is made of one large and two small pieces. The resources available are six large and eight small pieces (see **fig. 1**). Profit for a table is $20 and for a chair is $15. Your task is to select a product mix to maximize the company's profits, using the available resources.

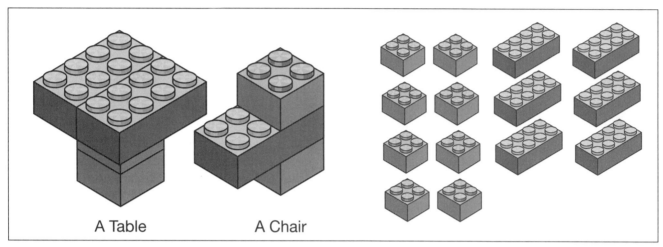

Fig. 1. A table, a chair, and the available resources

1. Use six large and eight small Lego pieces to explore this problem. What mix of tables and chairs that you were able to build with the available resources made the largest profit? To formulate this problem, if we let t represent the number of tables built and c the number of chairs, write an expression using t and c to represent the profit from building t tables and c chairs. _____

2. $P =$ _____

Recalling that only 6 large and 8 small pieces are available:

3. Every time you build a table, how many large pieces do you use? _____

4. Every time you build a chair, how many large pieces do you use? _____

5. Write an expression to represent the number of large pieces used if you build t tables and c chairs. _____
   _____

6. What can you say about the expression that you just wrote if you compare it with the number of available large pieces? Add that comparison to the expression you wrote in item 5. _____

7. Using similar reasoning, write a comparison between the number of small pieces used if you build t tables and c chairs and the number of available small pieces. _____
   _____

8. Could the numbers of either tables or chairs that you build be negative? Write expressions to represent these ideas. _____
   _____

High Step Sport Shoes:
A Linear Programming Problem

Mr. M. Jordan, director of manufacturing for the High Step Sport Shoe Corporation, wants to maximize the company's profits. The company makes two brands of sport shoe, Airheads and Groundeds. The company earns $10 profit on each pair of Airheads and $8.50 profit on each pair of Groundeds. The steps in manufacturing the shoes include cutting the materials on a machine and having workers assemble the pieces into shoes.

First we let A = the number of pairs of Airheads produced each week and let G = the number of pairs of Groundeds produced each week.

These variables are called *decision variables*, because Mr. M. Jordan's goal is to make the most money, or to *maximize* his profits. Using A and G, write a function to model the profit.

1.   Profit = _____ A + _____ G. This function is called the *objective function*.

The number of machines, workers, and factory operating hours put constraints on the number of pairs of shoes that the company can make. High Step Sport Shoe Corporation has the following constraints: 6 machines cut the materials, 850 workers assemble the shoes, and the assembly plant works a 40-hour week. Each hour, each cutting machine can do 50 minutes of work. How many minutes of work can 6 machines do in a 40-hour workweek?

2.   _____ machines × _____ min/hr × _____ hr/wk = _____ min/wk.

Each pair of Airheads requires 3 minutes of cutting time, whereas Groundeds require 2 minutes. Write an inequality representing the constraint of the cutting machines.

3.   _____ A + _____ G ≤ _____ minutes.

How many hours of work can 850 assembly workers do in a 40-hour week?

4.   _____ workers × _____ hr/wk each = _____ hr/wk total.

Each worker takes 7 hours to assemble a pair of Airheads and 8 hours to assemble a pair of Groundeds. Write an inequality representing the constraint of the amount of time it takes to assemble the shoes for the week.

5.   _____ A + _____ G ≤ _____ hours.

6.   Why do we use inequalities instead of equations to express the constraints? _____

The number of pairs of shoes that High Step manufactures is never negative but might be zero. Why? _____

_____

7.   Using A and G, write two more constraints to express the last idea. _____

High Step Sport Shoes:
A Linear Programming Problem

Sheet 3

The system of constraint inequalities is:

1) $3A + 2G \leq 12000$

2) $7A + 8G \leq 34000$

3) $A \geq 0$ and $G \geq 0$

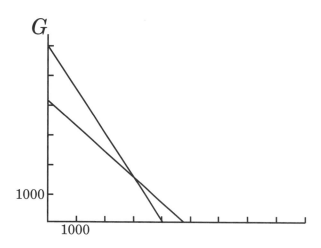

A graph of this system of constraint inequalities follows below. The shaded region represents the set of points that satisfy all the constraints. Values of A and G that satisfy all the constraint inequalities are called feasible, and the set of all such feasible points is called the *feasible region*.

8. *a.* Label each line with its equation.

   b. Label all points of intersection by solving the system of inequalities. These intersections are called *corner points*.

The best solution for this problem gives the High Step Sport Shoe Company a maximum profit. The process of determining this best solution is called *optimizing*, and the solution itself is called the *optimal solution*. To determine the optimal solution, you could use many strategies. One way is to try all the possible answers.

9. Pick three points inside the feasible region, and add them to the table. Then use the profit equation to calculate the profit for each of your points.

Compare your answers with those of other students and see who has the most profit. Who has the most profit in the class? How much profit was the highest?

10. Next, we are going to check the profit at the four corner points.

A = Airheads	G = Groundeds	Profit = $10A + 8.5G$
Example: 1,000	3,000	$35{,}500 = 10(1{,}000) + 8.5(3{,}000)$

11. Which point gives you the largest profit from both tables? _____
    This is an example of the *corner principle*.

A = Airheads	G = Groundeds	Profit = $10A + 8.5G$
0	0	$0 = 10(0) + 8.5(0)$

High Step Sport Shoes:
A Linear Programming Problem

Let's investigate why the corner principle works. The graph of the feasible region is on the second page of this exercise. Let's turn the profit equation around to yield $10A + 8.5G = P$, and substitute different quantities for the profit to see just how much money we can make for the High Step Company:

$10A + 8.5G = 25,000$

$10A + 8.5G = 30,000$

$10A + 8.5G = 40,000$

$10A + 8.5G = 45,000$

$10A + 8.5G = 50,000$

12. Solve each equation for G. What is the slope of each line? _____

    Draw each line on the graph of the feasible region on the second page of this exercise.

13. What do you notice about all these lines? _____

14. Are any of these values of P feasible? _____

15. Where will the line representing the *optimal solution* intersect the *feasible region*? _____

Geometrical Adventures in Functionland

Rina Hershkowitz, Abraham Arcavi, and Theodore Eisenberg

As long as algebra and geometry proceeded along separate paths, their advance was slow and their applications limited. But when these sciences joined company, they drew from each other fresh vitality and thenceforward marched on at a rapid pace toward perfection. — Joseph Louis LaGrange

This article discusses an approach in which algebra and geometry are interwoven in a series of problems that develop one from the other, forming an assignment of the kind described in Bruckheimer and Hershkowitz (1977). The two main concepts in this activity are the algebraic concept of the function and the geometric concept of the "family of quadrilaterals."

The geometric concept serves as the "real physical world" (Usiskin 1980) in which the student can develop the concept of function. Yet the algebraic way of thinking, necessarily involved in the concept of function and its application in the geometrical reality, stimulates the dynamic process that progressively extends geometrical ideas.

This assignment has been used with ninth-grade students of above-average ability and with teachers in in-service workshops. It has been particularly effective in eliciting the type of mathematical activity we believe should be at the core of the teaching of mathematics.

THE ASSIGNMENT

In the presentation of the activity, we shall describe the mathematical development of the problems and some of the strategies used by students in the different stages. In the description of students' behavior, we shall stress the effect of the algebraic-geometric interaction on the acquisition of concepts in the two domains.

Stage 1

From a given square of side length x, we create a new square whose side length is 5 cm longer.

1. Find an expression for the difference in area of the two squares in terms of x.

2. Specify the appropriate domain for the function whose expression you found in step 1.

3. Will the change of area be less than, equal to, or greater than the area of the original square?

The purpose of this stage is to introduce the main idea of a function being used to describe the change in area.

Some students begin geometrically, that is, they draw

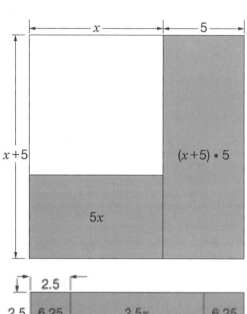

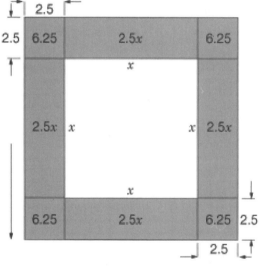

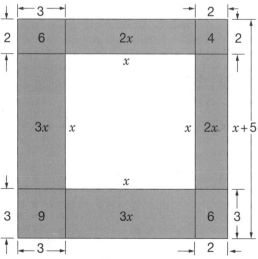

Fig. 1. Three students' diagrams in response to Stage I, question 1

145

an appropriate diagram to show the change in area and then describe it algebraically. **Figure 1** shows some responses. Other students start algebraically. In both approaches they obtain

$$f(x) = (x + 5)^2 - x^2 = 10x + 25.$$

And the domain? Here the geometry must be taken into account. Since x represents length, the "reality of the situation" requires that the domain be

$$\{x \in \mathrm{R} \mid x > 0\}.$$

In response to the third question, some students substitute numbers and obtain a partial solution by trial and error. Others try to solve it algebraically, by looking at

$$x^2 < 10x + 25,$$
$$x^2 = 10x + 25,$$

and

$$x^2 > 10x + 25.$$

Yet others, who regard the change of area and the area of the given square as functions of x, draw a graph and obtain the whole picture (see **fig. 2**). These students obtain not only the domains in which the values of one function are greater than (equal to, smaller than) the values of the other but also the geometrical meaning of these domains.

Stage II

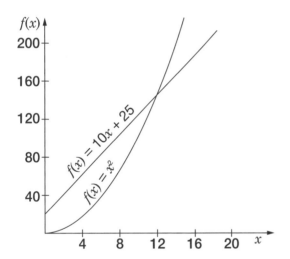

Fig. 2. The change of area and the area of the given square as functions of x

In the previous section the length of the given square was the variable and the change in the side length was constant. Here we turn the problem around, by having a square whose side is constant (e.g., 4 cm) and the change in side length is the variable. What can be said geometrically and algebraically about the change in the area?

A typical student's response is shown in **figure 3**. But in describing the change in area $f(x)$, in terms of the change in side length x, the algebraic expression obtained is this:

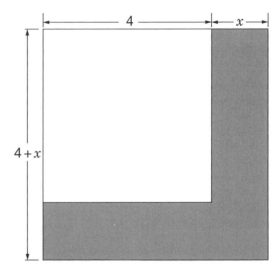

Fig. 3. Students' diagram in answer to Stage II

$$f(x) = (4 + x)^2 - 4^2 = x^2 + 8x.$$

And this time the change of area is not linear.

What about the domain of this function? As in the previous example, geometric constraints must be taken into account to define a meaningful domain. It is obvious that the domain includes all positive numbers and zero, but what about negative numbers? While searching for the largest meaningful domain, students soon realize that some negative numbers are certainly possible. In other words, the change in area can be negative, meaning the original square can be "pushed" inward (see **fig. 4**) until the resultant square has zero area.

The domain of the change in area function

$$f(x) = x^2 + 8x$$

can therefore be taken to be

$$\{x \in \mathrm{|R} \mid x > -4\};$$

its graph is shown in **figure 5**.

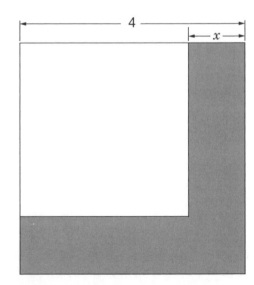

Fig. 4. A diagram illustrating negative change in area

146

Stage III: From a Square to ...

The main question remains the same. What is the resulting change in area if we change the length of the diagonals of a given square in one of the following ways:

1. Change only one diagonal.

2. Allow equal and symmetric changes in both diagonals.

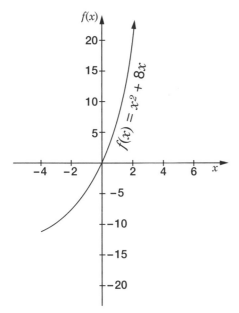

Fig. 5. The graph of the "change in area" function

3. Allow any change in both diagonals.

   - What are the geometric shapes obtained?
   - What are the functions that describe the change in area in each case?

1. One diagonal

When asked to change the length of one diagonal of the square, students seem to be more inclined to "pull" rather than to "push" (see **fig. 6**). For each value of $x > 0$ we obtain a specific trapezium. The lengths of its mutually perpendicular diagonals are a and $a + x$. Some students calculate the area of the trapezium as the sum of two triangles. Others conclude that the area of the trapezium is equal to half the product of the diagonals (since they are perpendicular). In both cases, the expression obtained is

$$\frac{1}{2}a(a + x).$$

Some calculated the change in area directly as the sum of the areas of two (four) obtuse-angle triangles (see the shaded areas in **fig. 6**) and obtain

$$f(x) = \frac{x \cdot \dfrac{a}{2}}{2} \cdot 2 = \frac{ax}{2}.$$

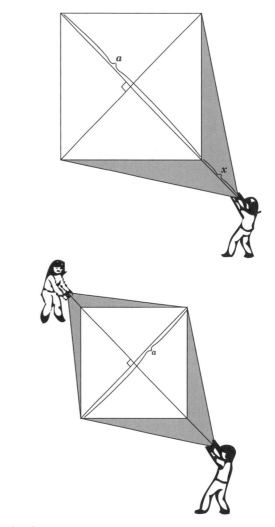

Fig. 6. "Change in area" of a given square by "pulling" one diagonal

As in previous stages, the largest possible domain has to be found, and, by now, it is natural to consider negative values of x. But the surprise arises in the corresponding geometry, when the students come to reconsider the new geometric "creatures" they have created from their algebraic intuition. These creatures could be regarded geometrically as being created by pushing the diagonals inward (for example, see **fig. 7**).

For

$$-\frac{a}{2} < x < 0$$

the trapezium flattens on one side.

For

$$x = -\frac{a}{2}$$

it disappears (**fig. 8**)!

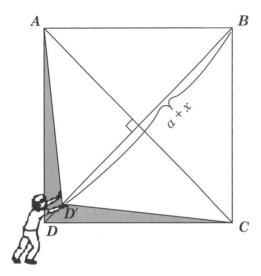

Fig. 7. "Change in area" of a given square by "pushing" one diagonal

The adventurous continue pushing for

$$-a < x < -\frac{a}{2}$$

and obtain the result shown in **figure 9**. The natural geometric transformations suggested by the algebra in

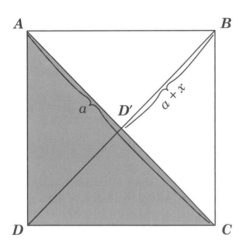

Fig. 8. "Pushing" the diagonal at $x = a/2$.

this situation bring the student to regard these concave quadrilaterals also as rhomboids.

The function describing the change of area is

$$f(x) \to \frac{ax}{2}, \; x > -a,$$

whose graph is given in **figure 10**. The graph stops quite unnaturally. What about $x < -a$? Is it possible

to push x beyond $-a$? It would appear not. We leave the geometrical meaning of that case for homework to our star pupil, Alice, and those who want to follow her through the looking glass.

2. Equal and symmetric change in both diagonals

In this case a set of isosceles trapezoids with perpendicular diagonals is obtained (**fig. 11**). The change in area is given by

$$g(x) = \frac{(a+x)^2}{2} - \frac{a^2}{2} = \frac{2ax + x^2}{2}.$$

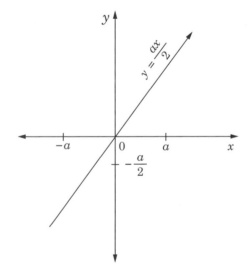

Fig. 9. "Pushing" the diagonal for $-a < x < -a/2$

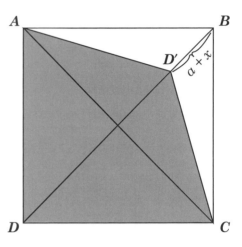

Fig. 10. Graph of the change in area function

Searching for the largest possible domain, we are again tempted to try to push. For a while,

$$-\frac{a}{2} < x < 0,$$

we obtain something quite to be expected (**fig. 12**).
Many students stop here and define

$$g(x) \rightarrow \frac{(a+x)^2}{2} - \frac{a^2}{2} = \frac{2ax+x^2}{2}, \ x > -\frac{a}{2},$$

whose graph is in **figure 13**.

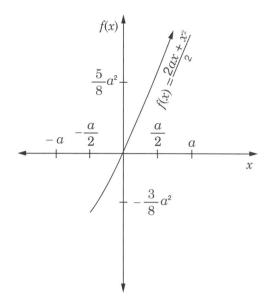

Fig. 13. Graph of the "change in area" function for $x > -a/2$

But the intrepid continue "pushing" to the point $x = -a/2$ and beyond. And lo and behold! Can these new shapes (see **fig. 14**) $A'D'BC$ be considered quadrilaterals? If, for example, we take a very general definition, such as *A quadrilateral is what you get if you take four points, A, B, C, and D* <u>*in a*</u> <u>*plane and join them with the*</u> *straight line segments $\overline{AB}, \overline{BC}, \overline{CD}, and \overline{DA}$.* (Galbraith 1981), then no problem exists.

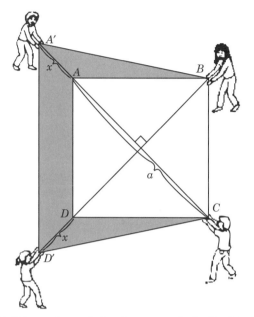

Fig. 11. "Change in area" of a given square by "pulling" both diagonals symmetrically

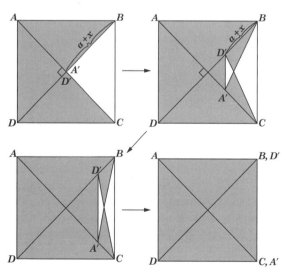

Fig. 14. The "change in area" when we "push" the two diagonals to the point $x = -a/2$ and beyond

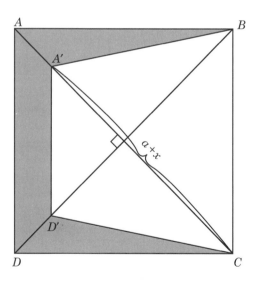

Fig. 12. "Change in area" of a given square by "pushing" both diagonals symmetrically

Can we say more than that? Are they still isosceles <u>trapezoids</u>? Why not? They have a <u>pair of parallel</u> sides $\overline{A'D'} \parallel \overline{BC}$ and two congruent sides $\overline{D'C} \cong \overline{A'B}$. In addition, the diagonals (yes! the diagonals), which are $\overline{D'B}$ and $\overline{A'C}$ (i.e., external), are still perpendicular. Nothing seems to have changed except the geometric figures.

But, what about the area? Is the area still equal to half the product of the diagonals? Unfortunately (or is it fortunately?), the answer is no! (For a full proof, contact the authors.) Here for the first time we have a major "break" in the regularity of the pattern. In fact, if other properties of trapezoids (or quadrilaterals) are checked for these new shapes, new surprises can be found. (For example, check the sum of the four internal angles.) At this point, inevitable discussion arises whether or not to adopt these trapezoids, that is, to consider them as members of the family (as in Crawforth [1967]).

Here, in our discussions with students and teachers, we add some general considerations relevant to extending a concept. For example, when extending do we have to, or even *can we*, retain all previous properties? When extending the natural numbers to the integers, we lose some properties, for example,

$$ac > ab \rightarrow c > b.$$

We have not, of course, finished; the case of any change in either diagonal still remains, and whatever else our imaginations suggest.

EPILOGUE

Our mental images of mathematical concepts are constructed in different ways, in and out of school. A "photograph" of the mental image of any particular mathematical concept will usually show a partial (and maybe even distorted) picture.

For instance, the concept of quadrilateral, with which we dealt in this article, includes for some students only squares (**fig. 15i**), for other it includes all convex shapes (**fig. 15i–ii**), for yet others it includes **figure 15iii**, and only a few develop the concept in its richest form (**fig. 15iv**)(Hershkowitz and Vinner 1983).

Among the other concepts in this paper was that of *function*. For some, the mental image may have been formed by the set approach (static), in which the function is a triad (domain, codomain, and a correspondence between their elements). This view may blur the dynamism involved in the idea of function as a dependence between two variable quantities. And the converse is also true: Where the "dependence" aspect is dominant, the domain of the function may have a minor role (Malik 1980). This project exemplifies a series of related questions in which the two aspects of the concept of function are interrelated. In addition, the function is given a geometric content, which enlarges and enriches the algebraic and geometric mental images.

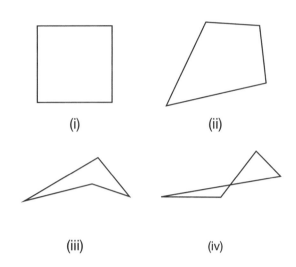

(i)   (ii)

(iii)   (iv)

Fig. 15. A collection of various quadrilaterals

SOLUTIONS

Sheet 1:

1) $f(x) = (x + 5)^2 - x^2 = 10x + 25$

2) $\{x \in \mathbb{R} \mid x > 0\}$;

3)

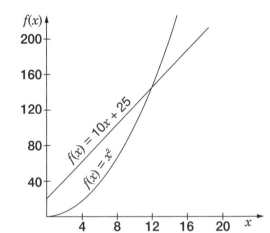

Fig. 2. The change of area and the area of the given square as functions of x

Students who regard the change of area and the area of the given square as functions of x draw a graph and obtain the whole picture as shown in **Figure 2** (reproduced above from page 146). These students obtain not only the domains in which the values of one function are greater than (equal to, smaller than) the values of the other but also the geometrical meaning of these domains.

Sheet 2:

1) $f(x) = (4 + x)^2 - 4^2 = x^2 + 8x$

2) $\{x \in \mathbb{R} \mid x > -4\}$

3)

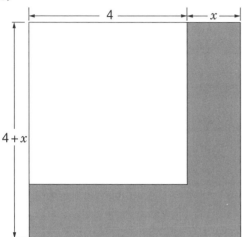

Fig. 3. Students' diagram in answer to Stage II.

A square whose side is constant (4 cm) and the change

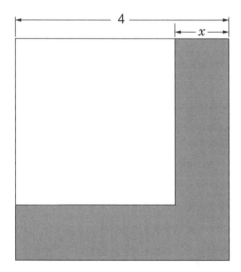

Fig. 4. A diagram illustrating negative change in area.

in side length is the variable.

The change in area can be negative, meaning the original square can be "pushed" inward (see **fig. 4**) until the resultant square has zero area.

Sheet 3:

1) For each value of $x > 0$ we obtain a specific rhomboid. The lengths of its mutually perpendicular diagonals are a and $a + x$.

2)

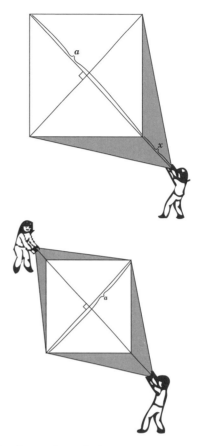

Fig. 6. "Change in area" of a given square by "pulling" one diagonal

For each value of $x > 0$ we obtain a specific rhomboid. The lengths of its mutually perpendicular diagonals are a and $a + x$.

Resulting shape if $x < 0$. Created by pushing the diagonals inward.

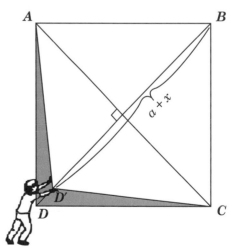

Fig. 7. Change in area" of a given square by "pushing" one diagonal

3)

$$f(x) = \frac{ax}{2}$$

3)

$$g(x) = \frac{(a+x)^2}{2} - \frac{a^2}{2} = \frac{2ax+x^2}{2}.$$

4)

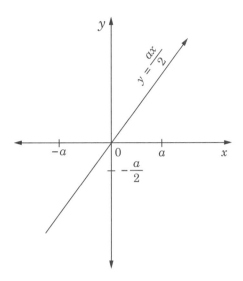

4) Graph of the function

$$g(x) = \frac{2ax+x^2}{2}$$

Graph of the function

$$f(x) = \frac{ax}{2}$$

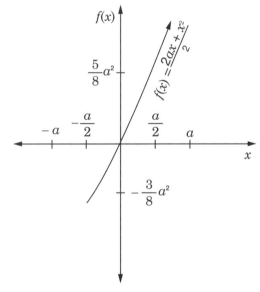

Fig. 13. Graph of the "change in area" function for $x > -a/2$

Sheet 4:

1) A set of isosceles trapezoids with perpendicular diagonal is obtained.

2)

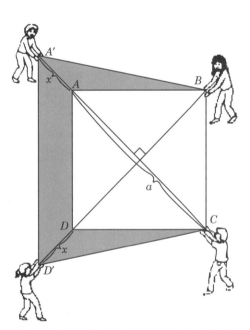

Fig. 11. "Change in area" of a given square by "pulling" both diagonals symmetrically

REFERENCES

Bruckheimer, Maxim, and Rina Hershkowitz. "Mathematics Projects in Junior High School." *Mathematics Teacher* 70 (October 1977): 573–78.

Crawforth, Denis. "What Is a Quadrilateral?" *Mathematics Teacher* 60 (November 1967): 778–81.

Galbraith, Peter L. "Aspects of Proving: A Clinical Investigation of Process." *Educational Studies in Mathematics* 12 (February 1981): 1–28.

Hershkowitz, Rina, and Shlomo Vinner. "The Role of Critical and Noncritical Attributes in the Concept Image of Geometrical Concepts." *Proceedings of the Seventh International Conference for the Psychology of Mathematics Education* (1983): 223–28.

Malik, M. A. "Historical and Pedagogical Aspects of the Definition of Function." *International Journal of Mathematics Education in Science and Technology* 11 (October 1980): 489–92.

Usiskin, Zalman. "What Should Be in the Algebra and Geometry Curricula of Average College-Bound Students." *Mathematics Teacher* 73 (September 1980): 413–24.

Geometrical Adventures in Functionland

Sheet 1

From a given square of side length x, we create a new square whose side length is 5 cm longer.

1. Find an expression for the difference in area of the two squares in terms of x.

2. Specify the appropriate domain for the function whose expression you found in step 1.

3. Will the change of area be less than, equal to, or greater than the area of the original square?

Geometrical Adventures in Functionland   Sheet 2

From a given square of side length 4, we create a new square whose side length is x cm longer.

1. Find an expression for the difference in area of the two squares in terms of x.

2. Specify the appropriate domain for the function whose expression you found in step 1.

3. What can be said geometrically and algebraically about the change in the area?

Geometrical Adventures in Functionland

Sheet 3

From a given square of side length x...

1. What is the resulting change in area if we change only one diagonal?

2. What are the geometric shapes obtained?

3. What is the function that describes the change in area?

4. Graph the function that describes the change in area.

Geometrical Adventures in Functionland

Sheet 4

From a given square of side length x...

1. What is the resulting change in area if we allow equal and symmetric changes in both diagonals?

2. What are the geometric shapes obtained?

3. What is the function that describes the change in area?

4. Graph the function that describes the change in area.

5. What is the resulting change in area if we allow any change in both diagonals? What are the geometric shapes obtained?

Helping Students Connect Functions and Their Representations

Deborah Moore-Russo and John B. Golzy

When we taught functions and their graphs in the past, our students frequently viewed the two as separate entities rather than recognizing that the graph is one of various representations of a function. They got so involved in finding ordered pairs using the equation of a function that they started to view a graph as a sketch through three or four points and never seemed to consider trends in families of functions or to recognize the role or significance of x-intercepts. They rarely considered the ways in which the graph provided a visual insight to the behavior of the function. They failed to understand the power of the graphical representation or to see how it connected to the other functional representations. This article describes how we changed our instruction to encourage student exploration of the graphical and then the algebraic representations of functions. By reversing the typical order of instruction, we enabled our students to see better how the graph, equation, and table of a function are related.

NCTM's *Curriculum and Evaluation Standards for School Mathematics* (1989) emphasizes the mathematical connections that need to be made between graphical analysis and algebraic processing. These standards also help bring the concept of function into the limelight, urging teachers to help students grasp the relationship among tables, graphs, and equations (NCTM 1989). In addition to seeing the connection between the algebraic and graphical representations of functions, students must be comfortable working with both representations of functions as related in NCTM's Representation Standard (NCTM 2000).

The new standards bring out an obvious but easy to overlook point: Besides seeing the connection between the two representations, students need to realize that each representation can offer different insight to the function it signifies. Clement and Sowder (2003) suggest that, to understand functions, students must be able to connect the various representations of a function. They point out that when students have more opportunities to make connections they will better understand these connections and will be more likely to search for connections on their own in the future.

As evidenced by the results of the Kerslake test (1981), many students traditionally have found it easier to recognize the graph of a linear equation than to produce the equation of a given straight line. Moreover, numerous studies (Kieran 1992) have reported that students struggle with graphical representation. In order to address these issues and to help students connect the

algebraic and graphical representations of a function, we have our students solve the following problem after they are comfortable graphing and working with linear functions:

> Given $y_1 = 2$ and $y_2 = 0.5x$, find the sum and product of the two functions.

The problem in no way specifies what approach the students should use. While students may solve it either algebraically or graphically, the order in which we introduce the two approaches affects how well they can connect the representations.

THE ALGEBRAIC SOLUTION FIRST

Students were asked to solve the problem algebraically by adding or multiplying the terms of the two functions:

$$y_3 = y_1 + y = 2 + 0.5x$$

$$y_4 = y_1 \cdot y_2 = 2(0.5x) = x$$

While they calculated y_3 and y_4 with little effort, these students had a hard time visualizing them. In order to help them with this process, we would have them graph y_1 and y_2. Then, we would ask them to predict and justify the way the graphs of y_3 and y_4 should look. Students usually hesitated to make any predictions, but we welcomed even incorrect solutions. Often after discussing only one incorrect solution someone would come up with the correct graph. Since the students had already solved the problem algebraically, they could justify the graphical solution. Once they did so, they had developed some connection between the equation and its graph.

THE GRAPHICAL SOLUTION FIRST

Based on our observation that students struggle with graphs as whole units, we now use graphs first. This switch seemed justified based on Kieran's (1992) position: Students in algebra courses often need significant prompting and instruction to use graphically represented information; students prefer working with numeric data in tables; and algebra teachers often encourage students very early to use efficient symbolic methods. Our experience had shown us that students use symbols before completely understanding the meanings behind the symbols. So, we decided to present the graphical solution before the algebraic one.

In this method, students begin by graphing y_1 and y_2 on the same set of axes (**fig. 1**). We then ask them to predict what their sum would look like. If they struggle

---
157

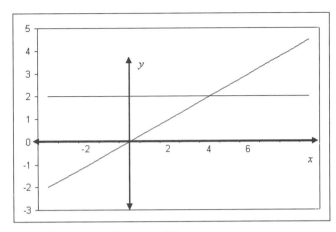

Fig. 1. Graph of $y_1 = 2$ and $y_2 = 0.5x$

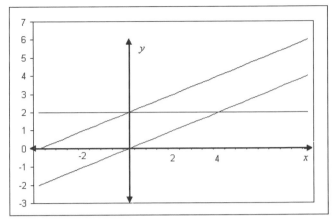

Fig. 2. Graph of $y_1 = 2$, $y_2 = 0.5x$, and $y_3 = y_1 + y_2$

to do so, we pose the following series of "routine" questions:

1. Find a point that will be in the sum of these two functions. How did you get it?

    If needed, we guide them to take a point from each graph—such as (4, 2), which happens to belong to both graphs—and have them find the sum of the y-coordinates. Since both functions go through the point (4, 2), their sum has to go through (4, 2 + 2) or (4, 4).

2. Where does the sum of these functions cross the x-axis? Why?

    In this example, the sum of the functions crosses at $x = -4$, $y_1 = 2$, and $y_2 = -2$. Therefore, the output for y_3 is 2 + −2 (or 0) when the input is −4. So, y_3 includes the point (−4, 2 + −2), or (−4, 0). We do not share this information with our students, but instead we guide a class discussion that leads them to this fact.

3. How many times, if any, will the resultant function go through the x-axis? Why?

4. Where do you think the resultant function is going to be positive or negative? Why?

After answering these questions, our students have a better understanding of what the sum of the two functions will look like. At this point, we use a graphing calculator to graph the functions $y_1 = 2$ and $y_2 = 0.5x$, as well as the points we have identified for y_3. With these points, the students can usually predict the shape of the graph. After that, we graph $y_3 = y_1 + y_2$ (**fig. 2**) using the calculator, and we ask the class to predict and justify the equation of y_3. The students must look at the general behavior of the graph, an increasing line with a y-intercept at 2. Using the form $y = mx + b$, they identify the y-intercept and, after some prompting, the slope as positive. This yields $y = mx + 2$, $m > 0$. In order to determine the specific slope, we take a point from y_3, such as (−4, 0), and substitute its coordinates into the

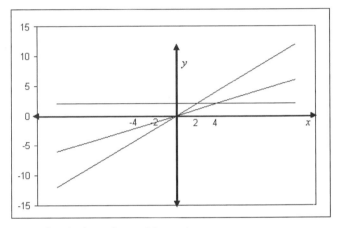

Fig. 3. Graph of $y_1 = 2$, $y_2 = 0.5x$, and $y_4 = y_1 \cdot y_2$

equation $y = mx + 2$ to solve for the slope, or we take two points and use them to calculate the slope as the ratio of the change in y-coordinates to the change in x-coordinates.

Once students have determined the correct equation, we move to the algebraic solution of the problem. Because they have already solved the problem graphically, they see and understand the algebraic solution better. By solving the problem graphically before algebraically, they seem to make a better connection between the equation and the graph than they did when they solved the problem algebraically first. We believe this difference occurs because the algebraic solution tends to be very quick and often seems easy. The students can do the algebraic approach with little thought about what the symbols actually represent. That is not the case when they start with the graphical solution. For practice, we often ask them to predict the graph then the equation for $y_5 = y_1 + y_3$ and $y_6 = y_2 + y_3$.

Next, we provide students with the graphs of y_1 and y_2 (**fig. 1**) and ask them to predict how the graph of the product will look. We help them with this by asking the same routine questions as before. We point out that the x-intercepts occur when either of the outputs of the two functions is equal to zero. Since y_1 is constant but y_2 crosses the x-axis once, the x-intercept for y_2 is also the

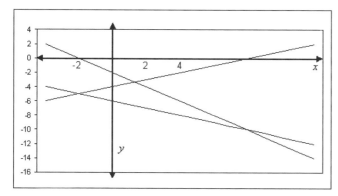

Fig. 4. Graph of $y_1 = 0.5x - 4$, $y_2 = -x - 2$, and $y_3 = y_1 + y_2$

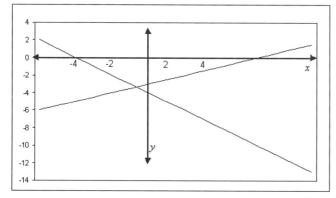

Fig. 5. Graph of $y_1 = 0.5x - 3$ and $y_2 = -x - 4$

only x-intercept for the product function y_4. After examining several points, students are able to see that the product $y_4 = y_1 \cdot y_2$ falls in the first and third quadrants (**fig. 3**). We then discuss the equation of y_4 in a manner similar to the discussion of y_3.

Once students are comfortable with this problem and others like it, we have them investigate the sum of other pairs of linear equations, such as $y_1 = 0.5x - 4$ and $y_2 = -x - 2$ (**fig. 4**), finding both the graph and then the equations of the resultant function. We then move to the product of two linear functions as a way to introduce quadratic functions.

INTRODUCING QUADRATIC FUNCTIONS

We give our students the graphs of two linear equations, such as $y_1 = 0.5x - 3$ and $y_2 = -x - 4$ (**fig. 5**), and ask them to predict what their product function would look like. We guide them through this process by looking for the x-intercepts of the product function, which occur when either y_1 or y_2 is 0, or at $x = 6$ and $x = -4$, respectively. Then we look at the areas to the left of, to the right of, and between the x-intercepts. To the right of $x = 6$, the y-coordinates of y_1 are positive and the y-coordinates of y_2 are negative, so their product function is negative. By substituting specific point values, students can tell that the graph's behavior is not linear, since the graph drops more rapidly the farther you move to the right. To the left of $x = -4$, the y-coordinates of y_1 are negative and the y-coordinates of y_2 are

positive, so their product function is negative. By plugging in specific values, students see the curved nature of the graph in this region. Finally, we explore the section between the two intercepts where $-4 < x < 6$. In this region, the y-coordinates of both functions are negative. Hence, the product function in this area yields positive outputs. We substitute some values and roughly sketch the curve by connecting all the points. Once the students have found the geometric solution (**fig. 6**), we ask them to find the algebraic solution using $y_3 = y_1 \cdot y_2$.

These two related activities, finding the geometric and algebraic solutions for the product function, require the students to engage in problem solving. The creation of a nonlinear function is something that is new, but within their reach. Students build quadratic functions by multiplying linear functions rather than seeing them as a new, unrelated topic. In a similar manner, this activity can be extended to introduce cubic functions either as a product of three linear functions or as the product of a linear and a quadratic function.

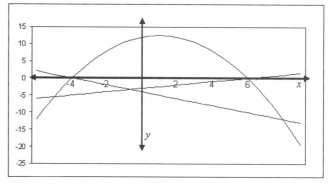

Fig. 6. Graph of $y_1 = 0.5x - 3$, $y_2 = -x - 4$, and $y_3 = y_1 \cdot y_2$

BENEFITS OF THE ACTIVITY

Once they have both the resultant graph and equation, students show the tabular representation of the functions in order to verify their results. By linking various representations we enable students to understand functions in a new way, one that could help students apply them more effectively (Kaput 1989). They also start to see the benefit of each functional representation, including the beauty of an efficient algebraic solution.

This approach also meets many of NCTM's *Principles and Standards* (2000). While the graphing calculator is used to enhance student learning and make the lesson more time effective, as suggested by the Technology Principle, the absence of this form of technology does not diminish the value of the approach itself. The inclusion of graphing calculators, however, affords more opportunities for "what if" questions and for exploration (Dunham and Dick 1994). A decided strength of the activity is that it helps students build on previous experience in a way that challenges them, by asking them to apply arithmetic operations and the distributive property to algebraic expressions. In this manner, the activity addresses both the Learning Principle and the Connection Standard. It also sets the stage for a more in-depth look at the role of linear factors in the graphs of quadratic functions and other polynomials. With this prior experience students should be able to grasp more easily the idea of the roots of a polynomial and their importance.

REFERENCES

Clement, Lisa, and Judith Sowder. "Making Connections within, among, and between Unifying Ideas in Mathematics." In *Integrated Mathematics: Choices and Challenges*, edited by Sue Ann McGraw, pp. 59–72. Reston, Va.: National Council of Teachers of Mathematics, 2003.

Dunham, Penelope H., and Thomas P. Dick. "Research on Graphing Calculators." *Mathematics Teacher* 87 (September 1994): 440–45.

Kaput, James J. "Linking Representations in the Symbol Systems of Algebra." In *Research Issues in the Learning and Teaching of Algebra*, edited by Sigrid Wagner and Carolyn Kieran, pp. 167–94. Reston, Va.: National Council of Teachers of Mathematics, 1989.

Kerslake, Daphne. "Graphs." In *Children's Understanding of Mathematics*: 11–16, edited by Kathleen M. Hart,
pp. 120–36. London: John Murray, 1981.

Kieran, Carolyn. "The Learning and Teaching of School Algebra." In *Handbook of Research on Mathematics Teaching and Learning*, edited by Douglas A. Grouws, pp. 390–419. New York: Macmillan, 1992.

National Council of Teachers of Mathematics (NCTM). *Curriculum and Evaluation Standards for School Mathematics*. Reston, Va: NCTM, 1989.

———. *Principles and Standards for School Mathematics*. Reston, Va.: NCTM, 2000.

Activity Sheet—Part One

Sheet 1

1. Use the axes below to graph $y_1 = 2$ and $y_2 = 0.5x$.

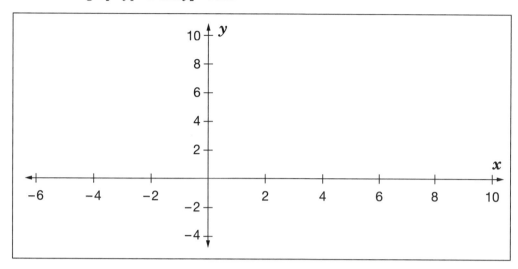

2. Just as you can add numbers, you can also add graphs of functions. What do you think the graph of $y_1 + y_2$ would look like? Explain you reasoning.

3. Using the graphs of y_1 and y_2 above, determine

   a. The value of y_1 when $x = 4$.

   b. The value of y_2 when $x = 4$.

   c. The value of $y_1 + y_2$ when $x = 4$. (Explain your answer.)

4. Using the graphs of y_1 and y_2 above, determine

   a. The value of y_1 when $x = 0$.

   b. The value of y_2 when $x = 0$.

   c. The value of $y_1 + y_2$ when $x = 0$. (Explain your answer.)

5. Using the graphs of y_1 and y_2 above, determine

   a. The value of y_1 when $x = -2$.

   b. The value of y_2 when $x = -2$.

   c. The value of $y_1 + y_2$ when $x = -2$. (Explain your answer.)

6. For what values of x would the graph of $y_1 + y_2$ cross the x-axis? Why?

7. How many times would the graph of $y_1 + y_2$ cross the x-axis? Why?

8. For what values of x would the graph of $y_1 + y_2$ be positive? Why?

Activity Sheet—Part One

9. For what values of x would the graph of $y_3 = y_1 + y_2$ be negative? Why?

10. Use the information above to sketch a graph of $y_3 = y_1 + y_2$ on the axes below.

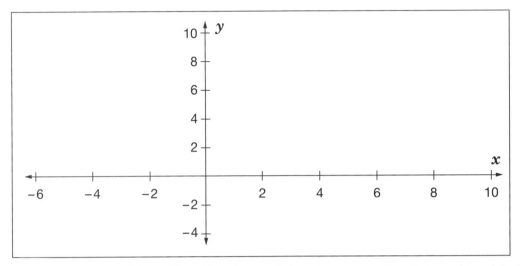

11. Look at the sketch you created in problem 10 and try to determine an equation that would match that graph. *Hint:* Since the graph should be linear, and you can find the y-intercept and then substitute in the coordinates of a point to determine the slope.

12. Sketch a graph and determine the equation of $y_4 = y_1 + y_3$. Explain your reasoning. Compare graphs of y_1, y_2, y_3, and y_4. How are the graphs similar and different?

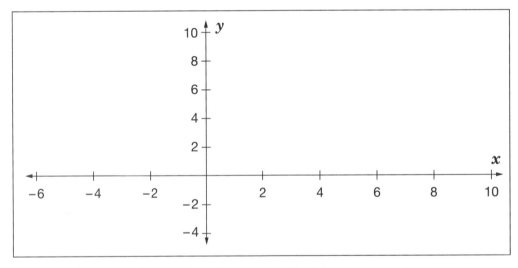

13. Sketch a graph and determine the equation of $y_5 = y_2 + y_3$. Explain your reasoning.

14. If $y_1 = 2$ and $y_2 = 0.5x$,

    a. Substitute in the values of y_1 and y_2 to algebraically determine the equation of $y_3 = y_1 + y_2$.

    b. Substitute in the values of y_1 and y_3 to algebraically determine the equation of $y_4 = y_1 + y_3$.

    c. Substitute in the values of y_2 and y_3 to algebraically determine the equation of $y_5 = y_2 + y_3$.

Activity Sheet—Part Two

Sheet 3

1. The graphs of $y_1 = 2$ and $y_2 = 0.5x$ are shown on the axes below.

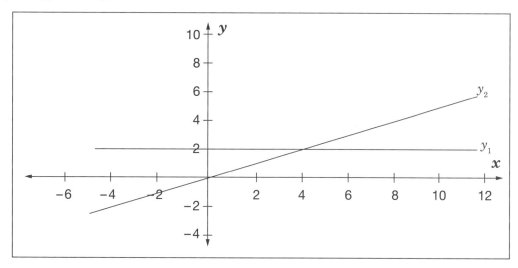

2. Just as you can multiply numbers, you can also multiply graphs of functions. What do you think the graph of $y_3 = y_1 \cdot y_2$ would look like? Explain your reasoning.

3. Using the graphs of y_1 and y_2 above, determine
    a. The value of y_1 when $x = 4$.
    b. The value of y_2 when $x = 4$.
    c. The value of $y_1 \cdot y_2$ when $x = 4$. (Explain your answer.)

4. Using the graphs of y_1 and y_2 above, determine
    a. The value of y_1 when $x = 0$.
    b. The value of y_2 when $x = 0$.
    c. The value of $y_1 \cdot y_2$ when $x = 0$. (Explain your answer.)

5. Using the graphs of y_1 and y_2 above, determine
    a. The value of y_1 when $x = -2$.
    b. The value of $y_2 \cdot x = -2$.
    c. The value of $y_1 \cdot y_2$ when $x = -2$. (Explain your answer.)

6. For what values of x would the graph of $y_1 \cdot y_2$ cross the x-axis? Why?

7. How many times would the graph of $y_1 \cdot y_2$ cross the x-axis? Why?

8. For what values of x would the graph of $y_1 \cdot y_2$ be positive? Why?

Activity Sheet—Part Two

9. For what values of x would the graph of $y_1 \cdot y_2$ be negative? Why?

10. Use the information above to sketch a graph of $y_1 \cdot y_2$ on the axes below.

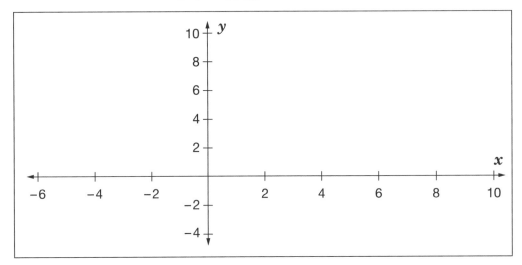

11. Looking at the sketch you created in problem 10, try to determine an equation for y_3 that would match that graph.

Activity Sheet—Part Three

Sheet 5

1. The graphs of $y_1 = 0.5x - 3$ and $y_2 = -x - 4$ are shown on the axes below.

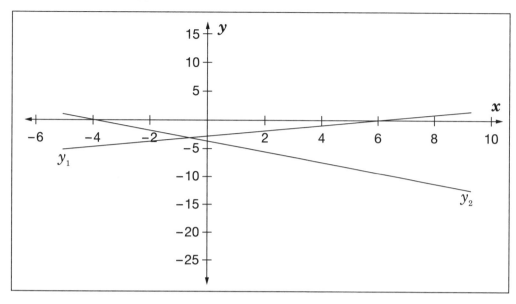

We are now going to consider the product function $y_3 = y_1 \cdot y_2$.

2. Using the graphs of y_1 and y_2 above, determine the x-intercepts of the product function when the value of $y_3 = 0$. (Explain your answer.)

3. For what values of x would the graph of $y_1 \cdot y_2$ cross the x-axis? Why?

4. How many times would the graph of $y_1 \cdot y_2$ cross the x-axis? Why?

5. For what values of x would the graph of $y_1 \cdot y_2$ be positive? Why?

6. For what values of x would the graph of $y_1 \cdot y_2$ be negative? Why?

Activity Sheet—Part Three   Sheet 6

7. Using the information above, try to sketch a rough graph of $y_3 = y_1 \cdot y_2$ on the axes below.

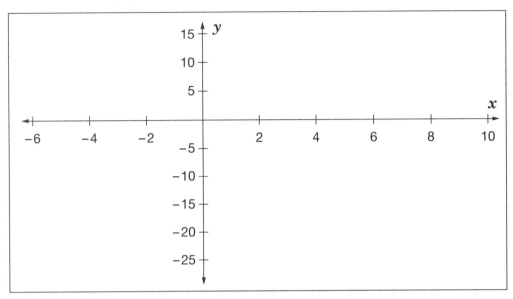

8. Now create an input-output table of values for y_3. Be sure to include input (x) values that are to the left of the left-most x-intercept, to the right of the right most x-intercept, and between the two x-intercepts. The ordered pairs in this table should correspond to the points of the graph of preceding problem.

x	$y_3 = y_1 \cdot y_2$

9. If $y_1 = 0.5x - 3$ and $y_2 = -x - 4$, substitute in the values of y_1 and y_2 to algebraically determine the equation of $y_3 = y_1 \cdot y_2$. The equation of y_3 should correspond with the graph and tables in the two preceding problems.

Nongeometry Mathematics and The Geometer's Sketchpad

Todd O. Moyer

The Geometer's Sketchpad (GSP) is a well-known interactive geometry software package. Its usefulness in geometry instruction has been well researched (Choi-Koh 1999; Dixon 1997; Groman 1996; Lester 1996; Moyer 2003; Weaver and Quinn 1999). Finzer and Jackiw (1998) recommend the use of GSP as the dynamic manipulative for geometric concepts. GSP allows students to construct a figure, to perform measurements of lengths and angles, and then to "click and drag" any part or parts of that figure to look for change.

But GSP is not limited to the geometry class. GSP Version 4 (Jackiw 1998) is equipped to make dynamic graphs beyond what is possible on a graphing calculator. A graphing calculator (such as the TI-83/84) permits changing the values of coefficients easily but only yields a snapshot of a graph. An example would be discovering the effects of the coefficients a, b, or c on the general quadratic $y = ax^2 + bx + c$. Students can see how the graph changes as a coefficient takes on different values. The activities described in this article can be adjusted to accommodate any type of function, even piecewise functions.

Using a graphing calculator, one must either manually change the leading coefficient or use the LIST function of the calculator. For example, letting Y1 = {2, 4, 6}X + 2 will graph all three lines at once. Setting Y1 = L1 • X + 2, with the values of 2, 4, and 6 in L1, would accomplish the same thing. But the student still has a sequence of snapshots of the graphs, not a "movie." In addition, the trace function does not allow the user to identify which graph goes with which leading coefficient. Animating the leading coefficient in GSP causes the graph to change. Students will be able to grasp the concept of slope more easily by watching a line change instantly as the leading coefficient changes. Instead of seeing snapshots, the student watches a movie.

To help students understand the transformations of the sine wave, GSP's Version 4 allows the user to manipulate any of the four coefficients of

$$f(x) = A \sin (Bx + C) + D$$

to show the graphical effects. Now, a teacher can create sketches before class, hide the appropriate information, and then ask students to determine what the equation is.

GSP can also be used for a better understanding of the composition of two functions. Students can guess whether a composition of a line and a parabola is always a parabola or is dependent upon the order of the functions. Using and animating parameters, students can investigate the composition of linear and linear, linear and quadratic, and quadratic and cubic functions much more easily with GSP than with a graphing calculator. Along with the composition of functions, we can consider inverses of functions. Students can play with functions and parameters to see if they can generate a function, other than $f(x) = x$, that is its own inverse.

GSP allows for the graphs of nonfunctions of the type $x = f(y)$. A function of this type would take two functions on a graphing calculator and would possibly require some algebraic manipulation. With GSP, changing equation types from $y = f(x)$ to $x = f(y)$ is easily done. Parameters can be entered and animated as well.

In a calculus class, the initial development of derivatives commonly begins with a tangent line to a curve at a point on the curve. Students sometimes have difficulty visualizing tangent lines. Using GSP, a dynamic tangent can be graphed effectively as a point travels along the path of the curve. A trace can also be set up so that as the point travels along the curve, the derivative is graphed at the same time. A student can see the relationship between an increasing function and a derivative above the x-axis and the relationship between an extreme point of a function and a derivative that equals zero.

During my fifteen-year career of teaching high school mathematics, I often incorporated Sketchpad (versions 1 through 3) into my instruction. Now, as an assistant professor specializing in mathematics education, I have used these activities as demonstrations in undergraduate methods classes and graduate courses with in-service teachers. Watching the enthusiastic response of these students and teachers has led me to believe that the activities are worthwhile. Concepts that do not translate well verbally are handled visually. Students can actually "see" that a derivative is the slope of the tangent line, for example, or how changing the value of A in $y = A \sin (Bx + C) + D$ affects the amplitude of the curve.

All the topics can also be presented on a graphing calculator. However, GSP adds a dynamic quality that emphasizes further the connection between an equation and its graph.

The following activities are intended for class explorations using a computer connected to an LCD projector. If students are familiar with GSP, these instructions can be emended easily for their use. Following the general graphing instructions below are suggestions for

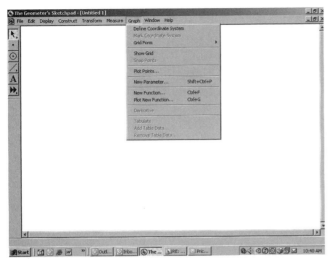

Fig. 1. To construct a graph, go to Graph and click on either Define Coordinate System or Show Grid.

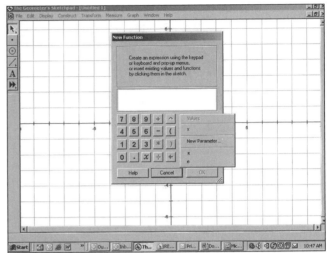

Fig. 2. To graph a function with parametric coefficients, go to Graph: New Function. Under Values, click on New Parameter.

six explorations dealing with quadratic equations, trigonometric functions, piecewise functions, derivatives, tangent lines, and the definition of a derivative.

GENERAL GRAPHING INSTRUCTIONS

1. Version 4 of GSP allows users to select multiple objects without using the shift key. Be attentive to what has been selected. When something does not work as it should, a first remedy is to deselect everything by clicking on blank space and then trying again.

2. To construct a graph, go to Graph and click on either Define Coordinate System or Show Grid (**fig. 1**). A coordinate system should be displayed.

3. GSP allows for functions to be graphed with parametric coefficients. For example, in order to graph $y = mx + b$, go to Graph: New Function. Under Values, click on New Parameter (**fig. 2**). In the Name box, type "m" and press OK. Then type ". x +" and go to Values, New Parameter, "b." By default, parameters are given the value of 1. You can change the value when creating a new parameter or by double-clicking on the parameter itself.

THE QUADRATIC EQUATION GRAPHING ACTIVITY

1. Go to Graph: New Function. Type the function "a • x^2 + b • x + c."

2. Once the function is entered, click OK and make sure that only "f(x) = a • x^2 + b • x + c" is selected. Then click on Graph: Plot Function. The graph should appear (**fig. 3**).

3. Reset the parametric values of b and c to 0. Select only parameter a, and click on Display: Animate

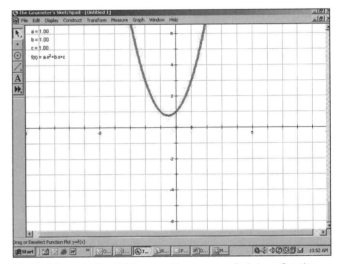

Fig. 3. After selecting "f(x) = a • x^2 + b • x + c," click on Graph: Plot Function. The graph of the quadratic should appear.

Parameter. A Motion Controller box should pop up, stating "Target: Parameter a." There are buttons for play, stop, reverse, and pause. The speed of the animation can also be adjusted. To start, a will increase. Press reverse and a will decrease. Set $a = 0$ for students to see why a 0 for the function to be quadratic.

4. Repeat the process for parameters b and c next.

5. Entertain student discussion about what effect each coefficient has on the graph. Two parameters can also be animated at the same time. For example, animating b and c will highlight the effects on the vertex of the parabola.

TRIGONOMETRIC FUNCTIONS

1. Go to Graph: New Function. Type "a · sin(b · x + c) + d." The sine function can be found under the Function pull-down menu. A dialogue box may appear,

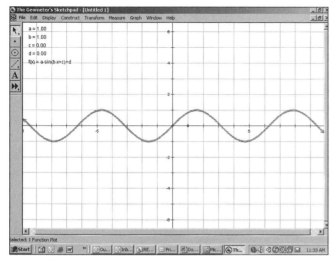

Fig. 4. After selecting "f(x) = a • sin(b • x + c) + d," click on Graph: Plot Function. The graph of the sine function should appear.

asking if you would like to change from degrees to radians. Radians are the preferred unit. Change to radians by going to Edit: Preferences and changing the angle measurement unit to radians.

2. Once the function is entered, press OK, and make sure that only "f(x) = a · sin(b · x + c) + d" is selected. Then click on Graph: Plot Function. The graph should appear (**fig. 4**).

3. Make sure the parametric values of a and b are 1 and c and d are 0. Select only parameter a and click on Display: Animate Parameter. A Motion Controller box should pop up, stating "Target: Parameter a." There are buttons for play, stop,

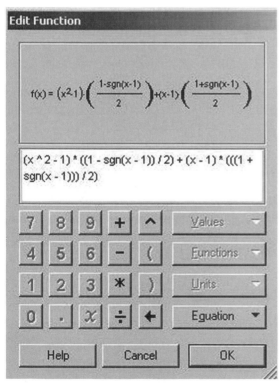

Fig. 5a

reverse, and pause. To start, a will increase. Press reverse, and a will decrease. The speed of the animation can also be adjusted as before.

4. Repeat the process for parameters b, c, and d.

5. Again, entertain discussion about how changing the coefficients of the equation affects the graph. Students should easily see how a affects the amplitude of the sine curve, for instance.

PIECEWISE FUNCTIONS

1. For a piecewise function, GSP suggests the use of the signum function. The signum function returns either a value of -1 or 1. The signum function, sgn, is found under Functions on the calculator.

2. In general, suppose the piecewise function is

$$f(x) = \begin{cases} p(x), x \le c \\ q(x), x > c \end{cases}.$$

3. For Plot New Function, type

$$(p(x)) \cdot \left(\frac{1 - \text{sgn}(x - c)}{2} \right) + (q(x)) \cdot \left(\frac{1 + \text{sgn}(x - c)}{2} \right).$$

4. For example, suppose

$$f(x) = \begin{cases} x^2 - 1, x \le 1 \\ x - 1, x > 1 \end{cases}.$$

Click on Graph: Plot New Function. Type the following (**figs. 5a** and **5b**):

$$(x^2 - 1) \cdot \left(\frac{1 - \text{sgn}(x - 1)}{2} \right) + (x - 1) \cdot \left(\frac{1 + \text{sgn}(x - 1)}{2} \right).$$

COMPOSITION OF FUNCTIONS

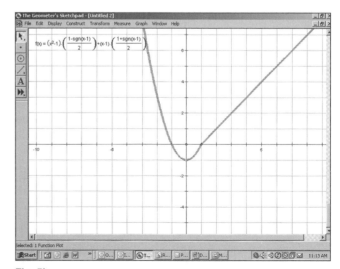

Fig. 5b

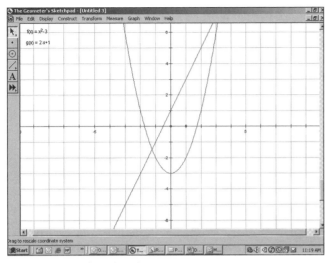

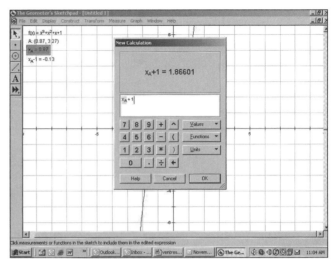

Fig. 6. A graph of two functions

Fig. 8. Constructing a tangent line

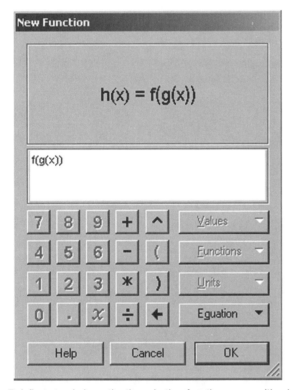

Fig. 7. A first step in investigating whether function composition is commutative

1. Type in any two functions. For example, let $f(x) = x^2 - 3$ and $g(x) = 2x + 1$. Once the functions are entered, make sure that only the functions are selected. Then click on Graph: Plot Function. The graphs should appear (**fig. 6**).

2. Go to Graph: Plot New Function. Click on the first function, $f(x)$, and $f(\ )$ should appear in the screen of the calculator. Click on the second function, $g(x)$, and $f(g(\ ))$ should appear. Type "x" inside the innermost parentheses, and click OK (**fig. 7**).

3. The opportunity now exists for students to investigate whether function composition is commutative by reversing the order of the functions. Repeat the instructions in #2, clicking on $g(x)$ first, then $f(x)$.

4. Enter functions in parametric form, as previously described, so students can animate and change the graph or graphs, looking for certain types of graphs.

5. Given a function f, ask students to find another function g such that $f(g(x)) = g(f(x)) = x$. In other words, ask students to find the inverse of f.

TANGENTS AND DERIVATIVES

1. Type in any function, although a cubic might be a good choice. Once the function is entered, make sure that only the function is selected. Then click on Graph: Plot Function. The graph should appear.

2. Select the point tool and construct a point on the graph. Click on Measure: Coordinates. Also click on Measure: Abscissa(x).

3. Click on Measure: Calculate. Select the abscissa, and then subtract 1. Click OK. (See **fig. 8**). Repeat, adding 1 instead of subtracting 1.

4. Evaluate the function at each of the two points, $x_A - 1$ and $x_A + 1$, using the calculator. Call up the calculator and click on your function. An $f(\ )$ should appear, with the cursor blinking inside the parentheses. Then click on $x_A - 1$; $f(x_A - 1)$ should now appear. Repeat the process using $x_A + 1$ to create $f(x_A + 1)$ (**fig. 9**).

5. Now plot $f(x_A - 1)$ and $f(x_A + 1)$. Select $x_A - 1$ and $f(x_A - 1)$, go to Graph: Plot as (x, y). Repeat using $x_A + 1$ and $f(x_A + 1)$. Select those two points and Construct: Line. Double-click on $x_A - 1$, and edit

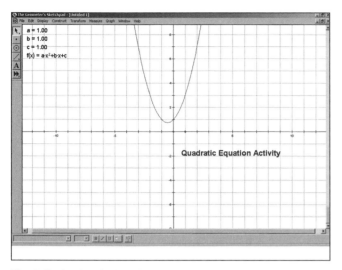

Fig. 9. Evaluating the function at $x_A - 1$ and $x_A + 1$.

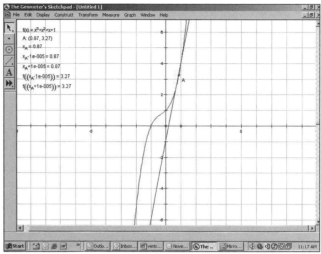

Fig. 10. Plotting $f(x_A - 1)$ and $f(x_A + 1)$.

it to be $x_A - .00001$. Do the same to $x_A + 1$. What started out as a secant line is now very close to being the tangent line to the curve at the original point (**fig. 10**).

6. Select the line and measure its slope under the Measure menu. Plot the abscissa and the slope using Plot as (x, y) under the Graph menu. With this new point selected, go to Display: Trace Plotted Point.

7. Click on the original point and go to Display: Animate point. A Motion Controller box will appear, which can be used to slow down, speed up, reverse, or pause the animation. As the point travels along the graph, the tangent line at that point should also be changing. The point, which is (abscissa, slope), should also be leaving a "trail" behind it as it moves. This trail measures the value of the derivative of the function.

8. *Optional*: Change the constant parameter to other values, and animate the original point. Note how the derivative is unchanged. This observation can lay the foundation for slopefields and the constant of integration.

THE DEFINITION OF A DERIVATIVE

1. Type in any function. Once the function is entered, make sure that only the function is selected. Then click on Graph: Plot Function. The graph should appear.

2. Select the point tool and construct a point on the graph. Relabel the point x. Click on Measure: Abscissa(x).

3. Under Graph, create a New Parameter, h.

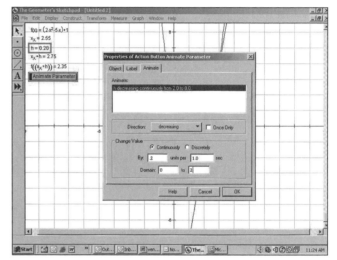

Fig. 11. Properties of Animation box.

4. Call up the calculator. Calculate the values of $x_x + h$ and $f(x_x + h)$.

5. Select $x_x + h$ and $f(x_x + h)$, then Graph: Plot as (x, y). Relabel the point as $x + h$.

6. Construct the line joining the points labeled x and $x + h$.

7. Select the parameter h. Then select Edit: Action Buttons: Animation. A Properties of Animation box will appear.

8. In the pull-down menu at Direction, choose "decreasing." For Change Value, choose 0.2 units per 1.0 seconds and a domain of 0 to 2. Click OK (**fig. 11**). An Animate Parameter button will appear.

9. Click on Animate Parameter. Clicking it a second time will stop the animation. Students will see that as h approaches 0, $x + h$ approaches x. The secant line determined by x and $x + h$ approaches the tangent line at x. With the animation stopped, drag x to another location on the function and restart the animation.

10. If this activity follows the previous one, students can make the connection that the derivative, the slope of the tangent at a point on the function, is the limit of the slope of the secant line determined by x and $x + h$.

Through these explorations, students will gain a better understanding of the illustrated mathematical concepts. For example, I used the tangents and derivatives exploration in a calculus course for certified elementary in-service teachers completing a master's degree to become middle school certified. I started with a cubic equation and graphed its derivative as described; I then reduced the equation to a quadratic by resetting the coefficient of x^3 to 0 and repeated the process of graphing the derivative. This process continued until a constant function was left. I had no need to introduce the power rule for derivatives—the students already saw it and knew it.

REFERENCES

Choi-Koh, Sang. S. "A Student's Learning of Geometry Using the Computer." *Journal of Educational Research* 92 (1999): 301–11.

Dixon, Julie. "Computer Use and Visualization in Students' Construction of Reflection and Rotation Concepts." *School Science and Mathematics* 97 (November 1997): 352–58.

Finzer, William, and Nicholas Jackiw. "Dynamic Manipulation of Mathematical Objects." Paper presented at Technology Conference: NCTM Standards 2000, Arlington, Va, June 1998. Available at www.keypress.com/sketchpad/sketchtalks.html.

Groman, Margaret W. "Integrating Geometer's Sketchpad into a Geometry Course for Secondary Education Mathematics Majors," pp. 61–65. Summer conference proceedings of the Association of Small Computer Users in Education, North Myrtle Beach, S.C., June 9–13, 1996.

Jackiw, Nicholas. The Geometer's Sketchpad, version 4. Berkeley, Calif: Key Curriculum Press, 1991.

Lester, Margaret. L. "The Effects of The Geometer's Sketchpad Software on Achievement of Geometric Knowledge of High School Geometry Students." PhD diss., University of San Francisco, 1996. *Dissertation Abstracts International* 57(06), AAT9633545.

Moyer, Todd O. "An Investigation of The Geometer's Sketchpad and van Hiele Levels." PhD diss., Temple University, 2003.

Weaver, Jennifer L., and Robert J. Quinn. "Geometer's Sketchpad in Secondary Geometry." *Computers in the Schools* 15, no. 2 (1999): 83–95.

Quadratic Equation Graphing Activity

Sheet 1

1. Go to Graph: New Function. Type the function $a \cdot x^2 + b \cdot x + c$.

2. Once the function is entered, click OK and make sure that only $f(x) = a \cdot x^2 + b \cdot x + c$ is selected. Then click on Graph: Plot Function. The following graph should appear:

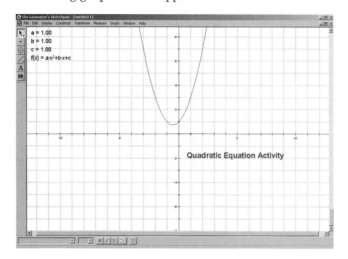

3. Reset the parametric values of b and c to 0. Select only parameter a and click on Display: Animate Parameter. A Motion Controller box should pop up, stating Target: Parameter a. There are buttons for play, ▶; stop, ■; reverse, ⇌; and pause, ‖. You can also adjust the speed of the animation. To start, a will increase. Press reverse and a will decrease. Set $a = 0$.

    What happened to the graph? _____

4. How did changing the value of a from negative to positive affect the graph? _____

    How would you make the graph become "skinny"? _____

5. Repeat the process for parameters b and c next.

    What effect does b have on the graph? _____

    What effect does c have? _____

6. You can also animate two of the parameters at a time. For example, animate b and c.

    What happened to the graph? _____

Trigonometric Functions

Sheet 2

1. Go to Graph: New Function. Type $a \cdot \sin(b \cdot x + c) + d$. You can find the sine function under the Function pull-down menu. A dialog box may appear, asking whether you would like to change from degrees to radians. Radians are the preferred unit. You can also change to radians by going to Edit: Preferences and changing the angle measurement unit to radians.

2. Once the function is entered, press OK, and make sure that only $f(x) = a \cdot \sin(b \cdot x + c) + d$ is selected. Then click on Graph: Plot Function. The following graph should appear:

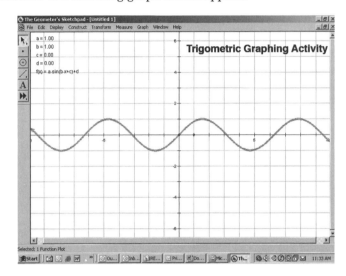

3. Make sure the parametric values of a and b are 1 and c and d are 0. Select only parameter a and click on Display: Animate Parameter. A Motion Controller box should pop up, stating Target: Parameter a. There are buttons for play, ▶; stop, ■; reverse, ⇌; and pause, ‖. You can also adjust the speed of the animation. To start, a will increase. Press reverse and a will decrease.

4. How did changing the values of a affect the graph?_____

    What happened when a was negative? _____

    What does the graph look like when a is between 0 and 1? _____

5. Repeat the process for the parameter b. How does the value of b seem to affect the graph?_____

    What does a negative value of b do to the graph?_____

    What if b is between 0 and 1? _____

6. Now change the values of c. What is changing on the graph?_____

    What is not changing? _____

7. Animate both b and c. What conclusion can you make? _____

8. Finally, change the value of d. What effect does d have on the graph?_____

Composition of Functions

Sheet 3

1. Type in any two functions that you wish. For example, let $f(x) = x^2 - 3$ and $g(x) = 2x + 1$. Once the functions are entered, make sure that only the functions are selected. Then click on Graph: Plot Function. The following graphs should appear:

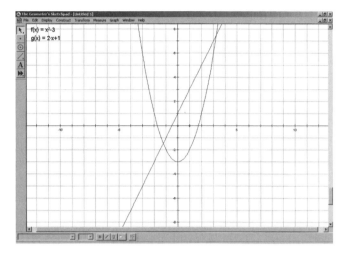

2. Go to Graph: Plot New Function. Click on the first function, $f(x)$, and $f()$ should appear in the screen of the calculator. Click on the second function, $g(x)$, and $f(g())$ should appear. Type x inside the innermost parentheses and click OK, as shown below:

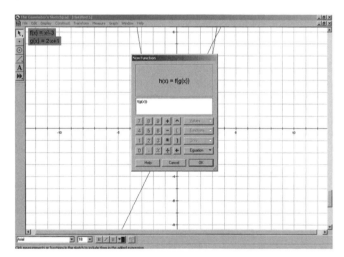

3. Go to Graph: Plot New Function, and enter the functions in the opposite order. Does the graph appear to be the same as the first one? Describe any differences.

Composition of Functions   Sheet 4

4. Experiment by typing in different types of functions. What is the result when the two functions are:

   a. Both linear? _____

   b. One linear and one quadratic? _____

   c. Both quadratic? _____

   d. One quadratic and one a square of a polynomial? _____

5. Use $g(x) = 2x + 1$, and find a function $h(x)$ such that $h(g(x)) = g(h(x)) = x$. Write down $h(x)$.

6. Can you find a another function $i(x)$ such that $i(f(x)) = f(i(x)) = x$? Use $f(x) = x^2 - 3$.

Tangents and Derivatives

Sheet 5

1. Type in the function $f(x) = ax^3 + bx^2 + cx + d$. Once the function is entered, make sure that only the function is selected. Then click on Graph: Plot Function. The graph should appear.

2. Select the point tool and construct a point on the graph. Click on Measure: Coordinates. Also Measure: Abscissa(x).

3. Next, construct the tangent line. Do this by clicking on Measure: Calculate. Select the abscissa, and then subtract 1. Click OK. Click on Measure: Calculate to call up the calculator (see below). On the calculator, select the abscissa again and add 1.

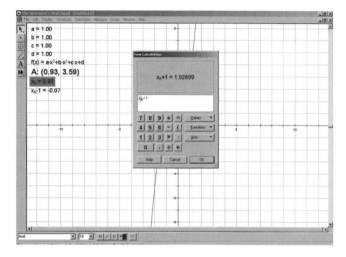

4. Using the calculator, evaluate the function at both points, $x_A - 1$ and $x_A + 1$. Call up the calculator and click on your function. An $f()$ should appear, with the cursor blinking inside the parentheses. Then click on $x_A - 1$. Then, f($x_A - 1$) should now appear. Repeat the process, using $x_A + 1$ to create $f(x_A + 1)$. See below.

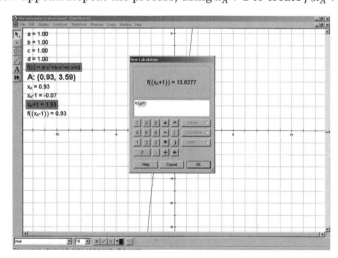

5. Now to plot $f(x_A - 1)$ and $f(x_A + 1)$. Select $x_A - 1$ and $f(x_A - 1)$, go to Graph: Plot as (x, y). Repeat using $x_A + 1$ and $f(x_A + 1)$. Select those two points and Construct: Line. What do we call this line?_____

Tangents and Derivatives   Sheet 6

6. Double-click on $x_A - 1$, and edit it to be $x_A - .00001$; likewise, do the same to $x_A + 1$ (see below). What kind of line is it now? _____

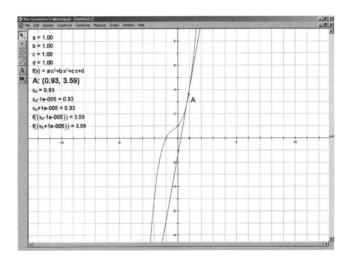

7. Select the line and measure its slope, under the Measure menu. Plot the abscissa and the slope, using Plot as (x, y) under the Graph menu. With this new point selected, go to Display: Trace Plotted Point.

8. Click on the original point and go to Display: Animate point. A Motion Controller box will appear, which can be used to slow down, speed up, reverse, or pause the animation. As the point travels along the graph, the tangent line at that point should also be changing. The point, which is (abscissa, slope), should also be leaving a trail behind it as it moves. The graph that is being traced is the graph of the derivative of the function.

9. When is the function increasing? _____

    a. When is the function decreasing? _____

    b. In terms of the derivative, when does the function appear to be at a maximum? _____

    c. In terms of the derivative, when does the function appear to be at a minimum? _____

10. Change the constant parameter to other values and animate the original point.
    Note any change in the derivative.

11. Go to Display: Erase Traces. Answer the following questions by resetting the appropriate parameter to 0 and allowing point A to move through the graph.

    a. What is the derivative of a cubic function? _____

    b. What is the derivative of a quadratic function? _____

    c. What is the derivative of a linear function? _____

    d. What is the derivative of a constant function? _____

Linear and Quadratic Change: A Problem from Japan

Blake E. Peterson

In the fall of 2003, I had the opportunity to conduct some research on the student teaching process in Japan. During my seven weeks of research at the junior high school affiliated with Ehime University in Matsuyama, Japan, I observed mathematics lessons taught by student teachers as well as many more lessons taught by experienced teachers. The basis for most of these lessons was wonderfully rich mathematics problems. In these lessons a problem was posed to students, time was given for them to explore it, and then a discussion of the solutions to the problem took place. A detailed description of similar problem-based lessons can be found in *The Teaching Gap* (Stigler and Hiebert 1999) and *The Open-Ended Approach: A New Proposal for Teaching Mathematics* (Becker and Shimada 1997).

Some of the assets of these problems were the connections students were able to make and the variety of representations they were able to employ in solving them. For example, connections were made between tabular, graphical, and symbolic representations, between linear and quadratic equations, and between geometric behavior and algebraic functions. Over time, I began to realize that the richness of these problems had a great deal to do with the connections and representations that were such a prominent part of these lessons.

Many teachers in the United States are making efforts to incorporate the Process Standards from the *Principles and Standards for School Mathematics* (NCTM 2000) into the teaching and learning of mathematics in their classrooms. If students are only presented with routine exercises that focus on a narrowly defined skill, connections are difficult to make. Solving broad, open-ended problems, however, allows students to see connections as part of the problem-solving process. Open-ended problem-solving situations also afford students the opportunity to use various representations as they solve the problem and communicate their solution to their peers.

Rich problems, like the ones I observed being used in Japan, are excellent sources about which to build lessons that incorporate the Process Standards of problem solving, connections, and representations. In this article, I begin by introducing a favorite from among the problems that I saw used by my colleagues in Japan and then go on to describe how this problem plays out in the classroom. Common student approaches to the problem will be presented, along with a discussion of where it might fit in the curriculum.

This problem is centered on the numerical and graphical behavior of linear and quadratic functions and is designed to encourage students to use tables and graphs on their way to describing geometric behavior symbolically. This problem asks students to describe the change in attributes of a sequence of geometric figures. Some attributes change linearly, and others change in a quadratic pattern. The conversations about the differences between linear and quadratic behavior allow students to make many of the types of connections described above.

LINEAR AND NONLINEAR GROWTH

This problem, which I saw used in a ninth-grade mathematics classroom in Japan, seems most appropriate for an algebra 2 class in the United States. Because it generates some linear and some quadratic solutions, it would be a nice problem for students after they have become familiar with both linear and quadratic equations. However, I used it in an algebra 2 class just before students began their study of quadratic equations and found it to be excellent for revisiting the concept of linear functions and also for motivating a discussion about nonlinear (quadratic) functions.

If this problem is introduced prior to a study of quadratic equations, students initially only see cases as linear and not linear. They are unsure what type of equation to use to describe the observed nonlinear situation. Because the nonlinear examples from this problem are quadratic, when I refer to nonlinear behavior, quadratic is implied.

Since most textbooks present students with equations and ask them to create a table and a graph from the equation, they think about linear and quadratic equations as first-degree equations with one variable and second-degree equations with one variable, respectively. Students may also visualize the graph of a line or a parabola, but they make connections between these representations and their corresponding tables far less often. In the problem presented here, students describe a geometric pattern by first building a table of values, then constructing the graph of those values, and finishing with an equation. This order, which differs from that of most textbook problems, allows students to make the connection between the table and other representations more readily. They can also compare the tabular values generated by a linear equation with those generated by a quadratic equation.

The following list gives some examples of responses to the preliminary question (see **fig. 1**) from students in both the Japanese ninth-grade classroom and the United States algebra 2 classroom:

> perimeter
> height
> width
> size of enclosing rectangle
> number of "toothpicks"
> number of interior toothpicks
> number of intersections
> number of corners
> number of convex corners
> number of squares
> number of nonadjacent squares
> number of right angles
> sum of the interior angles
> number of diagonals
> leftover space
> number of segments
> number of parallel lines
> length of longest line
> number of rectangles

As students generate a list of answers, questions of clarification need to be posed, such as "What do you mean by 'toothpicks'?" "What do you mean by 'number of squares'? Do you mean squares of any size?" "Are the diagonals only across one small square or can they cross multiple squares?" "Do intersections include points where two toothpicks meet or just where four toothpicks meet?" Each student or group of students may have their own answers to these questions, but it becomes their responsibility to define specifically what they are considering.

Once a list of changing attributes is identified, students are asked to describe the change in one attribute:

Problem: Using a table, a graph, and an equation, describe the step-by-step change observed in **figure 1**.

The solutions to this problem tend to fall into two general categories: linear and quadratic. As tables are constructed, the students are quickly able to identify patterns that are linear and patterns that are not linear. We will first look at some nonlinear (quadratic) examples and then some linear examples.

Quadratic

Many of the attributes in the list above generate quadratic growth. Those considered by students in the United States include total blocks or area, the number of inside right angles, leftover space, and the number of toothpicks.

Total blocks or area: The total number of blocks, or 1 × 1 squares, is the same as the area of the figure. One group of students saw this attribute as the total blocks and created the table, graph, and equation shown in

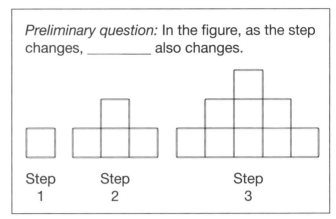

Preliminary question: In the figure, as the step changes, _____ also changes.

Fig. 1. What attributes change as the step increases?

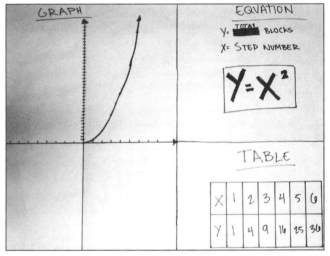

Fig. 2. One group counted the number of blocks (area).

figure 2. In this example, generating a table of values highlights a pattern that is readily recognized as perfect squares (see lower right-hand corner of **fig. 2**). Thus, the equation likely comes from the ability to recognize a numerical pattern rather than from any knowledge of quadratic functions.

Notice that although the problem situation has a domain of natural numbers, this group constructed a graph that is continuous with a domain of positive real numbers. All other groups, including groups from Japan, made the same generalization. My Japanese colleagues and I chose not to pursue this distinction, but it could easily become a rich point of discussion.

The number of inside right angles: If the number of right angles is investigated, the first question that must be resolved is whether to count the right angles on the outside of the figures. One group of students chose to consider only the interior right angles. Since each 1 × 1 square in the figure has four right angles in it, the total number of inside right angles is four times as big as the number of 1 × 1 squares.

The group that chose this approach has all three representations clearly displayed, as can be seen in **figure 3**. Based on observations, I believe that the students

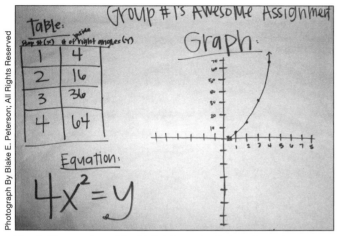

Fig. 3. The number of "inside" right angles is four times the number of blocks.

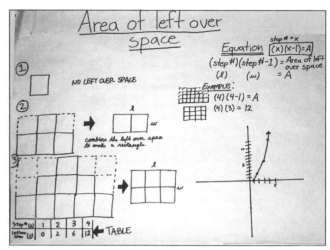

Fig. 4. "Leftover space" requires a slightly more sophisticated approach.

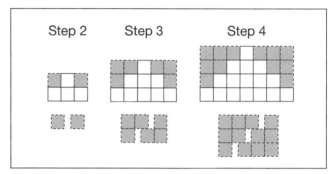

Fig. 5. Leftover space can be arranged to form rectangles.

were able to generate their equation by looking at the numerical pattern in the table. It is interesting to note that these students had not been formally introduced to quadratic equations at the time they were presented with this problem; thus, it is unlikely that the quadratic equations were generated from a knowledge of the behavior of quadratic functions.

The leftover space: Leftover space is determined by constructing a rectangle around each figure and computing the number of squares in the rectangle that are not part of the original figure. Although the change in this attribute is modeled by a quadratic function, the students are able to focus on the geometry of the problem in order to generate an equation without any formal knowledge of these nonlinear functions. The student work in **figure 4** highlights how the geometry of this attribute can be quite readily understood.

In the previous cases of the total number of blocks and the number of right angles, the function was easily generated by observing the numerical pattern in the table, but in this case the numerical pattern is more difficult to identify. The geometric pattern, on the other hand, can shed a great deal of light on the function. In the student work (**fig. 4**), the leftover space is rearranged to form a rectangle. This can be seen more clearly in **figure 5**.

As the step increases, the leftover space is in the shape of two inverted staircases. Each staircase height and width increases by one as the step increases. The staircases are placed together as shown in **figure 5** to form a rectangle. Thus, the leftover space for step 2 forms a 1×2 rectangle; the leftover space for step 3 can be rearranged to form a 2×3 rectangle; and the leftover space for step 4 can be rearranged into a 3×4 rectangle. The equation for the area of this new rectangle becomes the function that describes this attribute. Thus, the equation of the area of the leftover space is $A = x(x - 1)$ where x is the step number. After using geometry to generate the equation, connections can be made between the quadratic function and the numerical pattern in the

table by verifying that the equation does, in fact, generate the values in the table.

The number of toothpicks: If each of the original figures is viewed as being constructed with toothpicks, then generating a function that counts the number of toothpicks for a given step can be an interesting and challenging problem. In this case, the students created a table (see **fig. 6**) and quickly recognized that it was not linear because the differences between the values in the sequence 4, 13, 26, 43, 64, . . . are not constant. The pattern, however, is difficult to identify numerically as well as geometrically, and the students were unsure of the type of function that could describe the pattern. Since a linear function would have constant differences between successive terms, they concluded that the function must have some type of x^2 term in it. From that point, they used a guess-and-test method to create a function that would match the inputs and outputs in the table. In the guess-and-test process, they had a $2x^2$ term and changed it to a $3x^2$ term, only to realize that the function grew too quickly. Therefore, they went back to the $2x^2$ term and added an x term to make their equation of the form $y = 2x^2 + x + 1$. Next they used guess-and-test approach on the coefficient of the x and the constant term in order to find the appropriate equation of $2x^2 + 3x - 1$.

181

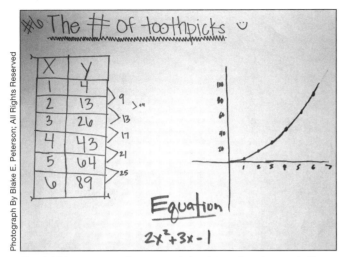

Fig. 6. This group used guess and check to arrive at a quadratic equation.

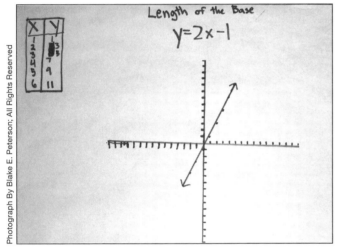

Fig. 7. The length of the base is a linear function of step number.

A slightly more sophisticated approach to this problem would be to count systematically the number of vertical toothpicks and the number of horizontal toothpicks. Some students began variations on this approach but struggled to put all the pieces together in the time allotted to find the resulting equation. For clarification of the solution to this problem, a summary of this counting method is shown in **table 1**. In the table, the vertical columns are counted left to right, and the horizontal rows are counted top to bottom.

Since the sum of the first n integers is

$$\frac{n(n + 1)}{2}$$

the number of vertical toothpicks becomes

$$2\left(\frac{n(n + 1)}{2}\right) = n(n + 1) = n^2 + n$$

It is also known that the sum of the first n odd integers is n^2, so the number of horizontal toothpicks is $n^2 +$

Step	Leftover Space
1	0
2	2
3	6
4	12
5	20
6	30

Fig. 8. Second differences are constant when computing leftover space.

$2n - 1$. Adding the number of vertical and horizontal toothpicks yields $(n^2 + n) + (n^2 + 2n - 1) = 2n^2 + 3n - 1$.

Summary: In each case that generated a quadratic equation, students constructed a table by looking at the geometric figure. They quickly recognized that the pattern in the table was not linear because the differences between consecutive outputs were not constant. Since they had not yet been introduced to quadratic equations, they did not know what to call the pattern other than "not linear." In the case of the total number of blocks and total right angles, perfect squares could be recognized in the table of values, and an equation followed. In the case of the leftover space, numerical patterns were not easily seen, but a pattern in the geometry of the shapes led to an equation. In the final case of the number of toothpicks, neither of the previous methods proved to be productive. Thus, students hypothesized that the equation contained a term to the second degree and used guess and test from that point.

In every case, students started with a physical situation and created a table on their way to writing an equation. Since this order is different from what is usually found in textbooks, it should help students move more flexibly, in either direction, between different representations.

Linear

The first of the two linear examples was generated directly from looking at the length of the base. The second example evolved from a group of students struggling to describe the leftover space because it was not linear; instead, they described the change in the leftover space.

Length of the base: When considering how the length of the base changes as the step changes, the table, graph, and equation shown in **figure 7** are generated. Students can clearly see from the table of values that the y values are increasing by 2 each time the x values increase by 1. Recognizing that the constant differences are indicative of the slope of a line is a valuable connection. This recognizable linear behavior implies

TABLE 1			
Counting Toothpicks			
Step	Vertical Toothpicks	Horizontal Toothpicks	Total
1	1 + 1	1 + 1	4
2	1 + 2 + 2 + 1	1 + 3 + 3	13
3	1 + 2 + 3 + 3 + 2 + 1	1 + 3 + 5 + 5	26
4	$1 + 2 + 3 + 4 + 4 + 3 + 2 + 1 =$ $2(1 + 2 + 3 + 4)$	$1 + 3 + 5 + 7 + 7 =$ $(1 + 3 + 5 + 7) + 7$	43
n	$2(1 + 2 + 3 + \cdots + n)$	$(1 + 3 + 5 + \cdots + (2n - 1)) + (2n - 1)$	

that the slope of the equation is 2. The corresponding equation, $y = 2x - 1$, easily follows.

Change of the leftover space: In the quadratic examples, one of the groups had considered the leftover space (see **fig. 4**). A second group also investigated the leftover space and, in their notes, generated the table shown in **figure 8**. When the students looked at the leftover space $(0, 2, 6, 12, 20, 30, \ldots)$ and differences in the leftover space $(2, 4, 6, 8, 10, \ldots)$, they realized that the leftover space was not changing linearly because the differences were not constant. When they looked at the difference of the differences $(2, 2, 2, 2, \ldots)$, however, they noticed a constant change, indicating a linear pattern. Thus, they created their poster (see **fig. 9**) to describe the behavior of the change in the leftover space instead of the behavior of the leftover space. In the poster, they consolidated their table to show only the step and the change in the leftover space, which would be described by the linear equation $y = 2x - 2$.

In the discussion of these solutions with students, a natural progression could be to start the group of linear solutions and then continue with the quadratics. The last case, the change in the leftover space, however, follows better after a discussion of the leftover space. For this reason, I reversed the order I might typically follow in a classroom.

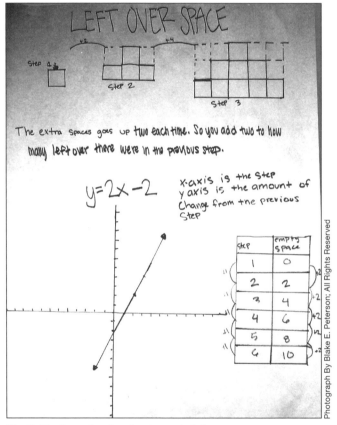

Fig. 9. The linear function for change in leftover space.

A JAPANESE COMPARISON

In Japan, I saw this problem taught in a ninth-grade classroom of forty students working in groups of four. The class worked on this problem for two and a half 50-minute periods. The students spent the first day deciding which changing attribute to investigate and formalizing their thinking. By the end of the first day, a few of the ten groups had presented their solutions. The remainder of the solutions were presented on the second day of the lesson. On the beginning of the third day, the teacher, Ms. Sunada, discussed some generalizations about linear and quadratic functions.

The Japanese student solutions were similar to those of the U.S. students and fell into the same two categories of linear and nonlinear functions. Japanese students

also looked at the total number of triangles created by cutting each square in half and at the total number of triangles in the leftover space. The triangle ideas came from a suggestion made by a student early in the discussion and got the whole class thinking about triangles. The reasoning needed to investigate these situations, however, is the same as what U.S. students had to use. Some of the other attributes the Japanese students investigated were the number of triangles with no exterior edge, the maximum number of nonadjacent squares, and the sum of interior angles around the perimeter of the polygon that was formed by the squares.

As the Japanese students presented their solutions, they were encouraged by the teacher to share how they found the equation. Ms. Sunada was very careful to push them to justify how each component of their equation related to the figure and numerical patterns. For example, when one group of students looked at the sum of the interior angles and generated the equation $y = 360 + 720(x - 1)$, she asked them why it was $x - 1$ instead of just x. The discussion of student solutions also focused on similarities and differences between the solutions. Ms. Sunada organized the presentations of the solutions so these similarities and differences would become more obvious.

CONCLUSION

The problem presented here is centered on the representations of tables, graphs, and equations. In addition, the problem was placed in a geometric context, which is yet another representation. The observation of patterns in the geometric situations and the subsequent conversion of these patterns to graphs and equations is also fertile ground for students to make connections.

For the purpose of making connections, "problem selection is especially important because students are unlikely to learn to make connections unless they are working on problems or situations that have the potential for suggesting such linkages" (NCTM 2000, p. 359). The problem presented here is particularly nice because the connections occur naturally in the problem-solving process, allowing students to make them without being told precisely what to look for. In Japan and the United States, the attributes that the students selected, the representations they used, and the connections they made were all similar. The learning that occurred was not an artifact of the language or the culture; it was a product of the rich mathematical problem in which they all engaged.

ACKNOWLEDGMENT

The author thanks Yutaka Kawasaki, Kyuoko Sunada, and Yoshiaki Umakoshi of Ehime University's Fuzoku Junior High School for graciously allowing him to watch, listen, and learn.

REFERENCES

Becker, Jerry P., and Shigeru Shimada, eds. *The Open-Ended Approach: A New Proposal for Teaching Mathematics*. Reston, Va.: National Council of Teachers of Mathematics, 1997.

National Council of Teachers of Mathematics (NCTM). *Principles and Standards for School Mathematics*. Reston, Va.: NCTM, 2000.

Stigler, James W., and James Hiebert. *The Teaching Gap: Best Ideas from the World's Teachers for Improving Education in the Classroom*. New York: Free Press, 1999.

Describe That Change

Sheet 1

The figures below have many attributes that are changing with each new step. For example, the number of unit squares changes as the step changes.

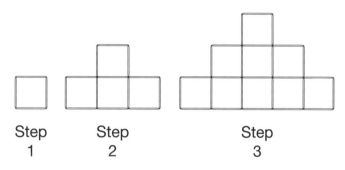

Step 1   Step 2   Step 3

1. Fill in the blank of the following question with at least 4 different attributes.

As the step changes, _____ also changes.

Attribute 1:_____   Attribute 2: _____

Attribute 3:_____   Attribute 4: _____

2. Select one of the attributes above or an attribute that one of your classmates identified and investigate it further.

3. Use a table, graph, and equation to describe the step-by-step change in the selected attribute.

Step	1	2	3	4	5	6
Attribute						

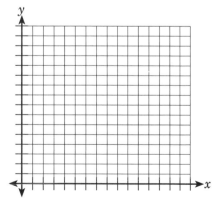

4. Explain how the elements of your equation are related geometrically to the figures in each step.

Using Graphs to Introduce Functions

Frances Van Dyke

The past decade has seen a shift toward functions as a central theme in beginning algebra. The advent of graphing calculators has meant that the graphical representation of functions is accessible and can be used in a meaningful way. Yet evidence indicates that students ignore graphs or resort to complicated algebraic expressions rather than read information from a graph. In "Understanding Connections between Equations and Graphs," Eric Knuth (2000) reported on a study that he conducted with 178 students from a suburban high school. He concluded that students may be missing the basic "Cartesian connection" and that they do not recognize a graph and an equation are two representations for the same set of points.

One idea for linking the representations and highlighting the importance of graphs is to introduce the concept of function using graphs, or pictorial representation of functions. A natural progression exists from qualitative graphs to quantitative graphs to tables to equations. Students become comfortable with functions that are introduced using this progression.

This article presents four activities that use this progression to introduce the theme of distance from an object as a function of time. The advantage of this application is that students are familiar with the idea, can walk the motion, and can use motion detectors to produce similar graphs. The disadvantage is that students may not understand the dependent variable and may think that it is speed or distance traveled, rather than distance from an object. They may also have a tendency to use iconic translation, that it, to impose an image of a scene onto the graph. An application in which these problems are avoided is one in which the value of the variable changes as a function of time, and teachers are encouraged to make up similar activities. After students have completed sheet 1, the teacher may want to discuss the answers in class. The class as a whole might discuss what each graph reveals about motion to underline the meaning of the dependent variable and to prevent students from using iconic translation.

Concentrating first on qualitative graphs allows students to think on an abstract level and to look at graphs globally. Exercises that connect the classroom with the real world may appeal to students who do not consider themselves mathematically oriented. By comparing qualitative graphs, a teacher can introduce important concepts from basic algebra without the burden of algebraic notation. The introduction of coordinates and points on the graphs allows students to interpolate and give specific information about the situation. Finally, when students are comfortable with the graphical and tabular representations, the teacher can introduce equations.

These activities can be used for grade levels 8–14. The activities challenge students who have a prealgebra background and are a good review for those who have basic algebra. The only materials needed to complete the activity are the activity sheets and blank paper for answers. A graphing calculator and a motion detector may also be used to complement the activity. The objective is to help students deepen their understanding of functions, graphs, and the underlying equivalence of the numeric, tabular, and algebraic representations of functions. Students complete the activity in groups. Each student can keep his or her own record of answers, so that the teacher can assess his or her work. After students complete all eight sheets, the teacher leads the class in a discussion of the notion of a function as a relationship between quantities that change. The activity sheets focus on different ways of representing this relationship. After completing the activity sheets, students write a paragraph on the aspect of the relationship that each representation highlights. Looking at the equation reveals initial distance, time, and direction. A table gives actual times with corresponding differences, whereas a graph is best for comparing two different motions. The progression can be developed for other functions—both linear and nonlinear—that students typically study.

SOLUTIONS

Sheet 1:

1) The correct answer is (*b*). A student who thinks of the dependent variable as speed might give (*a*) as the answer, and a student who thinks of the dependent variable as distance covered might think that (*c*) is the answer.

2) Answers will vary. For graph (*a*), Hugo might be standing a fixed distance from the object or moving in a circle around the object

3) The correct answer is (*c*). A student who thinks that the roller coast somehow gets pictured in the graph might give (*a*) as the answer; and if iconic translation occurs, that is, if the image of Mary, who is standing still, imposes itself on the graph, a person might give (*d*) as the answer.

4)

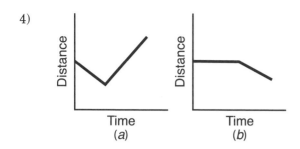

Time
(a)
Time
(b)

The sharper slope in the part of (a) nearest the vertical axis indicates that Chris's walk toward the tunnel was faster than his walk away from the tunnel.

Sheet 2:

5) Teachers may want to have students work on part (a) first, go over the answer, and then try parts (b), (c), and (d). Answers should indicate whether the people are moving toward or away from the ride, in addition to indicating the person who is moving faster. Sample answers are as follows:

a. Initially, Al was about twice as far from the ride as Bea was, and he took about one-quarter of the amount of time that she did to arrive at the ride. Al was therefore moving eight times as fast as Bea.

b. Initially, Clay was about four times as far from the ride as Di. Di moved about one-half the distance that Clay moved in three times the time, so she was moving at six times his rate. Di was moving away from the ride and Clay was moving toward it.

c. Eli was initially about three times as far from the ride as Faye was. They both reached the ride at the same time, so Eli was traveling three times as fast as Faye.

d. Hal was initially about half as far as Gus was from the ride. Hal moved away, and Gus stood still.

e. Although we can compare the two notions of the members of the pairs, we have no idea how far the people were from the rides or how fast they moved, since the graphs do not indicate units. Al, Clay, Eli, and Hal were all walking faster than Bea, Di, Faye, and Gus, respectively.

f. The starting point, or origin, of the walk for both is the ride. The x-intercepts indicate time taken for the walk; the y-intercepts indicate initial distance from the ride.

6) Jane's walk is depicted in (a), Angela's in (c), and Kathy's in (b). In part (a), equal distances are covered in equal time. As time passes in (b), less distance is covered in equal time intervals; whereas a greater difference is covered in (c) as time passes.

Sheet 3:

1) c

2) a. Alisa moved 24 feet in all.

b. She took 8 seconds to walk this distance

c. Alisa walked at a rate of 3 feet per second

d. After 1 second, she was 21 feet away from the pinball machine area.

e. After 2 seconds she was 18 feet away.

f. After 5 seconds, she was 9 feet away.

3)

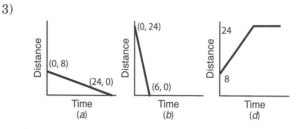

(0, 8)
(24, 0)
Time
(a)

(0, 24)
(6, 0)
Time
(b)

24
8
Time
(d)

Sheet 4:

4) c

5) a. Hugo was 9 feet away from a hot-dog stand and inched steadily toward it, reaching it after 21 seconds. He moved at the rate of 3/7 a foot per second

b. Hugo was 3 feet from a hot-dog stand and moved steadily away from it. He moved at a rate of 2 feet per second

c. Hugo waited for 3 seconds, then moved away at a rate of 3.5 feet per second. After 9 seconds, he was 21 feet away from the hot-dog stand.

6) a. The person started 15 meters from the exit and moved toward it at a rate of 3 meters per second, reaching it after 5 seconds. He or she then waited at the exit for 6 seconds and moved away at a rate of 2 meters per second. Since no indication of direction is given, he or she could be leaving the park, entering the park, or approaching and then backing away from the exit.

b. The individual started 9 meters from the exit, then moved away at a rate of 1.5 meters per second for 4 seconds. He or she then returned to the exit at a rate of 5 meters per second.

Sheet 5:

1) The answer is table (b). It is the only table that shows a constant rate of change.

2) c.

Sheet 6:

3) a.

Time	Distance
0	0
1	6
2	12
3	18
4	24
5	30
6	36

b.

Time	Distance
0	42
1	35
2	28
3	21
4	14
5	7
6	0

d.

Time	Distance
0	0
1	7
2	14
3	21
4	28
5	35
6	42

4) a.

Time	Distance
0	9
7	6
14	3
21	0

b.

Time	Distance
0	3
1	5
2	7
3	9
4	3
⋮	⋮
9	21

c.

Time	Distance
0	9
1	13
2	17
3	21

d

Time	Distance
0	0
1	0
2	0
3	0
4	3.5
5	7
6	10.5
7	14
8	17.5
9	21

5) For question 6 on *Sheet 4*, the tables are as follows:

a.

Time	Distance
0	9
2	6
3	3
5	0
6	3
⋮	⋮
11	0
12	2
13	4
14	6
15	8
16	10
17	12

b.

Time	Distance
0	9
1	10.5
2	12
3	13.5
4	15
5	10
6	5
7	0

Sheet 7:

1) The correct equation is (*d*). Each point in the table satisfies the equation. The initial distance is the *y*-intercept on the graph. In the table, the initial distance is the line with 0 as the time entry. In the equation, the initial distance is the constant term 40, corresponding to the *y*-intercept

2) a. The equation (*a*) $y = 42 - 6x$ corresponds to the table given.

  b. $y = 42 - 7x$ corresponds to table (*b*).

  c. $y = 6x$ corresponds to table (*a*).

  d. $y = 7x$ corresponds to table (*d*).

Sheet 8:

3) The answer is (*b*). Moving 20 feet in 8 seconds implies a rate of 2.5 feet per second, which is the coefficient of the variable standing for seconds elapsed. The initial distance corresponds to the constant term in the equation. The answers will vary. For (*a*), one possibility is, "Zarah was 4 feet away from a stand and walked away at 6 feet per second." For (*c*), one possibility is "Zarah was 24 feet away from a stand and walked away at 4 feet per second." For (*d*), one possibility is "Zarah was 8 feet away from a stand and walked away at 6 feet per second".

4) The answer is (*d*). The rate of 3 feet per second is the absolute value of the coefficient of the variable standing for seconds elapsed. The coefficient is negative when Josh is walking toward the bumper cars, so his distance from them decreases as time passes. The initial distance of 30 feet corresponds to the constant term in the equation. For (*a*), one possibility is "A man in a car was 3 feet from the sign, and he rode away at a rate of 30 feet per second." For (*b*), one possibility is "An object was 3 feet above sea level and dropped down at a rate of 30 feet per second." For (*c*), one possibility is "A man was 30 feet from a sign and walked away at a rate of 3 feet per second."

5) The rate of motion is positive when the individual is moving away from the object, and it is negative when he or she is moving toward the object. The distance is increasing (positive) when the individual is moving away from the object, and it is decreasing (negative) when the individual is moving toward the object. The rate of motion is the slope of the line on the graph. The rate of motion can be determined in the table by the difference in the distance for each increment in time, that is, difference in distance divided by difference in time.

6) a. $y = 9 - (3/7)x$

  b. $y = 3 + 2x$

  c. $y = 9 + 4x$;

  d. $y = -10.5 + (7/2)x$

7) 6*a* on *Sheet 4*: $y = 15 - 3x$, if $0 < x \le 5$; $y = 0$, if $5 < x \le 11$; $y = -22 + 2x$, if $11 < x \le 17$

  6*b* on *Sheet 4*: $y = 9 + (3/2)x$, if $0 < x \le 4$; $y = 35 - 5x$, if $4 < x \le 7$

REFERENCES

Knuth, Eric. "Understanding Connections between Equations and Graphs." *Mathematics Teacher* 93 (January 2000): 48–53.

Koellner-Clark, Karen, L. Lynn Stallings, and Sue A. Hoover. "Socratic Seminars for Mathematics." *Mathematics Teacher* 95 (December 2002): 682–87.

National Council of Teachers of Mathematics (NCTM). *Principles and Standards for School Mathematics.* Reston, Va.: NCTM, 2000.

Van Dyke, Frances. *A Visual Approach to Functions.* Emeryville, Calif.: Key Curriculum Press, 2002.

Using Graphs to Introduce Functions

Sheet 1

Each of the graphs on this page represents distance from an object as a function of time.

1. On the following graphs, distance, as labeled on the y-axis, refers to distance from an amusement park. Which graph best matches the following sentence?

   Hugo walked at a steady pace toward the amusement park.

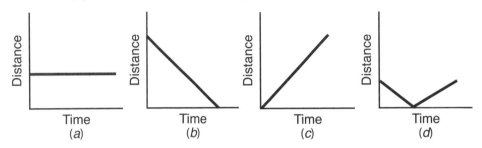

2. Describe a situation involving distance and time that could match each of the graphs you did not choose as the answer to question 1.

3. In this problem, distance on the y-axis stands for Mary's distance from the roller-coaster entrance. Which graph best matches the following sentence?

   While she was standing still, Mary looked at the roller-coaster track in the distance.

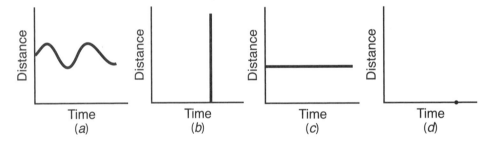

4. Draw a graph for each of the following situations. For graph (a), let the y-axis represent distance from the tunnel of love. For graph (b), let the y-axis represent distance from the entrance to the Ferris wheel.

   a. Chris walked quickly toward the tunnel of love and then slowly walked away from it.

   b. We stood for a while, and then the line in which we were waiting began to move slowly toward the Ferris wheel.

Using Graphs to Introduce Functions

Sheet 2

5. Each of the following graphs depicts the relationship between distance from a ride and time elapsed for two people. Each person walks at a steady rate toward or away from the ride or stands still. For each graph, describe the relationships and make observations and comparisons.

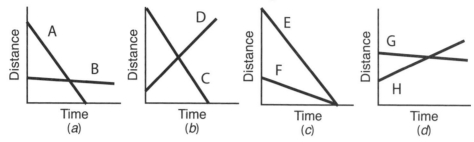

a.

b.

c.

d.

In addition to your observations, answer the following questions for each example:

e. Which person is walking faster?

f. What is the significance of the *x*-intercept? The *y*-intercept?

6. The graphs below show motion away from the park for three different mothers. Jane moves at a steady pace, Angela speeds up as she walks away, and Kathy slows down as she moves away. Which graph matches which woman's motion? Explain your reasoning.

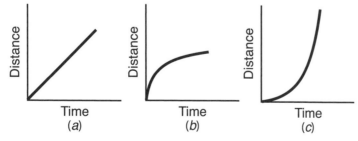

Using Graphs to Introduce Functions Sheet 3

1. Consider the graph below, in which time in seconds is graphed along the *x*-axis and distance in feet is graphed along the *y*-axis. The graph shows Alisa's distance from the pinball machine area as a function of time. Which sentence is a good match for the graph?

   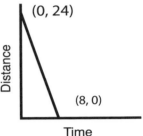

   a. Alisa stood 8 feet from the pinball machine area and moved toward it, reaching it in 24 seconds.

   b. Alisa stood 24 feet from the pinball machine area and moved toward it at a rate of 4 feet per second.

   c. Alisa stood 24 feet from the pinball machine area and moved toward it, reaching it in 8 seconds.

   d. Alisa stood 8 feet from the pinball machine area and moved away from it, stopping when she was 24 feet away.

2. *a.* How many feet did Alisa move in all? _____

   b. How long did she take to walk this distance? _____

   c. The straight line indicates a steady pace. How many feet did Alisa walk each second? _____

   d. How far was Alisa from the area after 1 second? _____

   e. How far was Alisa from the area after 2 seconds? _____

   f. How far was Alisa from the area after 5 seconds? _____

3. For each answer not chosen in question 1, sketch a graph that would correspond to that description. Write the description next to your graph.

Using Graphs to Introduce Functions

Sheet 4

4. Hugo was standing 9 feet from the hot-dog stand. He walked away, and after 3 seconds he was 21 feet away. Which of the following graphs corresponds to this situation?

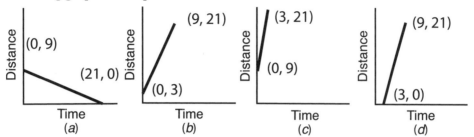

5. Write a description, similar to the one given in problem 1, that could correspond to each of the graphs not chosen in question 4. Determine the walking rate in each situation.

6. In the following two graphs, distance in meters from the main exit is graphed as a function of time in seconds for one individual. Describe the motion using the coordinates shown as endpoints of the line segments. Determine the rate at which the person walks for each segment. Assume that the person is moving directly toward or away from the exit or that he or she is standing still.

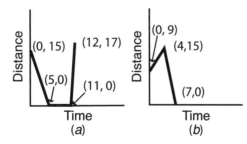

Using Graphs to Introduce Functions

Sheet 5

1. Consider again the graph at the right, where time in seconds is graphed along the x-axis and distance in feet is graphed along the y-axis. The graph shows Alisa's distance from the pinball machine area as a function of time. Which table describes the graph? Explain your decision.

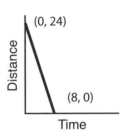

Time	Distance
0	24
2	16
4	8
6	0
8	0

a.

Time	Distance
0	24
2	18
4	12
6	6
8	0

b.

Time	Distance
0	24
2	18
4	11
6	3
8	0

c.

2. Consider the following table.

Time	Distance From Water Ride
0	42
1	36
2	30
3	24
4	18
5	12
6	6
7	0

Choose the sentence that best describes the table.

a. Kirk walked away from the water ride at a rate of 6 feet per second.

b. Kirk was 42 feet away from the water ride and walked toward it at a rate of 7 feet per second.

c. Kirk was 42 feet away from the water ride and walked toward it at a rate of 6 feet per second.

d. Kirk walked away from the water ride at a rate of 7 feet per second.

Using Graphs to Introduce Functions

3. For each answer that did not describe the table in question 2, make a table that could correspond to it. Give entries from each second from 0 to 6.

4. Make tables that correspond to the graphs given on sheet 4, question 4.

5. Make tables that correspond to the graphs on sheet 4, question 6.

Using Graphs to Introduce Functions

1. Consider the graph at the right where time in seconds is graphed along the *x*-axis and distance in feet is graphed along the *y*-axis. The graph shows a girl's distance from a spaceship ride as a function of time. A table of values that could accompany the graph is also shown. Use the table to help you decide which equation matches the graph.

Time	Distance
0	40
2	30
4	20
6	10
8	0

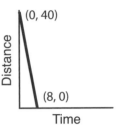

a. $y = 40x - 8$ *b.* $y = 40 - 8x$ *c.* $y = 5x + 40$ *d.* $y = 40 - 5x$

Explain why you made your choice.

How is the initial distance indicated on the graph? In the table? In the equation?

2. Consider again the sentences on sheet 5, question 2, and the tables created in question 3 on sheet 6.

Each of the following is an equation that describes one and only one of the sentences. Here, *x* corresponds to time in seconds and *y* to distance in feet. Write the appropriate sentence under each equation. Indicate the table that goes with each equation. Does each pair of values in the tables satisfy the equation?

a. $y = 42 - 6x$

b. $y = 42 - 7x$

c. $y = 6x$

d. $y = 7x$

Using Graphs to Introduce Functions Sheet 8

3. Zarah is standing 4 feet from a concession stand. She walks away at a steady pace. After 8 seconds, she is 24 feet away. Choose the linear equation that indicates her distance, y, from the stand as a function of time elapsed in seconds, x, since Zarah began walking. Explain your reasoning.

   a. $y = 4 + 6x$ b. $y = 4 + 2.5x$ c. $y = 24 + 4x$ d. $y = 8 + 6x$

   For each answer not chosen, give a scenario that describes the equation.

4. Josh is 30 feet from the bumper cars. He walks toward them at the rate of 3 feet per second. Choose the linear equation that describes the distance from the bumper cars as a function of time elapsed, x, in seconds. Explain your reasoning.

   a. $y = 30x + 3$ b. $y = 30 + 3x$ c. $y = 3 - 30x$ d. $y = 30 - 3x$

   For each answer not chosen, describe a situation that corresponds to the equation.

5. Notice that the equation can always be written in the following form: Distance after x seconds equals initial distance ± rate of motion times x. When is the sign positive, and when is it negative? Explain why. What indicates the rate of motion in the graph? In the table?

6. Determine the equations for the graphs given in question 4 on sheet 4. Be sure that the table entries that you gave on sheet 6, question 4, satisfy the appropriate equations.

7. Challenge: Find the equations for the graphs given in question 6 on sheet 4. These graphs consist of more than one line segment, so give the interval for each equation. (For example, a possible answer might be $2x + 1$ for $0 \le x \le 3$.)

CHAPTER 4

Geometry

Introduction

"Classically, geometry has been the subject in which students encounter mathematical proof based on formal deduction. Although proof should be naturally incorporated in all areas of the curriculum, attention to proof in the geometry curriculum is strengthened by a focus on reasoning and sense making." (NCTM 2009, p. 55).

This chapter's nine geometry activities from *Mathematics Teacher* capture the spirit of mathematics that *Focus in High School Mathematics: Reasoning and Sense Making* (NCTM 2009) advocates for. Mathematics for life, the workplace, and the scientific and technical community (NCTM 2009, p. 3) appear prominently in the activities. For example, Palmer (1946) provides a method for measuring neighborhood objects; Froelich (2000) presents a workplace application of optimizing space in soft drink packaging; and Hirsch (1974) investigates Pick's

TABLE 4.1			
Geometry Activities			
Author and title	Mathematical topic(s)	Context(s)	Materials
Edwards (2005), "Using Overhead Projectors to Explore Size Change Transformations"	Transformations, similarity	Overhead projector	Overhead projector, masking tape, transparencies, student activity sheets
Froelich (2000), "Modeling Soft Drink Packaging"	Volume	Soft drink packaging	The Geometer's Sketchpad (recommended), student activity sheets
Gernes (1999), "The Rules of the Game"	Deductive system	Sports and games	Student activity sheets
Hirsch (1974), "Pick's Rule"	Pick's rule	Mathematics	Transparencies, student activity sheets
Nelson and Williams (2007), "Sprinklers and Amusement Parks: What Do They Have to Do with Geometry?"	Bisectors	Amusement park	The Geometer's Sketchpad (recommended), student activity sheets
Palmer (1946), "Discovering the Tangent"	Trigonometry	Measuring neighborhood objects	Transit (homemade is fine), student activity sheets
Quinn and Ball (2007), "Explore, Conjecture, Connect, Prove: The Versatility of a Rich Geometry Problem"	Circles, locus of points	Path of a rowboat	Student activity sheets
Reys (1988), "Discovery with Cubes"	3-dimensional figures	Cube dropped in paint	Student activity sheets
Toumasis (1992), "The Toothpick Problem and Beyond"	Triangles	Toothpick triangles	Small box of flat toothpicks for each group, student activity sheets

rule to promote a deeper understanding of mathematics. **Table 4.1** summarizes the characteristics of the geometry activities.

NCTM (2009) suggests four key elements for geometry that, if present regularly, will make reasoning and sense making "a part of the fabric of the high school mathematics classroom" (p. 14):

1. *Conjecturing about geometric objects.* Analyzing configurations and reasoning inductively about relationships to formulate conjectures

2. *Construction and evaluation of geometric arguments.* Developing and evaluating deductive arguments (both formal and informal) about figures and their properties that help make sense of geometric situations

3. *Multiple geometric approaches.* Analyzing mathematical situations by using transformations, synthetic approaches, and coordinate systems

4. *Geometric connections and modeling.* Using geometric ideas, including spatial visualization, in other areas of mathematics, other disciplines, and real-world situations

Nearly every activity in this chapter addresses *conjecturing about geometric objects.* For example, Hirsch (1974) asks students to first predict the areas of polygons, next conjecture and test formulas for calculating areas as they move toward Pick's rule, and finally create polygons with a given number of sides with the smallest possible area. Similarly, Reys (1988) gives students an opportunity to search for patterns and make conjectures in an accessible geometric situation (e.g., painting a cube); therefore, the students are not struggling to understand the situation and can focus on the mathematics (i.e., identifying patterns).

Toumasis (1992) continues the emphasis on conjecturing found in Hirsch (1974) and Reys (1988) as students explore triangles. Brown (available on More4U) asks students to predict the number of triangles in the next iteration of a Sierpinski triangle and introduces geometric connections and modeling as students discover the exponential pattern of change. In Toumasis (1992), students hypothesize about the triangle inequality theorem, using toothpicks as construction tools. Toumasis (p. 543) stresses that "although many changes have occurred in the goals for teaching secondary school mathematics, the goal of motivating students to do mathematics and helping them to reason mathematically through problem solving have remained."

Several important mathematical topics are present in Edwards (2005) as students explore scale factor, area, and similarity in both activities. Edward's activity, which uses the overhead projector to investigate size

change, exposes students to constructing and evaluating geometric arguments as they must answer and defend their answers to "what if" questions at the end of the activity. The *geometric connections and modeling* element is also present as students discover the linear relationship between scale factor and projection distance and the quadratic relationship between scale factor and projected area.

Quinn and Ball (2007) address both the *construction and evaluation of geometric arguments* element, as students are asked to convince their classmates that the path of a boat takes on a particular shape given a set of conditions, and the *multiple geometric approaches* element, as they emphasize the consideration of a variety of proofs. Gernes (1999) also develops the *construction and evaluation of geometric arguments* element as students investigate the structure of a deductive system (i.e., undefined terms, definitions, postulates, and theorems) by using a variety of common games (e.g., Monopoly, soccer) in an effort to "bridge the gap between the student's world and the world of mathematics" (Gernes 1999, p. 424).

The *conjecturing about geometric objects, construction and evaluation of geometric arguments*, and *geometric connections and modeling* elements often go hand in hand. For example, Froelich (2000) presents the task of improving the efficiency of secondary soft drink packaging to students, asking them to hypothesize about the optimization of space, model the situation mathematically, and argue that they have found the best package. Similarly, Nelson and Williams (2007) offer an activity that addresses both *construction and evaluation of geometric arguments* and *geometric connections and modeling*. Students are asked to present their reasoning for the best location of a sprinkler and a soft drink stand at an amusement park given particular criteria. Finally, Palmer (1946) gives an oldie but goodie as students collaboratively investigate trigonometric values for common angles; this activity "takes the tangent from the textbook and gives it life and vitality" (Palmer p. 185).

These nine activities also, both implicitly and explicitly, give students opportunities to develop the four mathematical reasoning habits that *Focus in High School Mathematics: Reasoning and Sense Making* (NCTM 2009) describes: (1) analyzing a problem, (2) implementing a strategy, (3) seeking and using connections, and (4) reflecting on a solution. These habits reflect the components of problem solving. For example, Nelson and Williams (2007) expect students to *analyze a problem*. Students use their "own intuitions to guide their exploration and discovery" (Nelson and Williams 2007, p. 440) as they investigate bisectors in the interesting context of amusement parks. In Edwards (2005), students *implement a strategy*, making purposeful use of procedures including measuring, comparing measurements, calculating and comparing scale factors and areas, and constructing fit lines and curves. Gernes (1999) motivates students to develop a deductive system by *seeking and using connections* across a variety of contexts, including board games, sports, and a letter game. Finally, the open nature of the problem in Quinn and Ball's activity (2007) promotes the use of multiple strategies. The authors' suggestion that students share and discuss their strategies compels students to *reflect on a solution* to a problem. In particular, students reconcile different approaches to solving the problem.

REFERENCE

National Council of Teachers of Mathematics (NCTM). *Focus in High School Mathematics: Reasoning and Sense Making.* Reston, Va.: NCTM, 2009.

Using Overhead Projectors to Explore Size Change Transformations

Michael Todd Edwards

Through the study of *transformation*, a wide range of mathematical concepts may be introduced to secondary school students. *UCSMP Geometry* (Coxford, Usiskin, and Hirschhorn 1993) gives this definition of *transformation*:

> **Definition:** A transformation is *a* correspondence between two sets of points such that (1) each point in the preimage set has a unique image, and (2) each point in the image set has exactly one preimage. (p. 255)

First-year geometry students explore congruence as they apply reflection *transformations* to various planar objects. **Figure 1** illustrates the transformation of pre-image $\triangle ABC$ into image $\triangle A'B'C'$ by means of reflection over line *m*.

For instance, when one reflects a polygon over an arbitrary line *m*, a transformation is generated. As **figure 1** illustrates, reflecting preimage $\triangle ABC$ over line *m* results in image $\triangle A'B'C'$. A transformation exists between the two triangles.

As students compare angle measures and side lengths of preimage/image pairs, they are provided with opportunities to explore congruence within the context of transformation. **Figure 1** suggests that reflection preserves angle measure ($m(\angle A) = m(\angle A')$) and side length ($AB = A'B'$), creating congruent image/preimage pairs ($\triangle ABC \cong \triangle A'B'C'$).

As students investigate reflection in greater detail, they learn that reflections may be combined to create "new" transformations—notably rotations, translations, and glide reflections. See **figure 2**.

Because the transformations depicted in **figure 2** preserve length, they are referred to as *isometries*—from

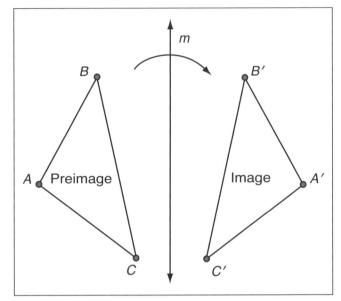

Fig. 1. Reflection over line *m* maps preimage $\triangle ABC$ onto image $\triangle A'B'C'$. We say that a transformation exists between the two triangles.

the Greek *isos*, meaning "equal," and *metron*, meaning "measure."

In the mathematics classes that I teach, students' first encounters with transformation involve reflection, translation, and rotation. After working exclusively with isometries for some time, it is natural for students to wonder if all transformations preserve length. What might a nonisometric transformation look like? Several years ago, while my students pondered this question, an insightful pupil noted that movie theaters employ principles of nonisometric transformation to project motion pictures. The comment led to a class discussion of *size change*. *UCSMP Geometry* (Coxford et al. 1993) defines *size change* transformations in the following manner:

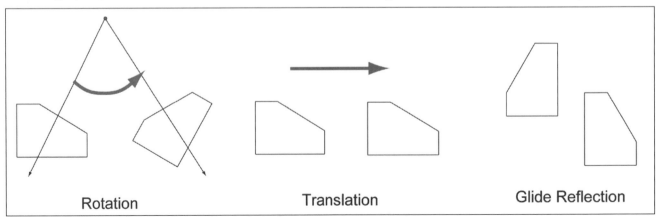

Rotation   Translation   Glide Reflection

Fig. 2. When reflections are composed, rotations, translations, and glide reflections are generated.

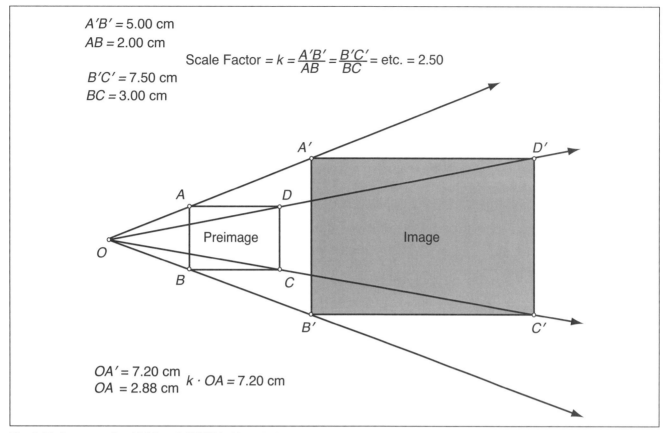

$A'B' = 5.00$ cm
$AB = 2.00$ cm

Scale Factor $= k = \dfrac{A'B'}{AB} = \dfrac{B'C'}{BC} =$ etc. $= 2.50$

$B'C' = 7.50$ cm
$BC = 3.00$ cm

$OA' = 7.20$ cm
$OA = 2.88$ cm $\quad k \cdot OA = 7.20$ cm

Fig. 3. Similar rectangles $ABCD$ and $A'B'C'D'$ with scale factor $= 2.5$

Definition: Let O be a point and k be a positive real number. For any point P let $S(P) = P'$ be the point on OP with $OP' = k \cdot OP$. Then S is the size change with center O and magnitude k.

Following up on the student's remark, I commented that our classroom's overhead projector creates *size change transformations* between objects placed on the projector surface (preimages) and projections on classroom walls (images). As we discussed *magnitude* of size change transformations, curiosity regarding the classroom overhead projector intensified. For instance, several students wanted to know the scale factor associated with the overhead. Many students asked questions that could not be answered without further investigation, such as "What is the relationship between magnitude of size change and an overhead's distance from the wall?" and "Is there a distance at which the scale factor of a projector is less than or equal to 1?" Questions such as these led to the creation of the classroom "experiment" discussed in this article.

I typically introduce the concept of *size change* to students within the context of similarity. Students learn that two polygons are similar when they satisfy the following conditions:

- corresponding angles are congruent
- corresponding lengths are proportional (Coxford, Usiskin, and Hirschhorn 1993, p. 587)

When two noncongruent objects are similar, a size transformation exists between the smaller object, considered the preimage, and the larger object, considered the image. The *scale factor* associated with the size change transformation is a ratio of corresponding lengths (from image to preimage). For instance, **figure 3** depicts two similar rectangles, $ABCD$ and $A'B'C'D'$. As calculations within the figure suggest, a size change transformation with scale factor $k = 2.5$ exists between the preimage $ABCD$ and image $A'B'C'D'$.

Using a series of lenses and mirrors, overhead projectors create similar shapes. Preimages are placed on the projector's glass top, creating magnified images on a projection surface—typically a wall or screen in front of the classroom.

CLASSROOM SETUP

Preparing my classroom for an investigation of overhead projector scaling properties requires setup time. Prior to class (typically the evening before), I dim the lights of my classroom and project a 6.6 cm × 4.6 cm cardboard rectangle (the preimage) to various locations in the room. I apply masking tape along the outline of each projected rectangle, centering the tape along the edge of each projection. The tape provides a rough outline of each image. To make each tracing more precise, I outline the borders of each projection with permanent marker.

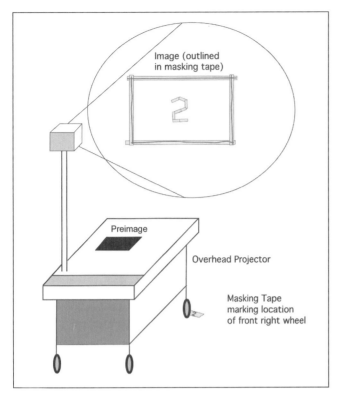

Fig. 4. Dilations of a cardboard rectangle by an overhead projector. The label "2" links the image on the wall with a specific projection location on the classroom floor.

For each projection, I also tape the floor, marking the location of the projector cart's right wheel. To facilitate student data collection, I label the masking tape from the first tracing with a "1," from the second tracing with a "2," and so forth. The setup for the second tracing is illustrated in **figure 4**.

I typically construct five or more rectangular projections on classroom walls from various distances throughout the room. Because a primary objective of the activity is to explore scale factor and its relationship to projection distance, I do not adjust the focus knob of the overhead projector. Adjusting the focus alters the size of images, needlessly complicating student interpretation of data. Although avoidance of overhead focusing may result in one or two blurred images, I've found that tracing images without focusing is not overly difficult.

During class, students explore scale factor and area while examining the rectangular projections they find on the classroom walls. Using a meter stick, each student begins by measuring the height and width of each projected rectangle and recording this data into a table similar to **table 1**. (See also sheet 1.) Next, each student uses knowledge of geometry, algebra, technology, and problem solving to complete the remaining entries of the table. Sample student calculations derived from the height and width of seven rectangles projected from the original 6.6 cm × 4.6 cm rectangle are provided in **table 1**.

INITIAL STUDENT WORK

To fill in the table, a student must make choices regarding measurement. For instance, one student might measure rectangles in terms of centimeters, while another might use inches. As students compare work with classmates, many are surprised to find that such choices do not affect calculations of scale factor.

Additionally, a student has the option of using rectangle width or height when performing calculations. I encourage each student to calculate scale factor for each rectangle twice—once using width, once using height. For instance, calculating the scale factor associated with rectangle 1 in **table 1** using width gives the following:

$$\text{scale factor} = \frac{\text{Width of Image}}{\text{Width of Preimage}} \approx \frac{63 \text{ cm}}{6.6 \text{ cm}} \approx 9.5$$

On the other hand, calculating scale factor using height provides different results:

$$\text{scale factor} = \frac{\text{Height of Image}}{\text{Height of Preimage}} \approx \frac{44 \text{ cm}}{4.6 \text{ cm}} \approx 9.6$$

Because the same dilation is applied to both width and height, one expects the two calculations to yield values that are equal (small errors in measurement often yield approximately equal values). For this reason, a comparison of values provides each student with a method of checking the reasonableness of his or her answers. Considerable difference in values may indicate significant calculation or measurement error.

Determination of scale factor (specifically choosing between height- and width-based calculations) is discussed more thoroughly in the following section.

EXPLORING RELATIONSHIPS BETWEEN SCALE FACTOR AND PROJECTION DISTANCE TO BUILD BASIC SKILLS

Plotting data

Students who find questions regarding the overhead projector intriguing complete curve-fitting and data modeling activities in an effort to find answers. As students explore data collected from various overhead projections, they unwittingly review basic ideas of algebra in realistic contexts. For instance, as students plot scale factor with respect to projection distance, they revisit topics such as slope and y-intercept. **Figure 5** illustrates a plot of scale factor with respect to projection distance generated by a handheld graphing calculator.

Many students are initially surprised by the apparent linearity of the plot. Examination of the plotted data provides teachers with opportunities to review linear concepts with students. Student recognition that scale factor and projection distance increase together may be strengthened with a demonstration using an actual overhead projector. See sheet 2 for some suggested questions.

TABLE 1						
Sample Data Collected from Projected Rectangles						
Image #	Height	Width	Scale factor (based on width)	Scale factor (based on height)	Projected area (i.e., area of image)	Distance from wall to front wheel of projection cart
1	44 cm	63 cm	$\frac{63 \text{ cm}}{6.6 \text{ cm}} \approx 9.5$	$\frac{44 \text{ cm}}{4.6 \text{ cm}} \approx 9.6$	42 cm × 42 cm $\approx 2800 \text{ cm}^2$	234 cm
2	14 cm	19 cm	$\frac{19 \text{ cm}}{6.6 \text{ cm}} \approx 2.9$	$\frac{14 \text{ cm}}{4.6 \text{ cm}} \approx 3.0$	14 cm × 19 cm $\approx 270 \text{ cm}^2$	42 cm
3	32 cm	45 cm	$\frac{45 \text{ cm}}{6.6 \text{ cm}} \approx 6.8$	$\frac{32 \text{ cm}}{4.6 \text{ cm}} \approx 7.0$	32 cm × 45 cm $\approx 1400 \text{ cm}^2$	165 cm
4	71 cm	101 cm	$\frac{101 \text{ cm}}{6.6 \text{ cm}} \approx 15$	$\frac{71 \text{ cm}}{4.6 \text{ cm}} \approx 15$	71 cm × 101 cm $\approx 7200 \text{ cm}^2$	393 cm
5	42 cm	59 cm	$\frac{59 \text{ cm}}{6.6 \text{ cm}} \approx 8.9$	$\frac{42 \text{ cm}}{4.6 \text{ cm}} \approx 9.1$	2500 cm²	210 cm
6	57 cm	81 cm	$\frac{81 \text{ cm}}{6.6 \text{ cm}} \approx 12$	$\frac{57 \text{ cm}}{4.6 \text{ cm}} \approx 12$	4600 cm²	307 cm
7	22 cm	32 cm	$\frac{32 \text{ cm}}{6.6 \text{ cm}} \approx 4.9$	$\frac{22 \text{ cm}}{4.6 \text{ cm}} \approx 4.8$	700 cm²	102 cm

Facilitating discussions of this sort enables me to clear up student misunderstandings regarding lines, slopes, and scale factor prior to subsequent curve-fitting activities.

Constructing fit lines

The apparent linearity of the plot provides each of my students with opportunities to explore linear regression within the context of size change and overhead projection. Rather than employing automated linear regression utilities to calculate fit lines (e.g., least squares or median-median regression), each student uses a graphing calculator in a step-by-step manner to build linear equations that fit data. Although a student-constructed fit equation may lack the predictive precision of those generated by the calculator, creation of the equations affords a more in-depth study of linear equations than possible with automated, "black box" calculator procedures.

Fig. 5. Projection distance and scale-factor values entered into calculator lists L1 and L2, respectively (left); appropriate window settings (center); plot of scale factor with respect to projection distance (right)

Fig. 6. First, a student approximates the slope of an initial fit line (left). Next, a linear equation is defined using this slope (center). The plot of the initial fit line is graphed along with data points (right).

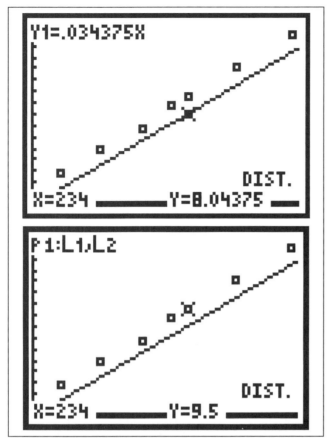

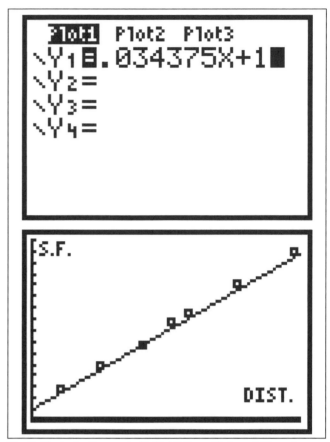

Fig. 7. The initial fit line predicts that scale factor will equal 8.04 when the projector is placed 234 cm from the wall (top). The actual scale factor at this distance is approximately 9.5 (bottom).

Fig. 8. Student refinement of initial fit line with the addition of a *y*-intercept

In introductory classes, uses of technology that mask underlying mathematics—those that provide "magic" answers by processes that students cannot understand—should be avoided. Calculator uses that illuminate rather than hide core mathematical concepts are highlighted throughout the remainder of this article.

Slope, y-intercept, and equation building

First, a student selects two data points and calculates the slope, *m*, of the line passing through them. The student uses this slope to build an initial fit equation for the projection distance/scale factor data. Using a graphing calculator, a student superimposes the graph of the line $y = m \cdot x$ into a scatterplot of his or her data. **Figure 6** highlights the calculation and graph of such an equation using data points (234, 9.5) and (42, 2.9).

Many students recognize that fit equations of the form $y = m \cdot x$ underestimate scale factor. This may be emphasized using tracing features of a graphing calculator. Tracing enables a student to examine scale factors (i.e., *y*-values) predicted by an initial fit line. Using the calculator's arrow keys, the student can jump back and forth between data points and the fit line—comparing predicted scale factors with those derived from actual data points. This process is illustrated in **figure 7**.

To translate the initial fit line upward, a student needs to include a non-zero *y*-intercept with the equation. To

encourage this, I ask students questions such as the following:

- What does the vertical axis (i.e., *y*-axis) represent in our plot?

- What is true about the distance of the overhead projector from the wall for any data point plotted on the vertical axis?

- Does it make sense for our fit line to cross the vertical axis at 0? Why or why not?

- Suppose our fit line crosses the vertical axis at a value less than 1. Explain what this would mean in terms of the overhead projector.

- What are some other values that we might expect the *y*-intercept to be?

To construct a reasonably good fit line, a student needs to recognize that the *y*-intercept represents the scale factor of the overhead projector when the cart's wheels are incident to the projection surface. Students can use the overhead projector to model this situation. Pushing the projection cart against the wall, my students find that preimage and image are approximately congruent, with the scale factor approximately equal to 1. In other words, when projection distance = 0,

$$\text{scale factor} = \frac{\text{Height of Image}}{\text{Height of Preimage}} \approx 1$$

In this way, the overhead projector provides students with a tangible example of y-intercept.

Incorporating improved estimates of y-intercept with initial fit equations enables one to construct an "improved" fit line. Because scale factor is approximately 1 when projection distance equals 0, a student typically chooses a y-intercept of + 1. An updated student fit line is depicted in **figure 8**.

As **figure 8** suggests, student estimates often fit the overhead projector data quite well. In addition, the process of thinking about fit coefficients provides a student with valuable insight regarding lines, y-intercepts, and slope. By building fit equations one parameter at a time and graphing results at each step, one is provided with an opportunity to see how parameter values affect the appearance of graphs. After students have successfully approximated a fit line, they have the ability to answer a host of mathematical modeling questions. For instance:

- If the projector were 400 inches away from the wall, what would one expect the scale factor of the enlargement to be?

- To triple the length and width of an object using the overhead projector, how far should the projector be placed from the wall?

EXPLORING RELATIONSHIPS BETWEEN SCALE FACTOR AND PROJECTION AREA

Initial student misconceptions

Many students falsely believe that a size change of magnitude 2 will double the area of a shape rather than enlarging area by a factor of $2 \times 2 = 4$. Analysis of overhead projector data provides students with a meaningful context in which to reconsider such beliefs. Initially, teachers may use rectangle data such as that in **table 1** to show students that (scale factor) • (preimage area) is not equal to image area, as seen in **table 2**. The fact that a plot of projected area with respect to scale factor is curved confirms that image area and scale factor are not linearly related. This curve is shown in **figure 9**.

The nonlinearity of the plot surprises many students, motivating further discussion of size change that pro-

TABLE 2			
(Scale Factor) • (Preimage Area) **Is Not Equal to Image Area**			
Preimage Area	Scale Factor	(Scale factor) • (Preimage area)	Image area
6.6 cm × 4.6 cm = 30 cm^2	9.5	9.5×30^2 cm = 285 cm^2	2800 cm^2
30 cm^2	2.9	87 cm^2	270 cm^2
30 cm^2	6.8	204 cm^2	1400 cm^2

vides a framework for subsequent curve-fitting activities. Construction of fit equations for the scale factor/projection area data provides teachers with opportunities to discuss linearity, square root, and squaring functions in meaningful ways with students.

Preliminary discussions of size change and area

A number of school textbook programs provide students with opportunities to study nonlinear functions in courses that precede high school geometry. For instance, texts such as *UCSMP Algebra* (McConnell, Brown, Eddins, Hackworth, and Usiskin 1991) and *Algebra I* (McDougal Littell 1991) devote several chapters to quadratic and exponential functions. Thus, when a student is asked to think of families of equations with curved graphs, he or she may recall quadratic and exponential functions. However, because experience with these forms is limited, a student is often unsure whether either form will provide an adequate model for the curved scale factor/projection area data. One gains a better understanding of relationships that exist between scale factor and area—and insight into possible fit equations—once one considers that *each dimension* of the preimage is multiplied by the scale factor. This observation leads to the derivation of image area illustrated in **figure 10**.

Because area of image = (scale factor)^2 · (area of preimage), students recognize that a quadratic equation appears to be the most appropriate model for the data. Properties of scale factor provide an additional clue regarding potential model equations. For instance, when the scale factor of a size change transformation

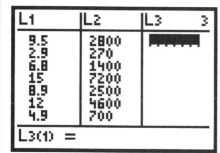
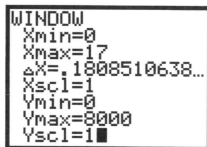
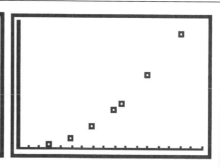

Fig. 9. Scale factor and projection area values entered into calculator lists L1 and L2, respectively (left); appropriate window settings (center); a plot of projection area with respect to scale factor (right).

Area of Image	= (Width of Image) • (Length of Image)
	= (Scale Factor • Width of Preimage) • (Scale Factor • Length of Image)
	= Scale Factor • Scale Factor • (Width of Preimage • Length of Image)
	= (Scale Factor)^2 • (Area of Preimage)

Fig. 10. Derivation of area of image as the product of the square of scale factor and area of preimage

equals 0, sides of the projected image have no length. This result suggests that a model equation should pass through the origin (0, 0). Hence, a quadratic model such as $y = a \bullet x^2$ is initially considered a potentially good model for the data.

Linearizing scale factor/projection area data

If, in fact, $y = a \bullet x^2$ is a suitable model for the plot of projection area with respect to scale factor, then a plot of the square root of projection area with respect to scale factor should result in a linear plot.

To test this hypothesis, a student can create a list containing square roots of all projection areas, then plot the square root of the projection area with respect to scale factor. Such a strategy is depicted in **figures 11** and **12**.

A process that converts nonlinear data into linear data through a series of algebraic transformations is informally referred to as *linearization* (Neter, Kutner, Nachtscheim, and Wasserman 1996, p. 126). Once data is linearized, a student may construct a linear equation that fits the transformed data. This process is highlighted in **figure 13**.

Delinearizing fit equations

In **figure 13**, the student found that the linear equation $y = 5.53x$ fit the plot of the square root of the projection area with respect to scale factor reasonably well. Recall that in the equation $y = 5.53x$, x represents scale factor and y represents the square root of the projection area. The satisfactory fit of equation $y = 5.53x$ to the data points suggests the following:

(1) $\qquad \sqrt{\text{projection area}} \approx 5.53 \cdot (\text{scale factor}).$

Furthermore, squaring each side of (1) suggests that

(2) $\begin{cases} \left(\sqrt{\text{projection area}}\right)^2 \approx 5.53 \cdot (\text{scale factor})^2 \\ \text{projection area} \approx 5.53 \ \cdot (\text{scale factor}) \ . \\ \text{projection area} \approx 30.58 \cdot (\text{scale factor})^2 \end{cases}$

Results from (2) suggest that $y = 30.58x^2$ will fit the scale factor/projection area data reasonably well. The reasonableness of $y = 30.58x^2$ as a model is supported by the graph shown in **figure 14**.

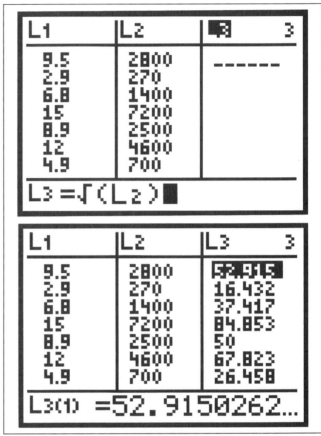

Fig. 11. Student defines list L3 to contain square roots of all areas in list L2 (top). After the list definition is entered, list L3 is filled with data values (bottom).

SUMMARY

In this article, I have illustrated ways in which ordinary classroom materials may be used creatively to enhance student understanding of concepts such as scale factor, fit equations, *y*-intercept, slope, square roots, and parabolic equations. The activity, which explores transformational properties of overhead projectors using cardboard, masking tape, graphing calculators, and permanent markers, is largely successful in entry-level algebra or geometry classrooms because it appeals to students' previous experience with overhead projectors in a manner that is actively engaging.

The activity helps students build connections among a wide range of mathematical topics—most notably transformation, measurement, and mathematical modeling—in a manner consistent with the vision of NCTM's *Principals and Standards of School Mathematics* (2000). While attempting to unravel surprisingly curious questions regarding the functionality of the classroom overhead projector, students actively collect and analyze data, perform algebraic calculations, and use technology to solve problems regarding size change transformations.

Fig. 12. Student sets up plot of square root of projection area with respect to scale factor, lists L1 and L3 (left); appropriate window settings are entered (center); resulting plot appears to be linear (right).

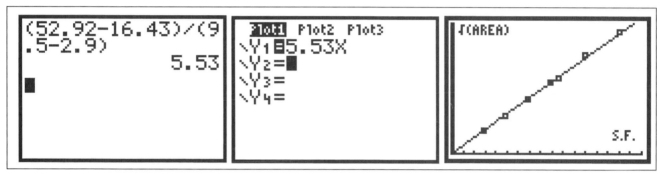

Fig. 13. Students utilize familiar techniques to construct a fit equation for the linearized scale factor/projection area data.

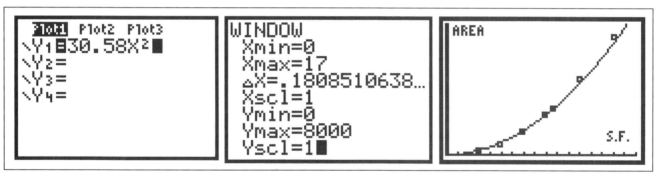

Fig. 14. Proposed model entered into Y= editor of calculator (left); appropriate window settings (center); a graph of model superimposed on plot of projection area with respect to scale factor (right).

EXTENSIONS

Certainly, the activities described here can be modified in a variety of ways. The list that follows provides a sampling of possible lesson extensions and modifications.

- Teachers (or students) may wish to project shapes other than rectangles onto classroom walls (e.g., circles, triangles, trapezoids, ellipses) and investigate relationships between area and scale factor. Such activities provide students with opportunities to review various area formulas in an active, exploratory fashion.

- Teachers (or students) may wish to project images using photographic slides or photographs copied to overhead transparency (e.g., aerial photographs of geographic regions such as continents or islands, faces of celebrities). This allows students to investigate size change transformations and areas of irregular regions using an even wider variety of images.

- Teachers (or students) may wish to focus images prior to tracing them onto classroom walls. Data collected might also include a "focus factor" to account for size change attributable to focus. Students may wish to compare analyses of "focused" data with "nonfocused" data. Does use of the focus knob of the overhead projector have a noticeable effect on models or conclusions drawn from data?

The author wishes to extend a warm thanks to everyone at Upper Arlington High School (Upper Arlington, Ohio) and Linworth Alternative School (Worthington, Ohio) who inspired the creation and development of the activities contained in this article.

REFERENCES

Algebra I: An Integrated Approach. Geneva, Ill.: McDougal Littell, 1991.

Coxford, Arthur, Zalman Usiskin, and Daniel Hirschhorn. *UCSMP Geometry*. Glenview, Ill.: Scott Foresman, 1993.

McConnell, John W., Susan Brown, Susan Eddins, Margaret Hackworth, and Zalman Usiskin. *UCSMP Algebra*. Glenview, Ill.: Scott Foresman, 1991.

National Council of Teachers of Mathematics (2000). *Principles and Standards for School Mathematics*. NCTM: Reston, Va., 2000.

Neter, John, Michael H. Kutner, Christopher J. Nachtscheim, and William Wasserman. *Applied Linear Regression Models*, 3rd ed. Chicago: Irwin, 1996.

Exploring Scale Factor and Area with Projected Rectangles

Use the images projected on the classroom walls to complete the chart.

Image #	Height	Width	Scale factor (based on width)	Scale factor (based on height)	Projected area (i.e., area of image)	Distance from wall to front wheel of projection cart
1						
2						
3						
4						
5						
6						
7						

Ideas for Further Study

Discuss the following questions and write your ideas.

1. What happens to scale factor as the overhead is moved away from the wall? How can this be determined from the plot?

2. Suppose scale factor is plotted with respect to projection distance and the data points appear to be horizontal. Is such an event likely or unlikely? Why or why not?

3. Suppose a line is constructed that passes as close to data points as possible. Would the slope of the line be positive, negative, or zero? Can you describe uses for projection devices that generate slopes less than or equal to zero? If so, describe these uses in more detail.

4. Complete your own idea for further study. "What happens when . . . ?"

Modeling Soft Drink Packaging

Gary Froelich

Mathematical modeling is the process of describing real-world phenomena in mathematical terms, obtaining mathematical results from the mathematical description, then interpreting and evaluating the mathematical results in the real-world situation. A simple diagram, shown in **figure 1**, depicts the process.

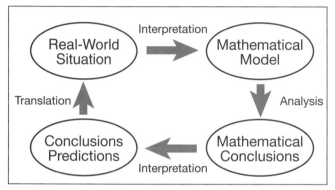

Fig. 1 The process of mathematical modeling

The mathematical-modeling process can also be described as a series of steps. For beginners, the number of steps should be minimal:

1. Identifying a real-world problem

2. Identifying important factors and representing those factors in mathematical terms

3. Using mathematical analysis to obtain mathematical results

4. Interpreting and evaluating mathematical results as they affect the real-world problem

Students who are new to mathematical modeling can better appreciate the diagram or the list of steps after they have experienced the process. Wait until the activity has been completed to formally discuss the process.

Emphasize the idea of process to your students. Encourage them to use *model* and *modeling* as verbs. Although a collection of equations or a computer program that implements the equations is often called a model, a model is really more—it includes assumptions about the phenomena being modeled, for example, and the modeler must consider those assumptions if the mathematical results are unacceptable.

In typical practice, at least a portion of the modeling process is repeated because initial results are inadequate or have become ineffective over time. Failure can occur because the process missed some important aspect of reality—a common occurrence, since capturing all aspects of a situation is difficult.

Modelers strive for simplicity. Students tend to resist simplification: questions that begin "But what about . . ." are common. Emphasize simplification, and ask students to delay discussing secondary concerns until they obtain results. Those concerns may need to be addressed when the mathematical results are tested against reality. Although the modeling process is only a shadow of reality, it produces results: mathematical modeling has helped take us to the moon, helped win wars, and saved lives by predicting natural disasters. Modeling has helped us accomplish these goals because it attempts to apply mathematics to only the essential elements of a situation.

Modeling can be theory driven or data driven. The results of data-driven modeling are often confirmed theoretically; conversely, theory-driven results are often confirmed by data collection—either in the real world or through simulation. For example, analyzing data on stopping distance of automobiles indicates that a quadratic function describes the relationship between automobile speed and stopping distance. A theoretical analysis shows that stopping distance has a linear component (reaction distance) and a component (braking distance) that varies with the square of the automobile's speed (Giordano, Weir, and Fox 1997, pp. 103–9).

In this activity, students conduct theory-driven modeling to improve the efficiency of secondary soft-drink packaging. The activity is adapted from *Mathematics: Modeling Our World: Course 2*, Unit 4: "The Right Stuff." Mathematics: Modeling Our World is a four-year high school mathematics program developed by the Consortium for Mathematics and Its Applications (COMAP; 1998) through a grant from the National Science Foundation.

The modeling process is accessible to students of various abilities. A good model is usually the product of more than just mathematical knowledge—insight and creativity are also important. COMAP conducts an annual High School Mathematical Contest in Modeling (HiMCM), in which teams of students research a real-world problem and develop a mathematical model.

For information on the Mathematics: Modeling Our World program or HiMCM, call (800) 772-6627 or send an e-mail to info@pop.comap.com.

TEACHER'S GUIDE

Prerequisites

- The ability to find areas of circles, triangles, and rectangles; the formula

$$\frac{s\sqrt{3}}{4},$$

  where s is the length of a side, for the area of an equilateral triangle is useful, but $(1/2)bh$ is sufficient.

- An understanding of 30°–60°, right-triangle relationships

- The ability to find surface area and volume of rectangular solids and cylinders

The preceding list of prerequisites is somewhat misleading, because portraying mathematical modeling as a process in which the modeler knows all the necessary mathematics is unrealistic. Often the modeler must research not only the context but also the mathematics. In fact, new mathematics must be developed in some situations. Thus, the teacher can use portions of this activity to encourage students to develop new mathematical ideas. For example, if the students do not know 30°–60°, right-triangle relationships, needing to evaluate such packages as the triangular six-pack or the hexagonal seven-pack can serve as motivation. The activity can be paused at the appropriate time, the mathematics developed, and then the activity resumed.

Whether or not modelers know all the mathematics needed in a situation, they often need to research the contextual setting. Since soft drinks and their packaging are familiar to students, you may think that research is unnecessary. However, a modest amount of research realistically depicts the modeling process and can enhance students' motivation.

An Internet search on "soft drink packaging" is likely to produce interesting results. Web sites of the American Beverage Association (**www.ameribev.org**) and the Canadian Beverage Association (**www.refreshments.ca**) are good starting points. Students are likely to discover that packaging in general is a major component of landfills; the volume and recyclability of packaging material therefore are important concerns. Students may also discover that effective use of space is important in package design. For example, soft drinks were formerly packaged in bottles that used space inefficiently; as a result, some retailers stocked soft drinks only in the summer.

If your students conduct research, have them report their findings to the class. Focus the discussion on research that relates to the use of space or to the nature of packaging material.

Although the activity, as presented here, assumes an ability to work with variables, it can easily be adapted for use with less sophisticated students. For example, you might give students six circles of radius 3.2 cm and have them build the various configurations by arranging the circles on a sheet of paper and then drawing the package boundary. They can calculate areas from direct measurements. Imprecise measurements and inaccurate drawings can serve as reasons to introduce an electronic drawing utility or an analytic investigation. Most students can benefit from this hands-on experience, and it is especially recommended as a first step in question 4 on sheet 4.

Sheets 1 and 2

Sheet 1 establishes secondary soft-drink packaging as the modeling context and gives students an opportunity to flex some of the mathematical muscles that they will need. In this situation, the muscles are geometric formulas.

Since students will find only slight differences among some package designs, question 1 shows that small differences can be worth pursuing. For example, an innovation that only slightly improves the efficiency of automobile engines could be worth millions of dollars to the inventor and to the automobile industry.

The goal of improving package design is too broad, so question 2 begins to define the problem. The primary concerns of this activity are maximizing the package space used by the cans, 2a, and minimizing packaging material per can, 2d. In the real world, the former might concern those engaged in storing, transporting, or selling soft drinks; the latter might concern soft-drink manufacturers or those public officials responsible for landfills.

As noted in question 3, modelers have at least two reasons to use simplification. One is for convenience—a simple problem is easier to tackle than a complicated one, and the knowledge gained can help with the more complicated problem. The other reason is to eliminate relatively unimportant factors. Simplification, however, carries risks, no matter what the modeler's motivation. Therefore, question 3c anticipates problems by showing that simplification to two dimensions does not compromise the results.

Question 3 assumes that packaging material has no thickness and that no overlap occurs. These assumptions may need to be reexamined.

In question 3a, students may need a hint: finding the percent of package space used by the cans requires taking the ratio of the total area of the tops of the cans to the area of the rectangular top of the package.

Question 4 makes an important point: for ease of comparison, the measure of efficiency may need modification.

After completing sheet 1, students should be comfortable with modeling criteria, particularly the percent material used per can. They should realize that in the former instance, optimization means maximization; and in the latter, optimization means minimization. As they progress in the activity, students should begin to realize what every modeler knows: optimizing two quantities simultaneously is difficult.

Sheet 3

The purpose of sheet 3 is to investigate the number of cans as a controlling factor when the modeling criterion is either maximizing the package space used or minimizing the packaging material used. Students see that the number of cans is not an important factor for the first criterion, but it is for the second.

Although this entire activity is well-suited to group work, question 1 of this sheet is especially good for this type of work. Groups can share their designs with the class by posting them in the classroom or by putting them on transparencies. Groups can be asked to defend their calculations. The previously described circular manipulatives can also be used.

Students can use a computer drawing utility to create their designs. If they do, the issue of accuracy is likely to arise. For example, measurements and calculations made with a drawing utility are not reliable if constructions are inaccurate. Transformation features of drawing utilities are often the simplest method to use to construct circles that are tangent to existing circles. If electronic sketches are accurate, measurements and calculations can substitute for a deductive investigation when such an investigation proves too difficult. **Figure 2** shows an accurately constructed "parallelopack." The measure of efficiency is the portion of package space used by the cans. Although the parallelopack is not an appropriate student design for this activity sheet, it is one that students might propose when they do sheet 3.

For this unit, The Geometer's Sketchpad files for several package designs can be downloaded from the COMAP website at **www.comap.com/highschool/ projects/mmow/weblinks_c2_g10_u4.htm**.

On the basis of the mathematical results that they describe in question 4, students should not conclude that the modeling process is complete. Mathematical results need interpretation and evaluation, as indicated

in question 5. The teacher might reinforce this point by asking students to build the package that they think is best and consider, for instance, the amount of additional material, or overlap, needed to hold the package together. Does, for example, a $3 \times 3 \times 2$ eighteen-pack need more overlap than a $3 \times 6 \times 1$ eighteen-pack? Or does an eighteen-pack require thicker packaging material than the standard twelve-pack?

This sheet presents opportunities for challenging better students. Students can algebraically analyze the formula that describes the efficiency of a rectangular package when the criterion is minimizing the package material used per can. For example, consider a rectangular package that holds eighteen cans, with m cans along one side. The surface area of each of the two long sides is $2rhm = (2m)(3.2)(12)$. If m cans are along one side, then $18/m$ cans are along the other side. The surface area of each of the two shorter sides is

$$2rh\frac{18}{m} = \left(2\frac{18}{m}\right)(3.2)(12).$$

The surface area of the top, as well as the bottom, is

$$lw = (2rm)\left(2r\frac{18}{m}\right)$$
$$= (2m)\left(2\frac{18}{m}\right)(3.2)(12).$$

The total surface area of the rectangular box is then

$$2[(2m)(3.2)(12)]+2\left[\left(2\frac{18}{m}\right)(3.2)(12)\right]$$
$$+ 2\left[(2m)\left(2\frac{18}{m}\right)(3.2)(3.2)\right]$$

The efficiency is the surface area divided by 18, the number of cans in the package:

$$e = \frac{2((2m)\bullet 3.2 \bullet 12)}{18}$$
$$+ \frac{2\left(\left(2\frac{18}{m}\right)\bullet 3.2 \bullet 12\right)+2\left((2m)\left(2\frac{18}{m}\right)\bullet 3.2 \bullet 12\right)}{18}.$$

A little simplification produces the equivalent formula

$$e = \frac{25.6 + \dfrac{460.8}{m} + 245.76}{3}.$$

A graph of this function, shown in **figure 3**, is interesting. Tracing the graph shows that the minimum amount of package material per can occurs when $m \approx 4.24$, which approximates the exact minimum $m = \sqrt{18}$. In other words, students who have concluded that the least packaging material occurs when the rectangular package is as close as possible to a square, in cross section, are indeed correct.

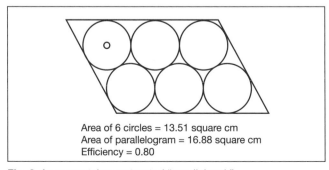

Area of 6 circles = 13.51 square cm
Area of parallelogram = 16.88 square cm
Efficiency = 0.80

Fig. 2 An accurately constructed "parallelpack"

When they complete sheet 3, students have had only partial success with the program of minimization and maximization. They have successfully identified the number of cans as an important factor when the criterion is minimizing the package material. But they have seen that changing the number of cans does not help improve efficiency when the criterion is maximizing the package space used by the cans, within the restriction of using rectangular packaging.

Sheet 4

Since associating the number of cans in the package with packaging material used per can has proved fruitful, sheet 4 is concerned only with maximizing the package space used by the cans. Before beginning this activity sheet, students should understand that their mathematical conclusions have shown that the identification of the number of cans as a controlling factor, when the criterion is maximizing the package space, is incorrect. Rejecting modeling results means that backtracking is necessary; other factors must be considered in this situation. The purpose of this sheet is to consider package shape as a controlling factor.

Although this sheet does not ask questions about minimizing packaging material, you may want to ask them of your students. If your students have not worked with the surface area of a cylinder, finding the amount of packaging material needed for the package in question 1 prepares them for the assessment problem.

Although the answers given for questions 1 and 2 are analytical, the questions themselves do not demand this type of approach. Less sophisticated students can construct the figures, perhaps using an electronic drawing utility, and calculate efficiencies from measurements. This approach is realistic from a modeling perspective; modelers often apply technology when mathematical analysis proves too difficult.

Question 4 of this sheet is an open-ended miniproject and is an excellent small-group activity. For ease of experimentation, consider having students use the previously described circular manipulatives. They can create posters or transparencies to share their designs and defend their calculations. Also consider asking students to confirm their theoretical calculations by using an electronic drawing utility to make accurate constructions.

Assessment

Because question 4 of sheet 4 is a miniproject, verbal or written reports discussing question 4 may furnish sufficient assessment for this activity. As an additional assessment, the teacher can show the photograph in **figure 4** or describe it to students and ask them to evaluate the two designs. Both coolers are designed to hold six soft-drink cans. The problem might be titled "The Tube or the Cube?" although students should understand that the rectangular design is not cubical.

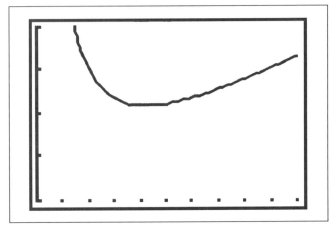

Fig. 3 [0, 10] × [100, 200]

Fig. 4 The tube or the "cube"?

A model answer to this assessment should indicate that when the criterion is the percent of package space used, the tube is perfect, since the cans use 100 percent of the available space. Students are likely to recall, without calculation, that the efficiency of the "cube" is about 78.5 percent.

A model answer should consider more than one criterion. Students may need to be reminded that a standard can has a radius of 3.2 cm and a height of 12 cm. Students should use those dimensions to calculate that the "cube" uses approximately

$$\frac{2(4 \cdot 3.2 \cdot 12) + 2(6 \cdot 3.2 \cdot 12) + 2(4 \cdot 3.2 \cdot 6 \cdot 3.2)}{6},$$

or 209.92, square centimeters of packaging material per can. The tube, which is a cylinder with a radius of 3.2 cm and a height of 72 cm, uses approximately

$$\frac{2\pi(3.2^2) + 72(\pi \cdot 6.4)}{6},$$

or 252, square centimeters of packaging material per can.

If you have emphasized that mathematical results need to be interpreted and evaluated in the real world, an exemplary answer might address other concerns. For example, since the tube has greater surface area than the "cube," it does not insulate as well. The tube may be inconvenient for such other reasons as its comparative inability to remain stationary.

SOLUTIONS

Sheet 1: **1a** 62.6 billion ÷ 12 ≈ 5.22 billion twelve-packs; 5.22 billion × $.001 ≈ $5.22 million; **b** A little over a half-million dollars ($522,000); **2a** One example is the percent of package space used by the cans; another appropriate measure would be cubic units per can;

b The space used might be measured in square inches per package if the packages all hold the same number of cans. Otherwise, it could be measured in square inches per can; **c** Attractiveness is subjective and is therefore difficult to measure. Perhaps a statistical experiment could be designed in which a random sample of consumers is asked to pick the most appealing of several designs; **d** The material used could be measured in cubic inches. If the thickness of the material is constant, square inches could be the measure. Measuring in cubic or square units per can is a better choice, since it allows comparing designs for different numbers of cans; **e** The cost could be measured in dollars or cents per can.

Sheet 2:

3a
$$\frac{12(3.2^2\pi)}{(6 \cdot 3.2)(8 \cdot 3.2)} \approx 0.785$$

The cans use about 78.5 percent of the package space.

b
$$\frac{12(r^2\pi)}{(6 \cdot r)(8 \cdot r)} = \frac{12\pi r^2}{48r^2}$$
$$= \frac{\pi}{4}$$
$$\approx 0.785;$$

c
$$\frac{12\pi r^2 h}{(6 \cdot r)(8 \cdot r)h} = \frac{12\pi r^2 h}{48r^2 h}$$
$$= \frac{\pi}{4}$$
$$\approx 0.785,$$

or about 78.5 percent.

4a 2(6 · 3.2 · 12) + 2(8 · 3.2 · 12) + 2(6 · 3.2 · 8 · 3.2) = 2058.24 cm²; **b** For comparison purposes, material per can is a better measure. The standard twelve-pack uses approximately 171.5 square centimeters per can; the competing design uses approximately 187.5 square centimeters per can. The standard twelve-pack is more efficient.

Sheet 3: **1** Sample answer: a nine-pack (3 cans × 3 cans)

Second sample answer: an eight-pack (2 cans × 4 cans);

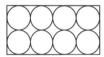

2 For the nine-pack:

$$\frac{9\pi r^2}{(6r)(6r)} = \frac{9\pi r^2}{36r^2}$$
$$= \frac{\pi}{4}$$
$$\approx 0.785;$$

For the eight-pack:

$$\frac{8(\pi r^2)}{(4r)(8r)} = \frac{8\pi r^2}{32r^2}$$
$$= \frac{\pi}{4}$$
$$\approx 0.785,$$

The number of cans is apparently not a controlling factor if the criterion is maximizing the package space used by the cans.

3 For the nine-pack, 2(6 · 3.2 · 12) + 2(6 · 3.2 · 12) + 2(6 · 3.2 · 6 · 3.2), or 1658.88, square centimeters; per can, 184.32 square centimeters. For the eight-pack, 2(4 · 3.2 · 12) + 2(8 · 3.2 · 12) + 2(4 · 3.2 · 8 · 3.2), or 1576.96, square centimeters; per can, 197.12 square centimeters. The amount of packaging material used per can varies, so the number of cans is a controlling factor when the modeling criterion is minimizing the packaging material used per can. **4** In general, increasing the number of cans decreases the amount of packaging material used per can. However, for two packages with the same number of cans, a configuration in which the ratio of length to width is closest to 1 is better. This ratio is sometimes called the aspect ratio. For example, a 4 × 4 sixteen-pack uses 158 square centimeters per can; a 2 × 8 sixteen-pack uses 177.92 square centimeters per can. **5** According to the modeling results, a 6 × 3 package will use less packaging material per can than a 2 × 9 package. Calculations can confirm this result.

For a 6 × 3 package:

$$\frac{2(12 \cdot 3.2 \cdot 12) + 2(6 \cdot 3.2 \cdot 12)}{18}$$

$$+ \frac{2(12 \cdot 3.2 \cdot 6 \cdot 3.2)}{18} = 158.72 \text{ cm}^2 \text{ per can}$$

For a 2 × 9 package:

$$\frac{2(4 \cdot 3.2 \cdot 12) + 2(18 \cdot 3.2 \cdot 12)}{18}$$

$$+ \frac{2(4 \cdot 3.2 \cdot 18 \cdot 3.2)}{18} = 175.79 \text{ cm}^2 \text{ per can}$$

However, a more insightful student should realize that stacking the cans in two 3 × 3 layers produces even better results:

$$\frac{2(6 \cdot 3.2 \cdot 24) + 2(6 \cdot 3.2 \cdot 24)}{18}$$

$$+ \frac{2(6 \cdot 3.2 \cdot 6 \cdot 3.2)}{18} = 143.36 \text{ cm}^2 \text{ per can}$$

When the modeling criterion is maximizing the package space used by the cans, stacking the cans in two layers has no effect: the cans still use 78.5 percent of the available space.

Sheet 4:

1
$$\frac{7\pi r^2}{\pi(3r)^2} = \frac{7\pi r^2}{9\pi r^2}$$

$$= \frac{7}{9}$$

$$\approx 0.778;$$

or about 77.8 percent. It is slightly less efficient than the standard twelve-pack.

2 In the figure, $CA = 2r$; $AB = r\sqrt{3}$. The base of the triangular package is $4r + 2r\sqrt{3}$.

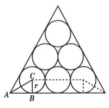

Since the triangle is equilateral, its area is

$$\frac{s^2\sqrt{3}}{4} = \frac{(4r + 2r\sqrt{3})^2\sqrt{3}}{4}.$$

The percent of space used by the cans is

$$\frac{6\pi r^2}{\frac{(4r + 2r\sqrt{3})^2\sqrt{3}}{4}} = \frac{6\pi r^2}{1} \cdot \frac{4}{4r^2(2 + \sqrt{3})^2\sqrt{3}}$$

$$= \frac{6\pi r^2}{(2 + \sqrt{3})^2\sqrt{3}}$$

$$\approx 0.781;$$

or about 78.1 percent. This design is slightly less efficient than the standard twelve-pack, but it is slightly more efficient than the cylindrical seven-pack.

3 On the basis of their experiences with rectangular packaging, students will probably expect that the number of cans is not a factor when the criterion is the percent of package space used by the cans. Although this expectation is a reasonable one, you can encourage students to be skeptical about whether this principle generalizes to packaging that is based on a triangle or on some other shape. Students should also realize that not every number of cans can be enclosed in an equilateral triangle.

4 Sample answer: Remove three small equilateral triangles from the triangular six-pack to form an irregular hexagon.

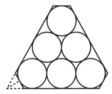

Each small equilateral triangle (one of which is shown at the lower left in the figure) has an altitude of length r. Therefore, the length of each side of the triangle is $2r\sqrt{3}$. The area of each small triangle is

$$\left(\frac{2r}{\sqrt{3}}\right)^2\left(\frac{\sqrt{3}}{4}\right) = \frac{r^2\sqrt{3}}{3}$$

The area of the irregular hexagon is

$$\frac{(4r + 2r\sqrt{3})^2\sqrt{3}}{4} - 3\left(\frac{r^2\sqrt{3}}{3}\right) = r^2((2 + \sqrt{3})^2\sqrt{3} - \sqrt{3}).$$

The efficiency of the package is

$$\frac{6\pi r^2}{r^2((2 + \sqrt{3})^2\sqrt{3} - \sqrt{3})} = \frac{6\pi}{(2 + \sqrt{3})^2\sqrt{3} - \sqrt{3}}$$

Second sample answer: A hexagonal seven-pack

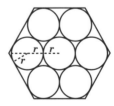

The hexagon can be divided into six equilateral triangles, each with sides measuring

$$2r + \frac{2r}{\sqrt{3}} \; .$$

Therefore, the area of the hexagon is

$$6\left[\frac{\left(2r + \frac{2r}{\sqrt{3}}\right)^2 \sqrt{3}}{4}\right] = \frac{\left(2r + \frac{2r}{\sqrt{3}}\right)^2 3\sqrt{3}}{2}$$

The package space used by the cans is

$$\frac{7\pi r^2}{\dfrac{\left(2r + \frac{2r}{\sqrt{3}}\right)^2 3\sqrt{3}}{2}} = \frac{14\pi}{3\sqrt{3}\left(2 + \frac{2}{\sqrt{3}}\right)^2}$$

$$\approx 0.851,$$

or about 85.1 percent.

REFERENCES

Carter, Claudia R. "Activities: Packing Them In." *Mathematics Teacher* 90 (March 1997): 211–14, 220–22.

Consortium for Mathematics and Its Applications (COMAP). "The Right Stuff." In *Mathematics: Modeling Our World: Course 2*, pp. 254–319. Cincinnati, Ohio: South-Western Educational Publishing, 1998.

Giordano, Frank R., Maurice D. Weir, and William P. Fox. *A First Course in Mathematical Modeling*. 2nd ed. Pacific Grove, Calif.: Brooks/Cole Publishing Co., 1997.

Hine, Thomas. *The Total Package*. Boston: Little, Brown & Co., 1995.

Modeling Soft Drink Packaging

In this activity, you will use mathematical modeling to describe and improve the efficiency of secondary soft-drink packaging. The primary packaging is the can; the secondary packaging is the container that holds several cans.

Before beginning the modeling process, consider the impact that an improvement in the design of soft-drink packages might have. Because the volume of soft drinks consumed is large, even a small savings on the cost of a secondary package could mean a lot to the soft-drink industry.

1. According to the National Soft Drink Association, 62.6 billion cans of soft drinks were consumed in the United States in 1995. Suppose that you find a way to save the soft-drink industry one-tenth of a cent ($.001) on the packaging of each twelve-pack sold.

    a. Estimate the total annual savings to the soft-drink industry. _____

    b. If you receive royalties worth 10 percent of the savings to the industry, estimate your annual income for the use of your innovation in the United States. _____

2. The first step in the mathematical-modeling process is to identify a real-world problem and define the modeling goal clearly. This example considers the efficiency of secondary packaging of soft drinks, and the modeler must decide how to measure efficiency. For example, if the primary concern of someone who is trying to improve the efficiency of automobile engines is economy, efficiency might be defined in terms of gasoline consumption and measured in miles per gallon. However, if the primary concern is the environmental impact of automobiles, efficiency might be defined in terms of emissions and measured in parts per million.

    Modeling often involves optimization—finding the best way to do something. For each of the following criteria for optimal packaging, give at least one example of how efficiency could be measured.

    a. The best soft-drink package is one in which the cans use as much of the available space in the package as possible. *Hint:* What numerical measure would indicate how well the cans use the space in the package?

    b. The best soft-drink package is one that uses shelf space well.

    c. The best soft-drink package is one that is most attractive to consumers.

    d. The best soft-drink package is one that uses the least packaging material.

    e. The best soft-drink package is one for which the packaging material is cheapest.

Modeling Soft Drink Packaging

Modelers use various mathematical tools to reach conclusions. Questions 3 and 4 show how geometric formulas can help calculate packaging efficiency. Questions 3 and 4 also introduce simplification, a necessary part of the modeling process.

3. Mathematical modeling always involves simplification. Simplification is often done to eliminate relatively unimportant information, but it is also done for convenience. The figure shown is a two-dimensional representation of the twelve-pack commonly used for secondary packaging of soft drinks.

 a. The radius of a soft-drink can is approximately 3.2 centimeters. Determine how well the cans use the space in this two-dimensional package, that is, calculate the percent of package space used by the cans.

_____

_____

 b. Show that your answer is independent of the size of the cans, that is, repeat your work, but use r to represent the radius. _____

_____

 c. Show that simplification to two dimensions does not affect the result, that is, find the volume of the package and the volume of the cans in terms of the radius r and height h of the cans, then calculate the percent of space used by the cans._____

_____

4. a. Determine the amount of packaging material used by the standard twelve-pack, as illustrated in the figure. For simplicity, assume that the material has no thickness and use the area of the packaging material as your measure. Also assume that the package requires no overlap of packaging material. Note that the height of a soft-drink can is about 12 centimeters._____

_____

 b. Suppose that a competing design uses 1500 square centimeters of packaging material to hold eight cans. Is this design more efficient than the standard twelve-pack? Explain.

_____

_____

_____

_____

Modeling Soft Drink Packaging

You have assessed the efficiency of a standard twelve-pack by two measures: the percent of package space used by the cans and the packaging material used per can. In the first case, the modeling objective is maximization; in the second case, the modeling objective is minimization.

After the goal of the modeling process has been defined, the modeler must identify important factors that control the chosen criterion, then conduct a mathematical investigation. For example, a modeler might identify temperature as a factor that controls the chirp rate of crickets and choose to ignore other factors until the effect of temperature is understood. If accurate predictions of the chirp rate of crickets can be made from temperature alone, no other factors need to be considered.

Your task here is to consider the number of cans in the container as a factor that controls a criterion for optimal packaging. To assess the effect of this factor, no others should be considered, that is, only the number of cans will vary.

1. Design a secondary package in which the number of cans is not twelve. Remember, everything else should remain the same. Do not change the size or the shape of the can. Do not change the shape of the package—keep it rectangular.

2. Find the percent of packaging space used by the cans in your design. Compare the percent with that used by other designs in your class. Is the number of cans a controlling factor if the criterion is maximizing the percent of packaging space used by the cans?

3. Find the amount of packaging material used per can in your design. Compare the amount with the amounts used in other designs that your class has created. Is the number of cans a controlling factor if the criterion is minimizing the amount of packaging material used per can?

4. Describe the effect of the number of cans as a controlling factor. Interpret the mathematical results that your class has produced, and recommend a package design.

5. Results produced by the mathematical-modeling process must be tested against reality. In the example of soft-drink packaging, the model may not address such considerations as convenience. For example, if you concluded that a package should contain a relatively large number of cans, some consumers may find the package too heavy or think that it requires too much storage space. Use your modeling results to develop a recommendation for a soft-drink company that has decided that a rectangular eighteen-pack is the most convenient.

Modeling Soft Drink Packaging

Your modeling efforts have been successful when the efficiency criterion is minimizing the packaging material used per can. But you have been unsuccessful in improving efficiency when the criterion is maximizing the package space used by the cans. When the mathematical-modeling process fails to produce results, part of the process must be repeated. You next examine another potential controlling factor: package shape.

1. Shown here is a cylindrical seven-pack. Determine the percent of package space used by the cans. How does this design compare with the standard twelve-pack?

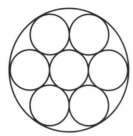

2. Shown here is a triangular six-pack. Determine the percent of package space used by the cans. How does this design compare with the standard twelve-pack?

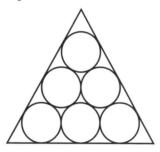

3. Would you expect a triangular three-pack or a triangular ten-pack to make better use of package space than the triangular six-pack? Explain.

4. Design your own package. Try to find a shape for which the percent of space used by the cans is greater than that of any of the designs that you have evaluated.

The Rules of the Game

Don Gernes

Concept development is both important and challenging; important because understanding a concept allows it to be used correctly and recognized in new situations, challenging because of the difficulty of enabling students to "see" clearly that which is confusing or vague.

One of the more difficult concepts to develop is that of a deductive system and what it means to "prove" a theorem. I think that one of the best ways to develop concepts is to start with situations with which the student is familiar and make a transition to the unfamiliar. In an effort to bridge the gap between the student's world and the world of mathematics, I developed a unique approach, one that uses ordinary games as a structure that is analogous to a deductive system. As with any analogy, it may have technical flaws in that subjective judgments sometimes creep into real-world events; but using board games, card games, and sports as models of deductive systems has worked very well in my classroom.

Games exhibit some characteristics similar to those of a deductive system, namely, undefined terms, definitions, postulates or axioms, and theorems. However, a different vocabulary is used—instead of "undefined terms," a game has "materials"; instead of "postulates or axioms," a game has "rules"; instead of "theorems," a game has "plays." Usually both situations have "definitions."

The rules of the game are accepted as being true, just like postulates. The interpretations of the plays of a game are not accepted but rather must be justified, that is, proved, by referring to the rules. To play or referee a game, a participant must thoroughly understand the rules and definitions. Likewise, to be able to write a proof in mathematics, the student must be familiar with "the rules of the game."

As an introductory activity, have students complete sheet 1.

When students realize that they are in fact familiar with the concept of a deductive system, they are ready to develop the concept of writing a proof. To prove a theorem in any game, whether Monopoly, football, or geometry, a good, clear understanding of the rules is necessary. To introduce students to the concept of proving a theorem, we need a simple deductive system where the rules are few and can be understood quickly, say, in five or ten minutes. One such system is the "letter game," patterned after the MIU system in *Gödel, Escher, Bach: An Eternal Golden Braid* (Hofstadter 1979).

THE "LETTER GAME"

Undefined terms

The letters M, I, and U

Definition

x means any string of Is and Us.

Postulates

1. If a string of letters ends in I, you may add a U at the end.

2. If you have M*x*, then you may add *x* to get M*xx*, by the doubling rule.

3. If three Is occur, that is, III, then you may substitute U in their place.

4. If UU occurs, you drop it.

In the "letter game," students change one string of letters into another string of letters using any of the four rules. Usually students are successfully writing proofs within ten to fifteen minutes. These proofs can be structured in whatever format that you want to promote. Sheet 2 uses the traditional two-column arrangement to introduce the format used later in a geometric minisystem. However, one could use a flow-proof format or a paragraph format. See the examples that follow.

Given: MI   *Prove:* MUIU

Flow-proof-format example:

MI → MII → MIIII → MIIIIU → MUIU

Paragraph-format example:

If MI, then MII, by rule 2. However, MII would imply MIIII, by rule 2. By using rule 1, MIIII would lead to MIIIIU. By applying rule 3 to MIIIIU, we end up with MUIU.

Next, have students complete sheet 2.

EUCLID'S GAME

After students have gained some confidence in proving theorems by playing the "letter game," they are ready for "Euclid's game," a game of geometry. Geometry has undefined terms, definitions, and postulates, just like basketball, Monopoly, and the "letter game." However, when we begin the study of geometry, students have a clear idea of what a deductive system is and what it means to prove a theorem. Students are then ready to be challenged by minisystems within the framework of Euclidean geometry, as recommended in NCTM's *Standards* document (1989). Such a minisystem would include undefined terms: point, line, and plane; and the definitions and postulates specifically needed to explore the topic, such as parallel lines, similar triangles, or congruent triangles.

One such minisystem is included as sheet 3.

SOLUTIONS

Sheet 1: Examples of acceptable answers follow; student responses may vary.
1) Monopoly, *Undefined terms:* Board, Playing pieces, Dice, *Defined terms:* Turn, Passing Go, Chance, Doubles, Bank, *Postulates:* If you roll three doubles in a row, then you go to jail, if you land on an unowned property, then you may buy it from the bank, If you pass Go, then you get $200.

2) Soccer, *Undefined terms:* Field, Ball, Players, Defined terms: Touch line, Goal, Penalty kick, Corner kick, Throw-in, *Postulates:* If the ball goes over the touch line, then the other team gets a throw-in, if a player is fouled, then that player's team gets a free kick, if the defensive team causes the ball to go over the goal line, then the other team gets a corner kick.

Sheet 2: The following are only one possible proof of each.

1)
MIII	Given
MIIIIII	Rule 2
MUU	Rule 3
M	Rule 4

2)
MIIIUUIIIII	Given
MIIIIIIIII	Rule 4
MIIUU	Rule 3
MII	Rule 4
MIIU	Rule 1

3)
MI	Given
MII	Rule 2
MIIII	Rule 2
MUI	Rule 3

4)
MII	Rule 2
MIIII	Rule 2
MIIIIIIII	Rule 2
MIUIU	Rule 3

5)
MI	Given
MII	Rule 2
MIIU	Rule 1
MIIUIIU	Rule 2

6)
MIIIUII	Given
MUUII	Rule 3
MUUIIU	Rule 1
MUUIIUUUIIU	Rule 2
MIIUIIU	Rule 4

Sheet 3: The decision to use "equal" angles instead of "congruent" angles was made so that the vocabulary and logic in this minisystem is clear and accessible.

1)
1	Lines l and m intersect	Given
2	$\angle 1 + \angle 2 = 180°$	Linear pairs postulate
3	$\angle 1 + \angle 2$ are supplementary	Definition of supplementary angles

2)
1	Lines l and m intersect	Given
2	$\angle 1 + \angle 2 = 180°$	Linear pairs postulate
3	$\angle 2 + \angle 3 = 180°$	Linear pairs postulate
4	$\angle 1 + \angle 2 = \angle 2 + \angle 3$	Substitution property
5	$\angle 1 = \angle 3$	Subtraction property of equality

3)
1	$a \parallel b$	Given
2	$\angle 1 = \angle 3$	Vertical angles theorem (problem 2)
3	$\angle 1 = \angle 2$	Corresponding angles "postulate"
4	$\angle 2 = \angle 3$	Substitution property

4)
1	$\overrightarrow{AB} \parallel \overrightarrow{EC}$ and $\overrightarrow{BC} \parallel \overrightarrow{EF}$	Given
2	$\angle 2 = \angle 3$	Corresponding angles "postulate"
3	$\angle 2 = \angle 1$	Corresponding angles "postulate"
4	$\angle 1 = \angle 3$	Substitution property

5)
1	$\overrightarrow{AB} \parallel \overrightarrow{DE}$ and $\angle 1 = \angle 2$	Given
2	$\angle 1 = \angle 4$	Alternate interior angles theorem (problem 3)
3	$\angle 4 = \angle 2$	Substitution property
4	$\angle 2 = \angle 3$	Vertical angles theorem (problem 2)
5	$\angle 4 = \angle 3$	Substitution property

6)
1	$\overrightarrow{AB} \parallel \overrightarrow{CD}$ and $\angle 4 = \angle 5$	Given
2	$\angle 2 = \angle 4$	Alternate interior angles theorem (problem 3)
3	$\angle 5 = \angle 1$	Corresponding angles "postulate"
4	$\angle 2 = \angle 5$	Substitution property
5	$\angle 2 = \angle 1$	Substitution property

CONCLUSION

The time spent developing concepts is time well spent. Games with which the students are familiar allow them to "see" an entire deductive system, and the "letter game" enables students to participate within a deductive system with a minimum of preparation. Thus, the key to developing the concept of a deductive system and the concept of proof is found in the "rules of the game."

REFERENCES

Hofstadter, Douglas R. *Gödel, Escher, Bach: An Eternal Golden Braid*. New York: Basic Books, 1979.

National Council of Teachers of Mathematics (NCTM). *Curriculum and Evaluation Standards for School Mathematics*. Reston, Va.: NCTM, 1989.

United States Department of Education, National Center for Educational Statistics. *Pursuing Excellence*. Washington, D.C.: United States Government Printing Office, 1996.

Usiskin, Zalman. *Van Hiele Levels and Achievement in Secondary School Geometry*. Final Report of the Cognitive Development and Achievement in Secondary School Geometry Project. Chicago: University of Chicago, Department of Education, 1982.

Structure of a Deductive System

Sheet 1

Deductive systems have the following structure:

1. Undefined terms—a few of the most basic terms

2. Definitions—the meaning of terms as used in the given system

3. Postulates—statements assumed to be true, usually expressed in if-then form

4. Theorems—statements that are proved

Assume for the moment that all games are deductive systems. Assume for simplicity that the postulates below always hold true.

Example: Basketball

Undefined terms:	Ball
	Player
	Court
	Baskets
Defined terms:	Field goal
	Foul
	Free throw
	Traveling
Postulates:	If a player is fouled, then the player gets to shoot a free throw.
	If a player travels, then the other team gets possession of the ball.
	If a player makes a field goal, then the player's team gets two points.
Theorems:	The referee objectively applies the rules of the game to each play.

Give examples of undefined terms, defined terms, and postulates for the following:

1. Monopoly

   Undefined terms:_____
   _____
   _____
   _____

   Defined terms: _____
   _____
   _____
   _____

   Postulates:_____
   _____
   _____
   _____

2. Soccer

   Undefined terms: _____
   _____
   _____
   _____

   Defined terms:_____
   _____
   _____
   _____

   Postulates: _____
   _____
   _____
   _____

The "Letter Game"

Sheet 2

Undefined terms:	Letters M, I, and U
Definition:	*x* means any string of Is and Us.
Postulates:	1. If a string of letters ends in I, you may add a U at the end.
	2. If you have M*x*, then you may add *x* to get M*xx*.
	3. If 3 Is occur, that is, III, then you may substitute U in their place.
	4. If UU occurs, you drop it.
Objective:	Given one string, you are to derive or prove some other string.
Example:	Given: MI
	Prove: MIIU

Statement	Reason
MI	Given
MII	Rule 2
MIIU	Rule 1

Prove the following theorems:

1. Given: MIII
   Prove: M

Statement	Reason

2. Given: MIIIUUIIIII
   Prove: MIIU

Statement	Reason

3. Given: MI
   Prove: MUI

Statement	Reason

4. Given: MI
   Prove: MIUIU

Statement	Reason

5. Given: MI
   Prove: MIIUIIU

Statement	Reason

6. Given: MIIIUII
   Prove: MIIUIIU

Statement	Reason

A Deductive Minisystem Involving Parallel Lines   Sheet 3

Undefined terms:	Point Line Plane
Defined terms:	*Supplementary angles:* Two angles whose sum is 180° *Vertical angles:* Two nonadjacent angles formed by two intersecting lines *Corresponding angles:* Two nonadjacent angles on the same side of a transversal where one is interior and one is exterior *Alternate interior angles:* Two interior nonadjacent angles on different sides of a transversal
Postulates:	*Linear pairs postulate:* If two adjacent angles have exterior sides forming a line, then the angles sum to 180°. *Subtraction property of equality:* For any numbers a, b, and c, if $a = b$, then $a - c = b - c$. *Corresponding angles "postulate":* If two parallel lines are cut by a transversal, then corresponding angles are equal. *Substitution property:* If $x + b = c$ and $x = y + 3$, then $(y + 3) + b = cz$.

1. *Given:* lines l and m intersect
   Prove: ∠1 and ∠2 are supplementary angles
   (Supplementary angles theorem)

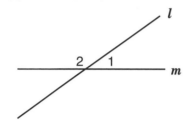

2. *Given:* lines l and m intersect
   Prove: ∠1 = ∠3
   (Vertical angles theorem)

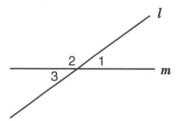

After theorems have been proved, they can be used to justify statements in a proof, just as definitions and postulates can. The supplementary angles theorem and the vertical angles theorem can be used in the following proofs.

3. *Given:* $a \parallel b$
   Prove: ∠2 = ∠3
   (Alternate interior angles theorem)

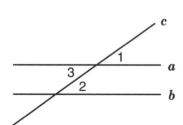

4. *Given:* the plane figure $\overleftrightarrow{AB} \parallel \overleftrightarrow{EC}$; $\overleftrightarrow{BC} \parallel \overleftrightarrow{EF}$
   Prove: ∠1 = ∠3

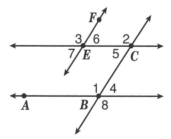

5. *Given:* $\overleftrightarrow{AB} \parallel \overleftrightarrow{DE}$; ∠1 = ∠2
   Prove: ∠4 and ∠3

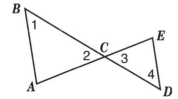

6. *Given:* $\overleftrightarrow{AB} \parallel \overleftrightarrow{CD}$; ∠4 = ∠5
   Prove: ∠1 = ∠2

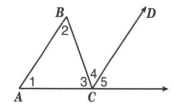

Pick's Rule

Christian R. Hirsch

TEACHER'S GUIDE

Grade level: 7–10

Materials: One set of worksheets for each student and a set of transparencies for class discussion

Objectives: The student will discover and apply Pick's theorem for finding the area of a polygon whose vertices are lattice points.

Directions: Distribute the worksheets one at a time. Be sure to allow sufficient time so that each student has an opportunity to discover the rule for himself or herself.

Have the students discuss their answers after they have completed sheet 2. A complete set of answers can be given on the transparencies.

Sheet 3 will provide opportunities for the students to apply their generalization and may suggest other areas of exploration. Students may enjoy forming more exotic polygons and computing their areas as well.

Supplementary activities: A formula for finding the area of a polygonal region with one or more holes, as in exercise 7, is closely associated with Pick's theorem, and students might be encouraged to experiment further to find such a formula. For discussion of this topic see Marshall (1970).

The results of exercise 8 may be used to suggest an investigation into whether for each natural number n, one can find a square on dot paper whose area is exactly n.

Some students may inquire if Pick's rule can be applied to three dimensions. The answer is yes, but not in the obvious manner. See Niven and Zuckerman (1967).

SOLUTIONS

1) a. 1/2, 1, 1 1/2, 2, 2 1/2

   b. iii

   c. 6 1/2

2) a. 2, 2 1/2, 3, 3 1/2

   b. $A = 1 + 1/2b - 1$

   c. 5

3) a. 4, 4 1/2, 5, 5 1/2, 6

   b. $2 + 1/2b - 1 = A$

4) 6 1/2

5) $1 + 1/2b - 1$, $2 + 1/2b - 1$, $3 + 1/2b - 1$, $4 + 1/2b - 1$, $8 + 1/2b - 1$, $i + 1/2b - 1$

6) a. 1

   b. 5

   c. 10

   d. 12

   e. 2

   f. 4

7) a. 20 1/2

   b. 18

   c. 22

BIBLIOGRAPHY

Harkin, Joseph B. "The Limit Concept on the Geoboard." *Mathematics Teacher* 65 (January 1972): 13–17.

Marshall, A. G. "Pick: With Holes." *Mathematics Teaching* 50 (1970): 67–68.

Niven, Ivan, and Herbert S. Zuckerman. "Lattice Points and Polygonal Area." *American Mathematical Monthly* 74 (1967): 1195–1200.

Sullivan, John J. "Polygons on a Lattice." *Arithmetic Teacher* 20 (December 1973): 673–75.

Dots and Area

This polygon has 6 dots on the boundary. We say $b = 6$. This polygon has 1 dot in the interior. We say $i = 1$.

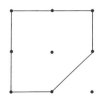

1. *a.* Below are some polygons with no dots in the interior ($i = 0$). Find the area of each of the polygons, and record your answers in the table provided. The figure labeled E has an area of 1.

Interior Dots i	Boundary Dots b	Area A
0	3	
0	4	
0	5	
0	6	
0	7	

   b. Circle the rule that shows how to find the area when the number of boundary dots (b) is known and the number of interior dots (i) is 0.

   i. $A = b + 1$ ii. $A = 2b - 2$ iii. $A = 1/2b - 1$ iv. $A = b^2 - 2$

   c. What do you think the area of a polygon would be if it has exactly 15 dots on its boundary and not dots inside?

   d. Form such a polygon and find its area.

2. *a.* The following polygons had one dot in the interior. Find the area, and enter it in the table.

Interior Dots i	Boundary Dots b	Area A
1	4	
1	5	
1	6	
1	7	

   b. When $i = 1$, write a formula relating b to A.

   c. Find the area when $b = 10$ and $i = 1$. Draw the polygon and check your result.

Dots and Area

Sheet 2

3. *a.* Exactly 2 dots are inside each polygon ($i = 2$) below. The number of dots on the boundaries are different. Find the area inside each of these polygons and record your answer in the table provided.

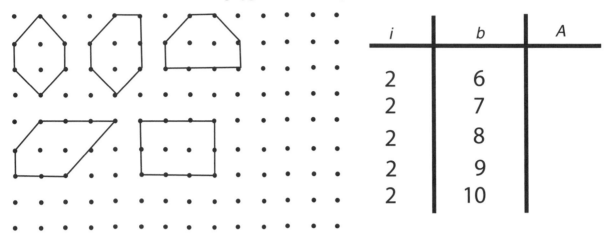

i	b	A
2	6	
2	7	
2	8	
2	9	
2	10	

b. State a rule relating b to A when $i = 2$ for these polygons. _____

4. Find the area when $b = 11$ and $i = 2$. Draw the polygon and check your results.

5. Using the results of questions 1 to 4, complete the following table. You will have to look for a pattern.

dots inside (i)	0	1	2	3	4	8	i
dots on boundary	b	b	b	b	b	b	b
Area	$0 + \dfrac{1}{2}b - 1$						

Dots and Area

The rule you discovered, $A = i + 1/2b - 1$, relates the area A of a polygon to the number of dots inside a polygon i and the number of dots in its boundary b. It is known as Pick's rule.

6. Find the area of each of the following polygons using Pick's rule.

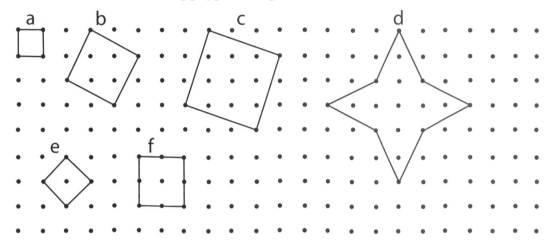

7. Use Pick's rule to find the area of each of the shaded regions.

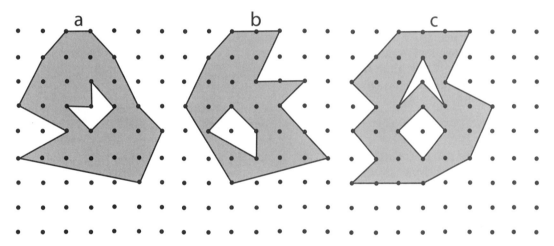

8. Use Pick's rule to help you draw polygons having 3, 4, 5, 6, 7, 8, 9, and 10 sides so that each polygon has the smallest possible area.

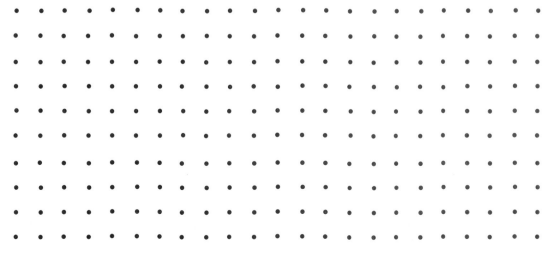

Sprinklers and Amusement Parks:
What Do They Have to Do with Geometry?

Cheryl Q. Nelson and Nicole L. Williams

Realistic Mathematics Education (RME), a theory of mathematics learning and instruction, was utilized in the development of the following activity. RME incorporates the two principles below, which were synthesized from de Lange and Treffers (Meyer 2001):

1. The starting point of instruction should be experientially real to students, allowing them to engage in meaningful mathematical activity.

2. The learning of a concept passes through various stages of abstraction. The initial stage should be a concrete example in which students can use their informal knowledge to construct their own personal meaning and connections. Students will get to greater abstract levels through various representations such as models, diagrams, and symbolic notation.

Two tasks presented in this activity were developed in accordance with these principles. Since most students have knowledge of sprinklers and amusement parks, the first principle is satisfied through experientially real activities. The second principle is achieved as students transfer the real-life situation into a geometric model, further analyzing this model to determine the appropriate constructions.

All too often, students in a geometry class learn terms and constructions with no idea of where they would be useful. Two examples are angle bisectors and perpendicular bisectors. Typically, students bisect angles and segments using a straightedge and compass. They then use these concepts to construct the incenter and circumcenter of triangles. Students primarily focus on these concepts in an abstract setting. This activity provides two real-world applications for bisectors that use the students' own intuitions to guide their exploration and discovery. With these applications, we found not only that students understood the geometric concepts better but also that the applications provided a foundation for them to develop a broader sense of the purpose of geometry.

TEACHER NOTES FOR THE SPRINKLER PROBLEM

Students are confronted with a problem in which a circular sprinkler is to be placed in a triangular park (see **fig. 1**). That is, students need to find a center—specifically, the incenter—of the triangle. Students can use a straightedge and compass to accomplish this task; but dynamic software such as The Geometer's Sketchpad

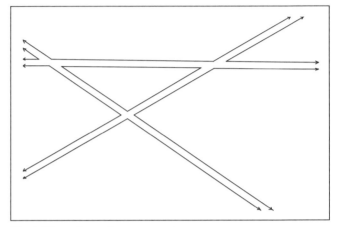

Fig. 1. Triangular park

(GSP; Jackiw 2001) is more effective in solving this problem as students can generate the incenter quickly and precisely. Students can then alter the triangle to generalize the solution to any triangle. They can even decide for which type of triangle it would make the most sense to use a circular sprinkler.

Extension: Many extensions are possible for this problem. For example, we could ask, "What percentage of the lawn is not getting water?" "What if the park were in the shape of a square? a rectangle? a diamond?"

TEACHER NOTES FOR THE AMUSEMENT PARK PROBLEM

Many of us have gone to amusement parks and searched for the closest vendor selling soft drinks. In this application (see **fig. 2**), students need to place a soft drink stand so that it is the same distance from the three most popular rides in the park (the vertices of a triangle). With this task, students need to think what it means to be equidistant from the vertices of a triangle and how they can find this point, called the circumcenter of the triangle. When first confronted with this problem, students can explore solutions using either software or a straightedge and compass. If they choose geometric software, however, they can further investigate what happens as the shape of the triangle changes.

Extension: Sometimes the location of a vendor can create difficulties in terms of crowds and long lines. In the early stages of developing a new amusement park, planners should explore where to place the three most popular rides so that a vendor is equidistant from all three

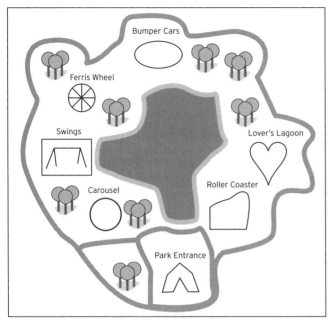

Fig. 2. Amusement park

but at the same time not within the triangle. What are some of the properties of this triangle?

REFLECTIONS AND CONCLUSION

This particular activity requires that students analyze the centers of a triangle in order to determine which ones would make sense in solving two real-world problems. If dynamic software is available, potential solutions are more readily tested and verified. As geometry concepts are examined through real-world situations, students make connections to more formal abstract mathematics.

SOLUTIONS

Sheet 1:

1) People walking the sidewalk would get wet. It would waste water.

2) No, the sprinkler cannot reach the corners unless the circle it makes sprays the sidewalks.

3) Answers will vary with students.

4) Answers will vary with students.

Using GSP to assist in solving the Sprinkler problem

12) a. The circle is inscribed in the triangle; this means that each side of the triangle is tangent to the circle.

b. No, the circle remains inscribed in the triangle.

c. It would pass through the center of the circle and the vertex of the angle.

d. It shows how to find the center of the circle, which is where the park's sprinkler should be located.

Sheet 3:

1) The location will provide easy access to the stand, and more potential customers will be attracted to it.

2) Students will probably choose a location within the triangle formed.

3) Answers will vary with students.

Using GSP to assist in solving the Amusement Park problem

8) a. The circle is circumscribed about the triangle. This means that every vertex of the triangle is on the circle.

b. No, the circle remains circumscribed about the triangle.

c. It would pass through the center of the circle and the midpoint of the side.

d. It shows how to find the point that is the same distance to each of the three rides at the park.

REFERENCES

Jackiw, Nicholas. The Geometer's Sketchpad. Berkeley, Calif.: Key Curriculum Press, 2001.

Meyer, Margaret R. "Representation in Realistic Mathematics Education." In *The Roles of Representation in School Mathematics*, edited by Albert A. Cuoco. Reston, Va.: National Council of Teachers of Mathematics, 2001.

National Council of Teachers of Mathematics (NCTM). *Principles and Standards for School Mathematics.* Reston, Va.: NCTM, 2000.

Bisectors in Geometry   Sheet 1

THE SPRINKLER PROBLEM

The Parks Department is installing a circular sprinkler (a sprinkler whose spray makes a perfect circle) to water the lawn at a park that is in the shape of a triangle and is surrounded by sidewalks (see **fig. 1.1**). The sprinkler should be placed so as to cover as much lawn as possible without spraying the sidewalks.

1. Why wouldn't you want water to spray the sidewalks?

2. Will all of the lawn be watered? Why or why not?

3. Indicate on figure 1 where you think the sprinkler should be placed.

4. Draw a rough sketch of the circle that the spray from the sprinkler would make.

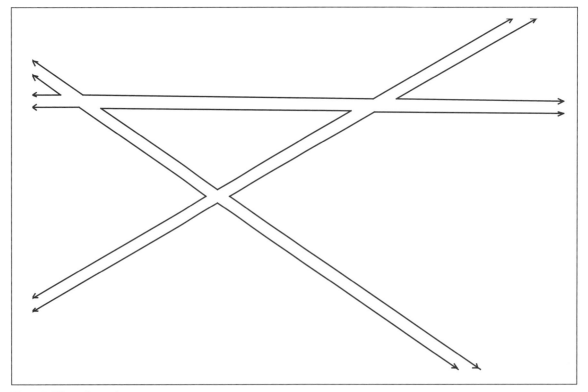

Fig. 1.1. Triangular park

Bisectors in Geometry

THE SPRINKLER PROBLEM

The Geometer's Sketchpad can assist in solving this problem.

How to use GSP to assist in solving the Sprinkler problem

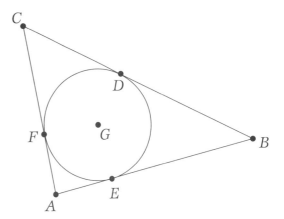

5. Construct a triangle.

6. Construct the bisectors of two angles of the triangle.

7. Construct a point at the intersection of the angle bisectors. This point is one of the centers of the triangle and is called the incenter of the triangle (point G in **figure 1.2**).

8. Construct a perpendicular line from the incenter to one side of the triangle.

9. Construct the intersection of the perpendicular line and the side of the triangle.

Fig. 1.2. Model of the incenter of a triangle

10. Create the segment from the center of the circle to this intersection. This segment can be used as the radius of a circle centered at the incenter of the triangle.

11. Using the incenter and the radius, construct the circle.

12. Locate the three points where the circle and the triangle intersect. Your construction should look like **figure 1.2**.

    a. Describe the relationship between the triangle and the circle.

    b. Move the vertices of the triangle to change its shape and size. Does this affect the relationship between the circle and the triangle?

    c. If the bisector of the third angle of the triangle were constructed, state two points that it would pass through.

    d. Explain how this construction helps you solve the worksheet problem.

Bisectors in Geometry

Sheet 3

THE AMUSEMENT PARK PROBLEM

Jo wants to open a new soft drink stand at an amusement park. She looks at the map of the amusement park (**fig. 2.1**) and notes the three most popular rides: roller coaster, swings, and bumper cars. Jo decides to locate the stand so that it is the same distance from all three of these rides.

1. Why do you think that Jo wants the stand located at a point equidistant from the three most popular rides?

2. Indicate on **figure 2.1** where you think the stand should be placed.

3. Do you think this location is a good place for a stand? Explain.

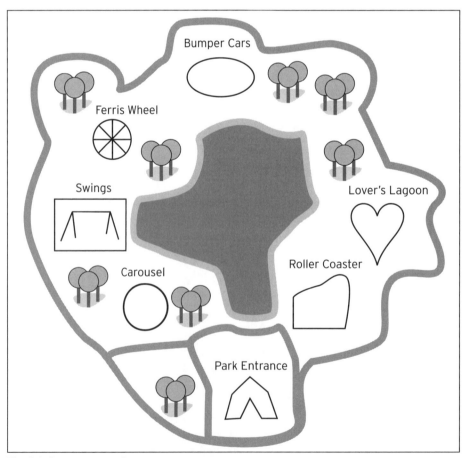

Fig. 2.1. Amusement park

Bisectors in Geometry

Sheet 4

THE AMUSEMENT PARK PROBLEM

The Geometer's Sketchpad can assist in solving this problem.

How to use GSP to assist in solving the Amusement Park problem.

4. Construct a triangle.

5. Choose any two sides of the triangle and construct the perpendicular bisector of each.

6. Construct a point at the intersection of the two perpendicular bisectors. This point will become the center of a circle and is called the circumcenter of the triangle.

7. Draw a segment from the intersection point to one of the vertices of the triangle. This segment will be the radius of a circle.

8. Construct a circle that has the intersection of the perpendicular bisectors as its center and the segment created above as its radius. Your construction should look like **figure 2.2**.

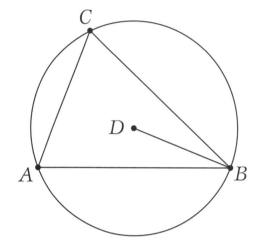

Fig. 2.2. Model of the circumcenter of a triangle

   a. Describe the relationship between the triangle and the circle.

   b. Move the vertices of the triangle to change its shape and size. Does this affect the relationship between the circle and the triangle?

   c. If the perpendicular bisector of the third side of the triangle were constructed, state two points that it would pass through.

   d. Explain how this construction helps you solve the worksheet problem.

Discovering the Tangent

Bertha Weir Palmer

Tangents, sines, and cosines furnish an excellent opportunity for creative learning for junior high school pupils. They also furnish the pupils an opportunity to have the thrill that comes to a discoverer if they are taught by the laboratory method.

The class can be divided into groups of three or four pupils to a group. Let the first pupil in each group draw a base line *CA* one inch long. Have the second pupils draw theirs two inches long, the third pupils three inches long and the fourth pupils four inches long. At the end *C* of the base line have them all erect with a protractor a line *CL* perpendicular to the base line. Have the right angle at *C* labeled 90°. Now give each group of four an acute angle to be made at point *A*. These might be 30°, 45°, 60°, and 64°. The pupils should extend the line which forms these angles to the perpendicular line *CL* at B and form a right triangle. Now the side *CB* should be labeled side opposite angle *A*. The side *CA* should be labeled the side adjoining angle *A*. The pupils should measure each of these lines very carefully and write the dimensions on their figures. A ruler graduated in tenths is convenient to use.

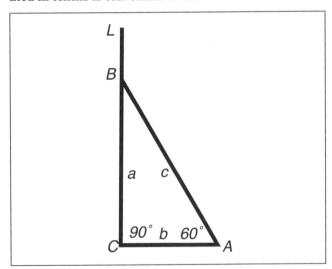

Fig. 1

Each pupil should set up this form with the data from his or her right triangle.

$$\frac{\text{Side opposite } \angle 60°}{\text{Side adjoining } \angle 60°} = \frac{BC}{CA} = \frac{3.46}{2} = 1.73$$

Use the blackboard to set up this form for each group of angles.

$$\frac{\text{Side opposite } \angle 60°}{\text{Side adjoining } \angle 60°} = \frac{BC}{CA} = \frac{1.7}{1} = 1.7$$

$$\frac{\text{Side opposite } \angle 60°}{\text{Side adjoining } \angle 60°} = \frac{BC}{CA} = \frac{3.46}{2} = 1.73$$

$$\frac{\text{Side opposite } \angle 60°}{\text{Side adjoining } \angle 60°} = \frac{BC}{CA} = \frac{5.2}{3} = 1.73$$

$$\frac{\text{Side opposite } \angle 60°}{\text{Side adjoining } \angle 60°} = \frac{BC}{CA} = \frac{6.9}{4} = 1.72$$

Have the pupils consider only the first decimal place and see what conclusion they can make from the results. They will readily observe that all the quotients are the same. When asked what other factor is the same in each of the four, they will respond that it is the acute angle. It should now be very carefully pointed out that in each case the sides opposite and the sides adjoining the acute angles are of different lengths. Use this same procedure for each of the acute angles assigned.

The class is now ready to make the conclusion that in each group the result of dividing the side opposite the acute angle by the side adjoining the acute angle is the same for all angles of 30°, also for all of 45°, and for all of 60°, and for all of 64°. Explain to the group that this result is in each case the tangent of the respective angle and that the relationship of these lines for each acute angle has been arranged for them in a tangent table.

This procedure will give the pupil a discoverer's feeling of ownership of the tangent table. He or she will realize that he or she has had a part in its construction and it will cease to be just a page to aid him or her in getting the answer to some problem. Many pupils, as a result of this experiment, will want to make tangent tables of their own to compare with the one in their text. If they construct three triangles and take the average, their result will be fairly accurate. This makes a good class project, and it gives the pupil a real understanding of what a tangent is.

The formal definition is no longer "formal" or, as has often been the case, "formidable." The fact that in a right triangle the tangent is the ratio of the side opposite the acute angle to the side adjoining the acute angle, has been learned through discovery and not through memorization.

Now have the class consider two triangles. Take one from the 60° group and the other from the 30° (**fig. 2**). Have the pupils in the angle 60° group consider the angle at Y. Have them determine its size and discover its tangent ratio. Have those in the 30° angle group consider in the same manner the angle at L. This will emphasize that all right triangles have two acute angles and serve to show further that the tangent ratio is always the same for angles of the same size regardless of their position in the right triangle.

The pupils are now ready to go out doors with a transit and measure heights of objects. That will make the meaning of the angle of elevation clear and will also teach them the reason for considering the height of the instrument.

The angle of depression can be taught from a classroom window. A second or third floor window is desirable for this project. They can find the distance from the base of the building to various trees and objects. They will also enjoy having an imaginary lake beside the building and measuring its width. They will be eager to make these measurements with a tape to see how accurate they were.

The pupils will have so much enthusiasm for this work that many of them will want to make transits of their own so that they can make measurements in their leisure time in their own neighborhoods. In our school we have had several made, and to date no two have been alike in all respects. Each has however measured satisfactorily horizontal and vertical angles. It has also given its creator keen satisfaction and enjoyment.

This same method can be used with sines and cosines to enable the pupils to discover for themselves what those ratios mean.

When the pupil has had an opportunity to make real applications and to learn through this laboratory method, the principles have more meaning and are more readily retained in his or her memory. This active learning will have given the pupils an experience which they can not readily forget. It will have taken the tangent from the textbook and given it life and vitality. Each pupil will have become a discoverer in his or her own right and will have had a learning thrill that is part of laboratory learning.

SOLUTIONS

NOTES TO TEACHER:

- The original **figure 1** did not have triangle BCA's right angle, C, marked "90°."

- The original article had errors in the five equations accompanying **figure 1**. We have corrected these, both in the article itself and on the student activity sheet.

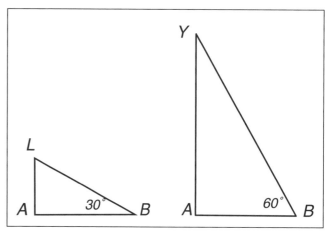

Fig. 2

1)

2) Construct a line, *L*, at *C* perpendicular to *CA*.

3) *TEACHER NOTE:* You may choose to assign each group a different angle. The article suggests using the angles 30°, 45°, 60°, and 64°. You could also choose to assign groups who finish early an additional angle to try after they have completed question 5.

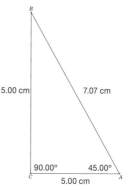

4) ***Table for 30°***

Triangle 1	side opposite 30°/side adjoining 30° = ????
Base length: **5 cm**	**= 2.9 cm/5 cm = 0.58**
Triangle 2	side opposite 30°/side adjoining 30° = ????
Base length: **7 cm**	**= 4.05 cm/7 cm ≈ 0.579**
Triangle 3	side opposite 30°/side adjoining 30° = ????
Base length: **10 cm**	**= 5.77 cm/10 cm = 0.577**
Triangle 4	side opposite 30°/side adjoining 30° = ????
Base length: **12 cm**	**= 6.93 cm/12 cm = 0.5775**

Table for 45°

Triangle 1	side opposite 45°/side adjoining 45° = ????
Base length: **5 cm**	**= 5 cm/5 cm = 1**
Triangle 2	side opposite 45°/side adjoining 45° = ????
Base length: **7 cm**	**= 7 cm/7 cm = 1**
Triangle 3	side opposite 45°/side adjoining 45° = ????
Base length: **10 cm**	**= 10 cm/10 cm = 1**
Triangle 4	side opposite 45°/side adjoining 45° = ????
Base length: **12 cm**	**= 12 cm/12 cm = 1**

Table for 60°

Triangle 1	side opposite 60°/side adjoining 60° = ????
Base length: **5 cm**	**= 8.66 cm/5 cm = 1.732**
Triangle 2	side opposite 60°/side adjoining 60° = ????
Base length: **7 cm**	**= 12.12 cm/7 cm ≈ 1.731**
Triangle 3	side opposite 60°/side adjoining 60° = ????
Base length: **10 cm**	**= 17.33 cm/10 cm = 1.733**
Triangle 4	side opposite 60°/side adjoining 60° = ????
Base length: **12 cm**	**= 20.8 cm/12 cm = 1.73**

Table for 64°

Triangle 1	side opposite 64°/side adjoining 64° = ????
Base length: **5 cm**	**= 10.25 cm/5 cm = 2.05**
Triangle 2	side opposite 64°/side adjoining 64° = ????
Base length: **7 cm**	**= 14.37 cm/7 cm ≈ 2.053**
Triangle 3	side opposite 64°/side adjoining 64° = ????
Base length: **10 cm**	**= 20.51 cm/10 cm = 2.051**
Triangle 4	side opposite 64°/side adjoining 64° = ????
Base length: **12 cm**	**= 24.63 cm/12 cm = 2.0525**

5) All the ratios are the same for the same angle measure.

6) Compare your findings with those of other groups with different values of α.

7) Given any right triangle with fixed acute angle α, the ratio of the length of the side opposite α to the side adjacent α will always be the same.

8) Yes, both triangles have an acute angle of 30°, and so I measured the lengths of the sides opposite and adjacent to the 30° angle and found that the ratios were the same.

9) The tangent tables show the ratios of the side lengths of the opposite side length divided by the adjacent side length for all right triangles and their acute angles.

Discovering the Tangent

1. Draw four line segments (with lengths of 5 cm, 7 cm, 10 cm, and 12 cm) horizontally on a blank piece of paper. Label each segment *CA*, as shown below.
   Hint: Leave enough room around the line segments so that you can construct triangles with these lengths as the bases.

$$C \qquad\qquad\qquad\qquad A$$

2. Construct a line, *L*, at *C* perpendicular to *CA*.

3. Use the acute angle your teacher gave you to draw an angle at *A*. Label this angle α. Extend the line created at *A* to the perpendicular line through *C*. This should form a right triangle, with *CA* as the base.

Triangle 1	side opposite α/side adjoining α = ????
Base length:	=
Triangle 2	side opposite α/side adjoining α = ????
Base length:	=
Triangle 3	side opposite α/side adjoining α = ????
Base length:	=
Triangle 4	side opposite α/side adjoining α = ????
Base length:	=

4. Measure and label the side lengths of all four of your triangles. Use this information to fill in the table.

5. What do you notice about the quotients?_____

6. Compare your findings with those of other groups with different values of α.

7. Write a conjecture about the relationship between the side lengths opposite and adjoining an acute angle in a right triangle.

8. Consider the following two triangles. Does your conjecture work for these two triangles?

9. The ratio that you have been exploring today is called the tangent ratio. Try searching the Internet for tangent table. What does this table represent? How does this table relate to your original conjecture? _____

_____

_____

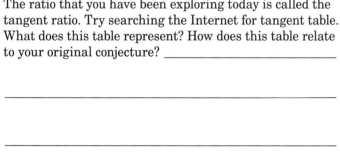

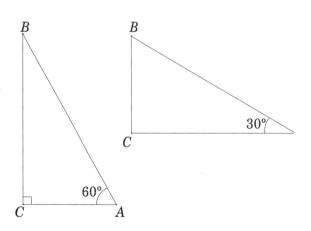

Explore, Conjecture, Connect, Prove:
The Versatility of a Rich Geometry Problem

Robert J. Quinn and Tom S. Ball

The National Council of Teachers of Mathematics suggests that "students will benefit from a rich and integrated treatment of mathematics content. Instruction that segregates the content of algebra or geometry from that of other areas is educationally unwise and mathematically counterproductive" (NCTM 2000, p. 213). Furthermore, it is important that students "have frequent encounters with interesting, challenging problems" (NCTM 2000, p. 211). Additionally, it remains critical that students be exposed to "problems that draw on a variety of aspects of mathematics, that are solvable using a variety of methods, and that students can access in different ways" (NCTM 2000, p. 289). Finally, "students need to develop increased abilities in justifying claims, proving conjectures, and using symbols in reasoning. They can be expected to learn to provide carefully reasoned arguments in support of their claims" (NCTM 2000, p. 288).

This article will present and discuss a rich mathematical problem that combines geometric concepts with those of algebra. This problem is well suited to having students work cooperatively. Initially, it can be viewed as an opportunity for them to make conjectures and explore an interesting mathematical scenario. As students become confident in their conjectures, they should be encouraged to delve more deeply into the integrated algebra and geometry embedded in this problem. Ultimately, even the most talented high school students will be challenged as they are pushed toward proving their conjectures rigorously. The diversity of solution strategies and the breadth of mathematical content both algebraic and geometric will be eye opening for students.

We have used this problem over the last decade or so with preservice secondary mathematics teachers enrolled in a secondary mathematics methods course. Though we are not aware of its source, the problem conforms to the recommendations of NCTM in all the ways described above. The lesson lends itself to a group activity that models the way these prospective teachers might use such a problem in their own classrooms. Students are initially asked to explore the problem in groups and make a conjecture regarding the nature of the path that the rowboat should follow. After the exploration stage, the groups are told to create a proof of their conjecture. After a variety of proofs have been completed, often with the help of appropriate hints, each group shares its method of proof with the class. The class ends with a discussion of the variety of purposes that an exploration of this rich problem can provide. This discussion focuses

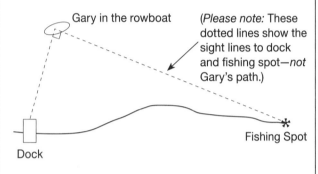

Gary rented a rowboat and took off from the dock as shown in the figure below. He decided to row the boat on a course so that his lines of sight back to the dock and forward to his favorite fishing spot on the shore were always perpendicular to each other.

Gary in the rowboat

(*Please note:* These dotted lines show the sight lines to dock and fishing spot—*not* Gary's path.)

Dock

Fishing Spot

What kind of path on the lake did he travel?
How did you come to that conclusion?
Develop an argument to convince someone else that your path is correct.

Fig. 1 Statement of the Rowboat problem

on how the problem could be adapted for use at a variety of levels and for numerous purposes.

THE PROBLEM

Gary rented a rowboat and took off from the dock as shown in **figure 1**. He decided to row the boat on a course so that his lines of sight back to the dock and forward to his favorite fishing spot on the shore were always perpendicular to each other. What kind of path on the lake did he travel? How did you come to that conclusion? Develop an argument to convince someone else that your path is correct.

INITIAL EXPLORATION

The simplicity of this problem allows students to explore the situation informally and to develop a reasonable conjecture regarding the nature of the path. Typically, we do not provide a compass, as that would be too strong a hint that the path is a portion of a circle. We do give students a ruler and encourage them to make sketches and try to estimate what the path would look like. Another helpful hint that we give, if the students do not come up with this on their own, is that they can use the corner of a piece of paper to represent a right angle. They can move this corner in such a way as to help trace the path that the rowboat must follow.

The students will often conclude that the path is some kind of curve. By using the right angle and attempting to construct the boat's path, they usually recognize that the path appears to be part of a circle.

A MISSTEP IN LOGIC

As students who are somewhat knowledgeable in geometry begin to seek a means of proving that the path will be a circle, they often recall the familiar theorem that states that an angle inscribed in a semicircle will be a right angle. Groups will often conclude that this theorem provides them with the cornerstone of the proof. At this point, we ask them what they are assuming if they are going to use this theorem. The answer is that they assume the path is a circle and then use properties of a circle to confirm that the angle formed by the sight lines to the fishing spot and the dock will be 90 degrees. We then ask them what it is they are trying to prove. The answer is that they are trying to prove that the path is a circle. These questions bring them to the realization that in using this theorem in their proof, they are assuming what they are trying to prove. Since that is not an appropriate logical means of deductive proof, they abandon this approach and look for another way to proceed.

A DEDUCTIVE PROOF

Although students' initial attempts at a deductive proof often fail, that does not mean that a simple yet elegant deductive proof does not exist. The key to one such proof is the familiar theorem that states that the midpoint of the hypotenuse of a right triangle is equidistant from each of the three vertices. This theorem can be applied since the position of Gary's rowboat is always the vertex of the right angle in the triangle. As the rowboat moves, the hypotenuse is always the segment connecting the dock to the fishing spot; therefore, all of the right triangles formed have the same hypotenuse. The theorem implies that the distance from the boat to the midpoint of the hypotenuse will always be half the length of the hypotenuse. Since the length of the hypotenuse remains constant, the position of the boat will be the set of points that are the same distance from the midpoint. These points will form a circle, proving the conjecture.

Unlike the assumption made in the misstep of logic in the previous section, in this instance we assume the right angle and show that this assumption leads to a circle. The dock and the fishing spot are points on this circle that represent the limiting cases in either direction as the appropriate acute angle of the triangle approaches zero. These two points satisfy the condition of being on the circle described, since they are each one-half the length of the hypotenuse from the midpoint of the hypotenuse.

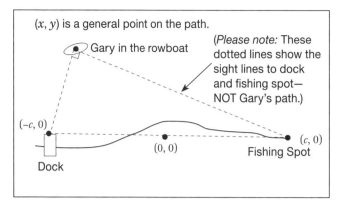

Fig. 2 Applying a coordinate geometry approach

A COORDINATE GEOMETRY PROOF

Coordinate geometry proofs provide another means by which the conjecture can be proven. Since students are often not familiar with the nature of a coordinate geometry proof, clues sometimes need to be provided. The most critical aspect of a coordinate geometry proof is the placement of the origin and the orientation of the axes. I suggest to students that they place the origin at the midpoint of the segment connecting the dock and the fishing spot. Given this configuration, the location of the boat is denoted by the general ordered pair, (x, y), the dock is located at $(-c, 0)$ and the fishing spot at $(c, 0)$ (see **fig. 2**). I then ask students to brainstorm for conclusions that can be drawn from the fact that the lines of sight are perpendicular to each other. Additional clues may or may not need to be provided. One such clue is to suggest that they use the relationship that the product of the slopes of lines that are perpendicular to one another is -1. Given this clue, students are usually able to determine that the slope of the segment from the boat to the dock is

$$\frac{y - 0}{x - (-c)}.$$

Similarly, the slope of the segment from the boat to the fishing spot is

$$\frac{y - 0}{x - (c)}.$$

Multiplying these slopes of the segment from the boat to the fishing spot yields the following equation:

$$\frac{y - 0}{x - (-c)} \cdot \frac{y - 0}{x - (c)} = -1,$$

which simplifies to

$$\frac{y^2}{(x + c)(x - c)} = -1$$

and

$$\frac{y^2}{(x^2 - c^2)} = -1.$$

Multiplying each side by $(x^2 - c^2)$ yields

$$y^2 = -1 \cdot (x^2 - c^2).$$

Thus,

$$y^2 = -x^2 + c^2.$$

Adding x^2 to each side yields $x^2 + y^2 = c^2$, which is the general equation for a circle of radius c whose center is at $(0, 0)$.

Thus, by assuming that the lines of sight to the dock and to the fishing spot are perpendicular, we have proven that the set of all points that meet this condition are the points that satisfy the equation of a circle whose center is at the midpoint of the segment connecting the fishing spot and the dock.

ANOTHER COORDINATE GEOMETRY PROOF

Given that the critical condition given in this problem is one of perpendicularity, students often conclude that the proof should include an application of the Pythagorean theorem. Indeed, such a proof can be completed using the same coordinate axes and origin used in the previous proof. To begin this proof, the distance formula is applied to find the distance from (x, y) to $(-c, 0)$ and the distance from (x, y) to $(c, 0)$. These distances are

$$\sqrt{(x+c)^2 + y^2} \text{ and } \sqrt{(x-c)^2 + y^2},$$

respectively. Since the length of the hypotenuse is $2c$, applying the Pythagorean theorem yields the following equation, which can be simplified as shown:

$$\left(\sqrt{(x+c)^2 + y^2}\right)^2 + \left(\sqrt{(x-c)^2 + y^2}\right)^2 = 4c^2$$

expanding:

$$x^2 + 2cx + c^2 + y^2 + x^2 - 2cx + c^2 + y^2 = 4c^2$$

combining like terms: $\quad 2x^2 + 2y^2 + 2c^2 = 4c^2$

Subtracting $2c$ from each side: $\quad 2x^2 + 2y^2 = 2c^2$

This, again, is the general equation for a circle of radius c whose center is at $(0, 0)$.

Thus, by assuming that the lines of sight to the dock and to the fishing spot are perpendicular, we have proven, using the Pythagorean theorem, that the set of all points that meet this condition are the points that satisfy the equation of a circle whose center is at the midpoint of the segment connecting the fishing spot and the dock.

A POSSIBLE EXTENSION

The use of an interactive geometry software package could be incorporated into the exploration of this problem. Each teacher should consider the advantages and disadvantages of such an approach based on the level of their students and the specific goals they hope to achieve during the lesson. We would recommend that students use the software after they have conjectured that the path would be a circle. The software would then allow students to make an accurate drawing of the situation. Once the drawing is completed, students can use the dynamic nature of the program and its measurement tools to help them gain confidence in the conjectures they have made. Of course, if they have drawn a circle and then measure angles, they are falling into the trap of assuming what they are setting out to prove, and they should be made aware of this. Further, they should be warned that even though all the cases they measure suggest that their conjecture is true, this does not constitute a formal proof.

CONCLUSION

The Rowboat problem is rich in mathematical rigor yet attainable for students at a variety of levels. It provides an exploration in the form of an interesting cooperative-learning problem that meets many of NCTM's recommendations. A complete treatment of the problem begins with a hands-on exploration leading to a conjecture, which can ultimately be proven by at least the three methods shown here. The variety of these proofs is illuminating to students, showing them the vastly different ways a problem can be viewed. Further, these proofs provide strong connections between algebra and geometry, reinforcing the importance of facility in the manipulation of algebraic terms and providing a useful application of the Pythagorean theorem and the relationship of the slopes of perpendicular lines.

REFERENCE

National Council of Teachers of Mathematics (NCTM). *Principles and Standards for School Mathematics.* Reston, Va: NCTM, 2000.

Gary in the Rowboat

Sheet 1

Gary rented a rowboat and took off from the dock as shown in the figure below. He decided to row the boat on a course so that his lines of sight back to the dock and forward to his favorite fishing spot on the shore were always perpendicular to each other.

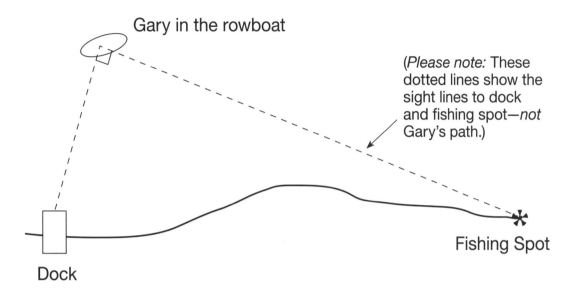

Gary in the rowboat

(*Please note:* These dotted lines show the sight lines to dock and fishing spot—*not* Gary's path.)

Fishing Spot

Dock

1. What kind of path did he travel on the lake?

2. How did you come to that conclusion?

3. Develop an argument to convince someone else that your path is correct.

Discovery with Cubes

Robert E. Reys

TEACHER'S GUIDE

Grade level: 6–12

Materials: Student worksheets

Objectives: Students will visualize three-dimensional figures, construct a table, discover patterns in the table, and use patterns to make predictions.

Directions: Make copies of the activities sheets for students. Divide the class into groups of two, and let them work together to solve this exercise. It would be quite helpful if the teacher had a set of cubes that were colored as stated in activities 1–3. In this way, students could verify their results.

After completing activity 4, students should record their results in the table (activity 6) found on sheet 3. Check the table with the class to insure that all students have the correct values, since predictions will be made on the basis of their data. Students should then sketch or construct a $6 \times 6 \times 6$ cube as indicated in activity 5 and add its data to the table.

Few students will be able to complete the table for a $10 \times 10 \times 10$ cube unless some patterns have been identified. Ask, "Are there any constants in a column? Any multiples?" Encouraging pupils to keep track of the factors used in the table aids pattern recognition. For example, 0, 6, 24, 54, and 96 are the first five values for one of the columns, A pattern is more discernible when these values are written as 0, 6×1, 6×4, 6×9, and 6×16.

Here is a question that might be used to culminate this activity: "Let the length of one side of the cube be n. When you complete this row of the table, is the sum of these values n^3?"

SOLUTIONS

Sheet 1:
1) 8
2) 8
3) 0
4) 0
5) 0
6) 8
7) equal
8) 27
9) 8
10) 12
11) 6
12) 1
13) 27
14) equal

Sheet 2:
1) 64
2) 8
3) 24
4) 24
5) 8
6) 64
7) equal
8) 125
9) 8
10) 36
11) 54
12) 27
13) 125
14) equal

Editorial comment: Additional answers for the table in activity 6 include the following:

$6 \times 6 \times 6$	64	96	48	8	Total	216
$7 \times 7 \times 7$	125	150	60	8	Total	343
$10 \times 10 \times 10$	512	384	96	8	Total	1000

Exploring the results for $n \times n \times n$ yields a key discovery:

$n \times n \times n$	$(n-2)^3$	$6(n-2)^2$	$12(n-2)$	8	Total n^3

The coefficients 6, 12, and 8 are the number of faces, edges, and vertices of a cube. These, of course, are the locations of those small cubes with one, two, or three faces painted, respectively.

Interesting modifications can be made on this activity by changing the painting process. How would the small cubes be painted if only the lateral faces of the large cube were painted? If just the top and bottom faces were painted? If three different colors were used, each on a different pair of opposite faces?

Discovery with Cubes

Sheet 1

Activity 1

1. How many cubes are in the large cube? _____

   If this large cube is dropped into a bucket of paint and completely submerged:

2. How many of the smaller cubes are painted on three sides?_____

3. How many on only two sides?_____

4. How many on only one side?_____

5. How many on zero sides?_____

6. What is the sum of your answers in 2, 3, 4 and 5?_____

7. How does your answer to 6 compare to 1?_____

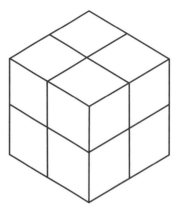

Activity 2

8. How many cubes are in the large cube? _____

9. How many of the smaller cubes are painted on three sides?_____

10. How many on only two sides?_____

11. How many on only one side?_____

12. How many on zero sides?_____

13. What is the sum of your answers in 9, 10, 11 and 12? _____

14. How does your answer to 13 compares to 8? _____

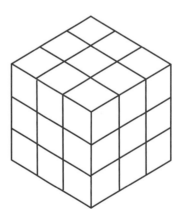

Discovery with Cubes
Sheet 2

Activity 3

1. How many cubes are in the large cube? _____

   If this large cube is dropped into a bucket of paint and completely submerged:

2. How many of the smaller cubes are painted on three sides?_____

3. How many on only two sides?_____

4. How many on only one side?_____

5. How many on zero sides?_____

6. What is the sum of your answers in 2, 3, 4 and 5?_____

7. How does your answer to 13 compares to 1?_____

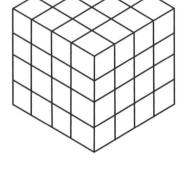

Activity 4

8. How many cubes are in the large cube? _____

9. How many of the smaller cubes are painted on three sides?_____

10. How many on only two sides?_____

11. How many on only one side?_____

12. How many on zero sides?_____

13. What is the sum of your answers in 9, 10, 11 and 12? _____

14. How does your answer to 13 compares to 8? _____

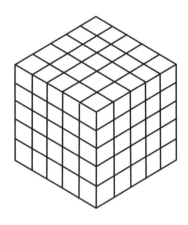

Activity 5

Suppose your cube was $6 \times 6 \times 6$. Complete this model by sketching a $6 \times 6 \times 6$ cube. Use it to determine the total number of cubes as well as the number of faces with zero, one, two, three, and four sides painted.

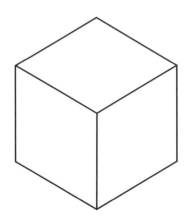

Discovery with Cubes

Activity 6

Now that you have solved several problems with the cubes, record this information in the table:

Length of Side of Cube	Number of Painted Sides					Total Number of Cubes
	0	1	2	3	4	
2						
3						
4						
5						
6						
7						
8						
9						
10						
n						

Do you observe any patterns? _____

If so, complete the table for a 7 × 7 × 7 cube. If not, sketch or construct a cube and then complete the table.

Have you got the idea? If you think so, try to complete the table for a 10 × 10 × 10 cube.

Let the length on one side of the cube be n. try to complete the table for an n × n × n cube.

The Toothpick Problem and Beyond

Charalampos Toumasis

- The mathematics curriculum should include the study of geometry of one, two, and three dimensions in a variety of situations so that students can identify, describe, compare, and classify geometric figures (NCTM 1989, p. 112).

- The mathematics curriculum should include numerous and varied experiences that reinforce and extend logical reasoning skills so that all students can make and test conjectures (NCTM 1989, p. 143).

TEACHER'S GUIDE

Introduction: Although many changes have occurred in the goals for teaching secondary school mathematics, the goals of motivating students to do mathematics and helping them to reason mathematically through problem solving have remained. NCTM's *Curriculum and Evaluations Standards* (1989) focus on these goals. All students should be encouraged to explore, to guess, to conjecture, to test, and to build arguments about the validity of their conjectures.

Secondary school geometry is a visual subject and may lead to more surprises than other branches of mathematics. Many students have more spatial intuitions than numerical ones and find it easy and enjoyable to work with figures and drawings. Therefore, geometry is an appropriate subject in which to implement the central goal of involving our students in the process of discovering mathematics and doing mathematics through exploratory activities.

This article was inspired by the "toothpick problem" discussed in the *Curriculum and Evaluations Standards* (NCTM 1989, p. 113). Basic properties of triangles can be developed and strengthened through investigating this problem. The extension problem gives students opportunities to explore and describe patterns and to make and test conjectures.

While my eighth-grade geometry class was engaged in the original toothpick problem, the discussion among

students generated the second problem, which challenged both me and my students. Modifying an existing problem to generate other problems for investigation is a method frequently used by research mathematicians.

Grade levels: 7–10

Materials: A set of activity sheets for each student and a box of toothpicks for each small group of students. (Because they don't roll, flat toothpicks may be preferable to rounds ones for this activity.)

Prerequisities: Students need to be able to identify and classify scalene, isosceles, and equilateral triangles.

Directions: These activities require two or three class periods to complete, depending on the amount of class discussion generated and the number of triangles that students produce.

Sheet 1: Divide the class into groups of four of five students. Distribute one copy of sheet 1 to each student and of box of toothpicks to each group. Use the figures at the top of the page to make sure students know how to use toothpicks to make triangles and count the number of toothpick on each side to identify their triangles. These triangles could be identified as (3, 4, 5) and (1, 5, 5). Make sure that students understand that (3, 4, 5) and (4, 3, 5) describe the same triangle but (3, 4, 6) describes a different triangle. Read the problem together and make sure that students understand the problem before they continue.

Have students work individually to answer questions 1 and 2. Be very clear about which definition of *isosceles triangle* you are using—a triangle with at least two equal sides or a triangle with exactly two equal sides. Suggest that the students in each group work together to complete the chart in question 3. After students have filled in the chart in question 3, they should have some ideas about what side lengths produce different types of triangles. In a class discussion, ask students to verbalize their ideas. Have students explain what they did to

TABLE 1							
Number of toothpicks	3	4	5	6	7	8	9
Is a triangle possible?	Yes	No	Yes	Yes	Yes	Yes	Yes
Number of triangles	1	0	1	1	2	1	3
Number of scalene triangles	0	0	0	0	0	0	1
Number of isosceles triangles	0 or 1	0	1	0 or 1	2	1	1 or 2
Number of equilateral triangles	1	0	0	1	0	0	1

TABLE 2

Measure of the longest side, α	1	2	3	4	5	6	7	8	9
Number of triangles	1	2	4	6	9	12	16	20	25
Number of scalene triangles	0	0	0	1	2	4	6	9	12
Number of isosceles triangles	0 or 1	1 or 2	3 or 4	4 or 5	6 or 7	7 or 8	9 or 10	10 or 11	12 or 13
Number of equilateral triangles	1	1	1	1	1	1	1	1	1

make their triangles and guide them in a discussion of their initial observations before they answer questions 4 and 5. Try to get students to formulate their own statement of the triangle inequality.

Sheet 2: Distribute a copy of sheet 2 to each student. Read the problem together and make sure that students understand the problem. They will collect and record data as they work through question 1 through 5 to help them answer questions on sheet 3 and solve the problem.

Students can work individually on questions 1 through 3 and then compare their answers with those of the rest of their group. Assign two numbers to each group to answer question 4 and 5. Each of the numbers, 4 through 9, should be assigned to two groups.

As the value of α increases, finding all the triangles becomes laborious and students may encounter some difficulties. For the most part, however, students enjoy this activity and can proceed independently. Some of them can be asked to continue the search at home for some greater values of α. When they have completed their investigations, perhaps during the next class period, record the data collected by all the groups on a transparency of the chart in question 6.

Sheet 3: Distribute copies of sheet 3 to each student. Let students work in pairs to complete sheet 3. When students have completed the sheet, lead a class discussion about the patterns students discovered and the predictions they made. Give students the opportunity to discuss any other observations they may have made.

The questions in this sheet are a little more advanced than those on sheets 1 and 2. Students must study the data in their tables and look for patterns. On the basis of these patterns, students will formulate generalizations about all cases. Gathering data, searching for patterns, and making conjectures are vital aspects of the problem-solving process that students should experience.

SOLUTIONS

Sheet 1:
1) a. 1
   b. 0;
   c. 0 or 1
   d. 1

2) a. 0; b 0; c 0; d 0

3) See table 1. The lower number given for isosceles triangles refers to isosceles triangles having exactly two equal sides; the high number refers to isosceles triangles having at least two equal sides.

4) The sum of the lengths of any two sides is greater than the length of the third side.

5) $a + b + c = n$, $a < b + c$, $b < a + c$, $c < a + b$.

Sheet 2:
1) a. 1
   b. 0
   c. 0 or 1
   d. 1

2) a. 2
   b. 0
   c. 1 or 2
   d. 1

3) a. 4
   b. 0
   c. 3 or 4
   d. 1

4)–6) See table 2.

Sheet 3:
1) 36, 49, n^2

2) 30, 42, $n(n + 1)$

3) 20, 30, $(n-2)(n-1)$

4) 16, 25, $(n-1)^2$

5) 15, 18, $3(n-1)$

6) 13, 16, $3n - 2$

7) One equilateral triangle results for any positive integral value of α.

REFERENCE

National Council of Teachers of Mathematics (NCTM). *Curriculum and Evaluations Standards for School Mathematics.* Reston, Va.: NCTM, 1989.

Original Toothpick Problem

Sheet 1

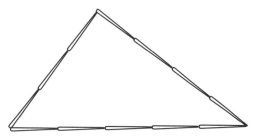 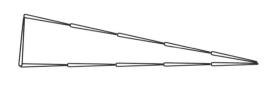

You are given a pile of toothpick all the same size. Take a number, $n \geq 3$, of toothpicks. How many different triangles can you make using all n toothpicks placed end to end in the same plane? What kinds of triangles are possible?

1. First use three toothpicks and answer these questions.

    a. How many different triangles can you form? _____

    b. How many are scalene? _____

    c. How many are isosceles? _____

    d. How many are equilateral? _____

2. Next use four toothpicks and answer the same questions.

    a. Number of triangles _____

    b. Number of scalene triangles _____

    c. Number of isosceles triangles _____

    d. Number of equilateral triangles _____

3. Repeat with more toothpicks and complete the chart below.

Number of toothpicks	3	4	5	6	7	8	9
Is a triangle possible?							
Number of triangles							
Number of scalene triangles							
Number of isosceles triangles							
Number of equilateral triangles							

4. Consider the ways in which your triangles were formed. Find the sum of measures of two sides of each triangle, and compare it with the measure of the third side.

5. Suppose that you have n toothpicks, $n \geq 3$, and that you are going to use those toothpicks to make triangles. Under what conditions does (a, b, c) represent one of your triangles? Remember that a, b, and c are the lengths of the sides of the triangle.

Extending the Toothpick Problem

Sheet 2

You are given a pile of toothpicks of the same size. Assume that this size is one unit. If the longest side of a triangle you can build is α units in length, then what kinds of triangles are possible whose other two sides have length less than or equal to α? How many of each type are possible?

1. First, take $\alpha = 1$.

   a. How many different triangles can you form? _____

   b. How many are scalene? _____

   c. How many are isosceles? _____

   d. How many are equilateral? _____

2. Next take $\alpha = 2$ and answer the same questions.

   a. Number of triangles _____

   b. Number of scalene triangles _____

   c. Number of isosceles triangles _____

   d. Number of equilateral triangles _____

3. Repeat with $\alpha = 3$.

   a. Number of triangles _____

   b. Number of scalene triangles _____

   c. Number of isosceles triangles _____

   d. Number of equilateral triangles _____

4. Repeat with $\alpha =$ _____. *(This number will be assigned to your group.)*

   a. Number of triangles _____

   b. Number of scalene triangles _____

   c. Number of isosceles triangles _____

   d. Number of equilateral triangles _____

5. Repeat with $\alpha =$ _____. *(This number will be assigned to your group.)*

   a. Number of triangles _____

   b. Number of scalene triangles _____

   c. Number of isosceles triangles _____

   d. Number of equilateral triangles _____

6. Use the data collected from all the groups to fill in the chart below.

Measure of the longest side, α	1	2	3	4	5	6	7	8	9
Number of triangles									
Number of scalene triangles									
Number of isosceles triangles									
Number of equilateral triangles									

Extending the Extension

Use the information that you recorded on sheet 2 to help you answer the following questions.

1. Fill in the chart below when α is an odd number. Do you notice a pattern? How many triangles do you think you could make when $\alpha = 11$?_____ When $\alpha = 13$? _____Use the pattern you found to fill in the table for $\alpha = 2n - 1$. For any natural number n, $2n - 1$ is a common way of expressing an odd integer.

α	1	3	5	7	9	$2n - 1$
Number of triangles						

2. Fill in the chart below when α is an even number. How many triangles do you think you could make when $\alpha = 10$? _____When $\alpha = 12$?_____ Use the pattern you found to fill in the table for $\alpha = 2n$.

α	2	4	6	8	$2n$
Number of triangles					

3. Fill in the chart below when α is an odd number. Do you notice a pattern? How many scalene triangles do you think you could make when $\alpha = 11$? _____When $\alpha = 13$?_____ Use the pattern you found to fill in the table for $\alpha = 2n - 1$.

α	1	3	5	7	9	$2n - 1$
Number of scalene triangles						

4. Fill in the chart below when α is an even number. How many scalene triangles do you think you could make when $\alpha = 10$?_____ When $\alpha = 12$? _____ Use the pattern you found to fill in the table for $\alpha = 2n$.

α	2	4	6	8	$2n$
Number of scalene triangles					

5. Fill in the chart below when α is an odd number. How many isosceles triangles do you think you could make when $\alpha = 11$?_____ When $\alpha = 13$? _____ Use the pattern you found to fill in the table for $\alpha = 2n - 1$.

α	1	3	5	7	9	$2n - 1$
Number of isosceles triangles						

6. Fill in the chart below when α is an even number. How many isosceles triangles do you think you could make when $\alpha = 10$?_____ When $\alpha = 12$? _____ Use the pattern you found to fill in the table for $\alpha = 2n$.

α	2	4	6	8	$2n$
Number of isosceles triangles					

7. How many equilateral triangles are possible for each value of α?

Statistics and Probability

Introduction

"In our increasingly data-intensive world, statistics is one of the most important areas of the mathematical sciences for helping students make sense of the information all around them, as well as for preparing them for further study in a variety of disciplines (e.g., the health sciences, the social sciences, and environmental science) for which statistics is a fundamental tool for advancing knowledge.... The common thread throughout the statistical problem-solving process is the focus on making sense of, and reasoning about, variation in data. The goal is not only to solve problems in the presence of variation but also to provide a measure of how much the variation might affect the solution" (NCTM 2009, p. 73).

Statistics and probability activities did not appear in *Mathematics Teacher* until the 1970s. Therefore, most of the activities in this chapter were developed relatively recently (seven of the nine activities were published within the past ten years). Supporters of statistics and probability as key mathematical topics have established its rightful place in *Mathematics Teacher*, evidenced by the fact that 20 percent of all activities in the past decade focused on statistics and probability. Here, we present activities from a wide range of topics, contexts, and levels of difficulty to engage and challenge students as they tackle this content. Four of the activities (Lanier and Barrs 2003; Lappan et al. 1987; Richardson and Gabrosek 2004; and White 2001) use games ranging from Plinko to basketball as their context, and one activity (Groth and Powell 2004) guides students through a statistical investigation allowing students to choose their own context. **Table 5.1** summarizes the characteristics of the statistics and probability activities.

According to *Focus in High School Mathematics: Reasoning and Sense Making* (NCTM 2009), the following are four key elements in statistics and probability:

1. *Data analysis.* Gaining insight about a solution to a statistics question by collecting data and describing features of the data through the use of graphical and tabular representations and numerical summaries.

2. *Modeling variability.* Developing probability models to describe the long-run behavior of observations of a random variable.

3. *Connecting statistics and probability.* Recognizing probability as an essential tool of statistics, understanding the role of *probability in statistical reasoning.*

4. *Interpreting designed statistical studies.* Drawing appropriate conclusions from data in ways that acknowledge random variation. (Interpreting

results from designed statistical studies involves statistical inference and other more formal levels of statistical reasoning.)

Franklin and Mulekar (2006) provide an experience for students to investigate global warming through an analysis of the temperature in Central Park during the last century. The students *analyze the data* by using a variety of representations (e.g., time series plot, histogram), *model variability* by simulating the situation using either a coin or a computer program, *connect statistics and probability* through the use of *p*-values, and *interpret statistical studies* by using hypothesis testing and the binomial distribution.

Even though these articles were written before *Focus in High School Mathematics* was published, the authors of the activities make statements in support of the key elements. For example, Shaughnessey and Pfannkuch (2002, pp. 254–55) emphasize the importance of the *modeling variability* element, saying, "we have concentrated heavily on measures of central tendency and we have neglected variation ... the central element of any definition of statistical thinking is variation." Similarly, Bryan (1988, p. 659) supports the *interpreting designed statistical studies* element in a suggestion to teachers: "Be sure to emphasize the fact that 'one' correct answer does not always occur. The idea is to have students explore the data, ask questions, and make observations. They should discuss their individual interpretations and attempt to support them when differing conclusions are reached."

The *data analysis* element is particularly well represented in Groth and Powell (2004) because the authors expect the students to carry out the entire statistical process, including formulating a researchable question, collecting appropriate data, analyzing the data, and interpreting and communicating results (cf. Franklin et al. 2007). Students also collect or analyze data in several other activities. For example, in McGivney-Burelle, McGivney, and McGivney's article (2008), students analyze data from an atlas or the Internet to determine the relationship between latitude and average

TABLE 5.1

Statistics and Probability Activities

Author and title	Mathematical topic(s)	Context(s)	Materials
Bryan (1988), "Exploring Data with Box Plots"	Box-and-whisker plots	Automobiles	Transparencies, student activity sheets
Franklin and Mulekar (2006), "Is Central Park Warming?"	Theoretical and experimental probabilities, hypothesis testing	Global warming	Coins, Minitab software (recommended), student activity sheets
Groth and Powell (2004), "Using Research Projects to Help Develop High School Students' Statistical Thinking"	Statistical process	Various	Fathom software (recommended), student activity sheets
Lappan, Phillips, Fitzgerald, and Winter (1987), "Area Models and Expected Values"	Expected values, experimental probability	Games of chance	Transparencies, bobby pins or paper clips (for spinners), student activity sheets
McGivney-Burelle, McGivney, and McGivney (2008), "Investigating the Relationship between Latitude and Temperature"	Line of best fit, correlation coefficient	Geography	TI-83/84 calculators, atlas or Internet access, student activity sheets
Perry and Kader (1998), "Counting Penguins"	Sampling, central limit theorem	Penguins in Antarctica	Sampling board and penguin cards (directions in article), random-number generator (e.g., icosahedron die, calculator), student activity sheets
Richardson and Gabrosek (2004), "Activities: A - B - C, 1 - 2 - 3"	Linear, quadratic, and cubic regression	Scrabble	Newspaper or magazine articles, Scrabble game (to show), graphing calculators, student activity sheets
Shaughnessy and Pfannkuch (2002), "How Faithful Is Old Faithful? Statistical Thinking: A Story of Variation and Prediction"	Statistical thinking, variation	Geysers	Student activity sheets
White (2001), "Connecting Independence and the Chi-Square Statistic"	Chi-square statistic	Basketball	Number cubes, calculators, student activity sheets

temperature. In a second correlation activity, Richardson and Gabrosek (2004) ask students to collect and analyze data from a sample text to examine the relationship between the frequency of the use of particular letters and various aspects of a Scrabble game (e.g., point value for each letter, number of tiles for each letter).

Lappan and colleagues (1987) address, among other elements, the *modeling variability* element in activities that use the context of games of chance. Lappan and colleagues (1987, p. 650) state, "Students have a natural affinity for games that involve chance. These activities build on this interest to reinforce understanding of dependent probabilities and to introduce the concept of expected value in an experimental way." Both activities develop the relationship between experimental probability and the long-run behavior of a random variable (i.e., expected value or theoretical probability). Lanier and Barrs (2003, p. 629), available on More4U, suggest that their students "experienced mathematics by simulating real-world phenomena. They developed an understanding of the importance of experimental probabilities and their relationship to theoretical probabilities. Perhaps most important, they had fun with mathematics. We hope that they will continue to develop a positive attitude about mathematics."

Perry and Kader (1998), in which students estimate the number of penguins in Antarctica, and White (2001), in which students determine whether playing basketball on a team's home court confers a statistically significant advantage, address both the *connecting statistics and probability* element (connecting probability, statistics, and the real world) and the *interpreting designed statistical studies* element as students are required to draw conclusions from their results. Perry and Kader (1998) address random and spatial sampling and the central limit theorem (i.e., as sample size increases, the sampling distribution moves toward a normal distribution). White (2001) uses the chi-square statistic to test goodness of fit, homogeneity of proportions, and independence.

In the context of automobiles (e.g., speed, repairs), Bryan (1988) provides experiences with both the *data analysis* and *interpreting designed statistical studies* elements. Students are expected to construct and compare data representations (e.g., box plots compared to stem-and-leaf plots, comparing a series of box plots) and make decisions based on their interpretations of the results. Finally, Shaughnessey and Pfannkuch (2002) use the fascinating phenomenon of Old Faithful to develop students' statistical thinking. The activity addresses all four key elements: the students collect and use data to answer the question "How faithful is Old Faithful?", investigate the pattern of eruption times with a small data set and then a larger data set, and discuss the amount of data needed to make reasonable conclusions for a given question.

Also well represented in these activities are the mathematical reasoning habits described in *Focus in High School Mathematics: Reasoning and Sense Making* (NCTM 2009): (1) analyzing a problem, (2) implementing a strategy, (3) seeking and using connections, and (4) reflecting on a solution. For example, White (2001) guides students through the *analysis of a problem* to determine whether home court advantage is indeed an advantage by asking students to first conjecture about the results, next to use the chi-square statistic to test their hypothesis, and finally to revisit their conjectures and revise their conclusion accordingly. Lappan and colleagues (1987) offer students an opportunity to implement a strategy as students calculate expected probabilities by using a given formula and compare these expected probabilities with their experimental probabilities. The emphasis on seeking and using connections is apparent in Franklin and Mulekar's activity (2006) as students connect different representations of the Central Park data and connect probability and statistics. Finally, Shaughnessy and Pfannkuch (2002) emphasize *reflecting on a solution* by providing a set of thought-provoking questions for discussion at the end of the activity.

REFERENCE

National Council of Teachers of Mathematics (NCTM). *Focus in High School Mathematics: Reasoning and Sense Making.* Reston, Va.: NCTM, 2009.

Exploring Data with Box Plots

Elizabeth H. Bryan

TEACHER'S GUIDE

Grade levels: 7–12

Materials: One set of activity sheets for each student; a set of transparencies for class discussion

Objectives: Students will organize and display data with a box-and-whiskers graph. They will use the plot as a summary display to detect patterns and to highlight the important features of the data for purposes of comparison.

Procedures: These activities assume that the students understand how to determine the median of a data set and how to construct stem-and-leaf plots. This background is useful for computing the statistics needed to construct the box plot and can be found in a previously published "Activities" (Landwehr and Watkins 1985) related to these concepts.

Distribute sheet 1 first. You may find it helpful to do sheet 1 together with your class and discuss the answers thoroughly before before assigning sheets 2 and 3. The procedures for drawing a box-and-whiskers plot are described on sheet 1 and include the following:

1. Draw and label a number line.

2. Find the median, lower quartile (LQ), upper quartile (UQ), lowest value (LV), and highest value (HV).

3. Draw the box.

    a. The length of the box is the interquartile range (from LQ to HQ).

    b. The width of the box can be anything.

    c. The median is marked with a line widthwise across the box.

4. Draw the whiskers—

    a. From LV to LQ;

    b. From HV to HQ.

You should stress that the box plots are appropriate when dealing with large sets of data because they focus attention on only a few characteristics of the data. However, students often have a difficult time interpreting them, and they are not as useful as a stem-and-leaf graph for showing details.

The major objective is to help student learn how to interpret data. Constructing the graphs and plots is only a secondary goal. You should encourage the students to make conjectures on the basis of examining the plots. Be sure to emphasize the fact that "one" correct answer does not always occur. The idea is to have students explore the data, ask questions, and make observations. They should discuss their individual interpretations and attempt to support them when differing conclusions are reached.

Finally, the students should organize their results into a paragraph describing what they have learned. Writing interpretations of statistical data is often difficult for students. It is important for you to require them to complete the activity with a written summary, since this approach teaches the students to follow a model for data analysis in the same way a good statistician would.

SOLUTIONS

Sheet 1:

1)

4	9
5	2 3 3 3 4 6 6 7 7 8 8
6	0 0 1 1 2 4 9
7	1 1 3 5 8
8	2

2) a. 60 mph
   b. 55 mph
   c. 70 mph
   d. 49 mph and 82 mph

3) and 4)

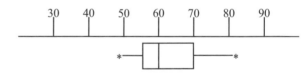

5) a. 50 percent (48% actual)
   b. 75 percent (76%)
   c. 50 percent (52%)
   d. 25 percent (24%)

6 Yes. It means that the faster speeds (those in the top 25%) are more spread out than those in the bottom 25 percent.

7) It is not in the center, because the values in the second and third quarters are not equally spread out.

8) 25 percent (24%)

9) Answers will vary.

> *Example:* The average car in the sample was traveling at a rate between 55 mph and 70 mph. The six fastest cars (those in the top 25%) were exceeding 70 mph, whereas the six slowest were traveling less than 55 mph. The rates of the cars traveling in excess of 60 mph displayed much more variability that those traveling below 60 mph.

Sheets 2 and 3

1) *New cars:* Median = 2, LQ = 1, UQ = 2.5,; LV = 1, HV = 6

   Three- to four-year-old cars: Median = 6; LQ = 5; UQ = 9; LV = 3; HV = 14

   Six-year-old cars: Median = 11; LQ = 10.5; UQ = 15.5; LV = 5; HV = 18

2)
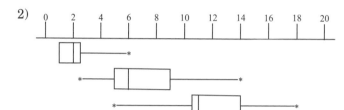

3) It has only one whisker because more than 25 percent of the categories of new car data displayed a required repair rate of 1 percent (which was the lowest percentage)

4) a. New: IQR = 1.5 percent, three to four years old,IQR = 4 percent, six years old: IQR = 5 percent

  b. Older cars evidence greater variability of reliability. This fact may be attributed to the quality of the product, the treatment the vehicle has received since its purchase, or a combination of both.

5) The new-car age group has the smallest percentage of required repairs and would, based on that criterion, be the most reliable,

6) The six-year-old cars have the largest median, and would, based on that criterion, be the least reliable.

7) Most of the detail of individual values is omitted in box plots, and we can therefore compare relative reliability ratings better. In addition, it is not possible to compare three sets of data on a back-to-back stem-and-leaf plot.

8) No. When you use a box plot to display information, only the five summary measures remain. Box plots do offer a way to focus on the relative positions of different sets of data for the purpose of making comparisons, and they are especially useful when data set are very large.

REFERENCES

Burrill, Gail. "Statistical Decision Making." *NCTM Student Math Notes*, May 1988.

Landau, Irwin, ed. "Frequency-of-Repair Records, 1982–87." *Consumer Reports* 53 (April 1988): 248–49.

Landwehr, James M., and Ann E. Watkins. *Exploring Data*. Palo Alto, Calif.: Dale Seymour Publications, 1987.

———. "Stem-and-Leaf Plots." *Mathematics Teacher* 78 (October 1985); 528–38.

Moser, James M., ed. *Teaching Quantitative Literacy: A Manual for Workshop Leaders*. Madison, Wis.: Wisconsin Department of Public Instruction, 1987.

Exploring Data with Box Plots

Sheet 1

The speed (in mph) of a sample of twenty-five cars checked by radar on an expressway is listed below:

1. Make a stem-and-leaf plot for these data.

2. Use your stem-and-leaf plot to find each of the following:

    a. The median speed

    b. The median of the lower half (the lower quartile)

    c. The median of the upper half (the upper quartile)

    d. The extremes (the lowest and highest values)

3. Mark dots for the median, quartiles, and extremes beneath the number line below question 4.

4. Draw a box between the two quartiles. Mark the median with a line vertically across the box. Draw two "whiskers" from the quartiles to the extremes.

Speed of Twenty-five Cars				
57	53	53	71	73
54	69	56	58	49
56	53	52	82	62
61	60	71	75	60
57	61	58	78	64

```
30      40      50      60      70      80      90
 |       |       |       |       |       |       |
```

5. About what percent of the speeds are—

    a. below the median? _____

    b. below the lower quartile? _____

    c. above the lower quartile? _____

    d. in the box? _____

    e. on each whisker? _____

6. Is one whisker longer than the other? What do you think this result means?

7. Why isn't the median in the center of the box?

8. If an officer is writing tickets to each driver in the sample whose speed is more than 70 mph, about what percentage of the drivers will be ticketed?

9. Write a description summarizing the information in the box plot.

Exploring Data with Box Plots

Sheet 2

Consumer Reports (Landau 1988) generated a reliability report designed to demonstrate how the "average" car ages. The data below give the percentage of problems requiring repairs in seventeen different categories for new cars, for cars three to four years old, and for cars six years old. The figures in this table are based on data obtained from over 544,000 responses to Consumer Union's annual questionnaire. The study included 252 cars for the model years 1982–1987.

How the "Average" Car Ages			
	Percentage of Cars Needing Repairs		
	New	3–4 Years Old	6 Years Old
Electrical system, chassis	6	14	18
Body integrity	6	9	11
Body hardware	5	9	12
Fuel system	3	11	17
Transmission, automatic	2	6	10
Transmission, manual	2	4	5
Body exterior, paint	2	6	11
Brakes	2	9	16
Air-conditioning	2	7	17
Suspension	1	8	11
Ignition system	1	5	8
Exhaust system	1	5	15
Engine mechanical	1	6	11
Engine cooling	1	5	12
Drive line	1	3	5
Clutch	1	6	13
Body exterior, rust	1	3	11

1. Find the median, the quartiles, and the extremes for each set (column) of data from activity sheet 2.

2. Follow the same procedure you used in activity sheet 1 to draw box plots for each of the three sets of data using the number line below.

```
   0     2     4     6     8    10    12    14    16    18    20
   |  |  |  |  |  |  |  |  |  |  |  |  |  |  |  |  |  |  |  |  |
```

Exploring Data with Box Plots   Sheet 3

3. How many whiskers does the box plot for new cars have? What do you think the missing whisker tells us about new-car data?

4. The length of the rectangle in a box plot is called the *interquartile range* (IQR).

    a. Compute the IQR for each data set.

    b. What does the IQR tell us about the variability of these data sets?

5. If the most reliable age group is the one with the smallest percentage of repairs, which age of car is the most reliable?

6. If the least reliable age group is the one with the largest median, which age of car is the least reliable?

7. Why do box plots furnish a better way to compare the relative reliability rating of the three age groups of cars than stem-and-leaf plots?

8. If you wanted to compare the reliability of the fuel systems for the three age groups of cars, would the box plot give you the information you needed? What does your answer indicate about what box plots tell us and what they do not tell us?

Is Central Park Warming?

Christine A. Franklin and Madhuri S. Mulekar

Many students in introductory statistics courses, in high schools as well as colleges and universities, find probability to be a difficult topic to understand and apply when presented in a formal mathematical way. Statistical concepts such as *p*-value, which involve probability, become even more difficult for students to comprehend. Students also have difficulty differentiating between the distribution of a variable and the sampling distribution of a statistic. Simulation is an alternative approach that may be used to demonstrate concepts involving probability prior to the formal introduction of such concepts. Quinn and Tomlinson (1999) give a lesson on simulating random variables to study the relationship between theoretical and experimental probabilities.

Here we present a simulation activity that is useful in demonstrating the concepts of hypothesis testing, *p*-value, and a sampling distribution. This activity provides an opportunity for students to develop a better understanding of random variables and their probability distributions (NCTM 2000).

BACKGROUND AND GOALS

This activity is useful in a non-calculus-based introductory statistics course in colleges and universities. It is also appropriate for high school students taking their first statistics course. Easily available temperature data is employed to demonstrate the use of simulation for decision-making through hypothesis testing. The major goals of this activity are to help students—

- Deal with a large data set;

- Simulate an event;

- Estimate probability of occurrence of an event using repeated simulations;

- Make a decision using estimated probability from a simulated sampling distribution; and

- Understand the difference between the distribution of a variable and the sampling distribution of a statistic.

The data on mean annual temperatures (in degrees Fahrenheit) recorded over the last century in New York City's Central Park is used in the activity. The data were extracted from the U.S. Historical Climatology Network (USHCN) database. (Also, from Franklin's homepage, **www.stat.uga.edu/people/faculty/chris-tine-franklin**, under Other Information, click on the "Centpark Dataset" link to download the Central Park

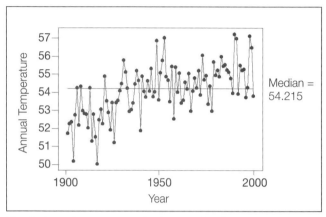

Fig. 1. Annual temperatures (in degrees F) in Central Park, 1901–2000

Table 1	
Numerical Summary of Annual Temperatures	
	Temp (°F)
n	100
Mean	54.097
Median	54.215
StDev	1.447
Min	50.040
Max	57.220
Q1	53.055
Q3	55.113

dataset.) Minitab, software commonly used in introductory statistics courses at colleges and universities, is used to demonstrate the simulation. Many high schools use such programs as well.

The temperature data for different sites are easily available on the Web. The National Climatic Data Center (**www.ncdc.noaa.gov/oa/ncdc.html**) is a particularly good place to look. Teachers may use local temperature data for this activity, providing more association for students.

ACTIVITY

The file CENTPARK.XLS, available via Franklin's homepage as described above, contains data on mean annual temperatures in Central Park for the years 1901 to 2000. Henceforth the mean annual temperatures

---

265

are referred to as annual temperatures. This data can be described as a time series, that is, the measurements are taken periodically over time. A time series plot, appropriate for investigating the behavior of temperatures over time, is a scatter plot in which one axis describes the time period over which the data are collected while the other axis describes the measurements, in this case temperatures. **Figure 1** shows such a time series plot of Central Park annual temperatures from years 1901 to 2000.

Figure 1 shows numerous fluctuations in the annual temperatures, but the annual temperatures at Central Park were comparatively lower during the first half of the century and higher during the latter part of the century. The numerical summaries in **table 1** show that the mean temperature in Central Park for the last century was about 54.1° F with standard deviation 1.4° F. The median annual temperature was 54.2° F; that is, during at least fifty years the annual temperatures were 54.2° F or higher. How do annual temperatures of the last century compare with the median temperature of the century?

Comparison of annual temperatures with median temperature (the horizontal line in **fig. 1**) shows that for most years early in the century the annual temperatures were below the median temperature of the century, whereas for most years during the latter part of the century the temperatures were above the median temperature. Suppose we look at the most recent three decades (the last thirty years of the century). How were annual temperatures relative to the median of 54.2° F during this time period? Of these thirty years, twenty-three years recorded annual temperatures above the median temperature. Could this reasonably be attributed to variation due to chance (or randomness) in a process such as weather, or are we experiencing an unusually large number of warm years?

To answer this question, one would need to know the likelihood of observing at least twenty-three out of thirty consecutive years with above-median annual temperatures. Let us simulate the situation assuming the temperatures from year to year are independent and estimate this probability. We will discuss the validity of this assumption later. If there is no trend, and the relatively warm weather of recent years is just part of the variability expected in the process of weather, then we can assume that the probability of observing an above-the-median annual temperature in a given year is equal to .5. This situation can be easily simulated using a coin or a computer program.

Simulation using a coin. Let "heads" represent a year with above-the-median annual temperature (i.e., annual temperature above 54.2° F) and "tails" represent a below-the-median temperature.

Toss a fair coin 30 times and note the number of heads in 30 tosses. This completes one simulation.

Repeat this process 100 times, recording the number of heads each time.

Count the number of simulations with twenty-three or more heads.

Estimate the probability of observing at least twenty-three years out of thirty with above-median annual temperatures as

$$\frac{\text{Number of simulations with 23 or more heads out of 30}}{100}$$

If the class or teacher does not have access to a computer, then the simulation using coins is useful. The teacher may have each student do the necessary number of simulations, and combine the results. For example, if the class has only 25 students, then each student would perform 4 simulations of 30 tosses each. Then the simulations from all the students in the class can be combined to give 100 simulations. However, simulation using coins is time-consuming and should be used only to develop the idea of simulating a situation. In the current technological world, simulation using programmable calculators or some software is strongly recommended.

Simulation using computer software. This simulation can be easily done using instructional statistical software. Here it is described using Minitab.

Open a blank worksheet in Minitab.
Click on Calc → Random Data → Integer

A window for Integer Distribution will open. Fill in the following information:

Generate 100 rows of data
Store in column(s): C1–C30
Minimum value: 0
Maximum value: 1

Click OK.

Notice that 100 rows and 30 columns are filled with 0s and 1s. Each row represents one simulation of thirty years, and 100 such rows represent 100 simulations of the situation. Define 0 = year with temperature below the median (similar to getting "tails" on a coin toss) and 1 = year with temperature above the median (similar to getting "heads" on a coin toss). The values 0 and 1 each have a .5 probability of occurrence. This is equivalent to assuming that the occurrence of a warmer or cooler year is due simply to random variation in the annual temperatures, that is, the probability that in a given year the temperature is above the median temperature is equal to the probability that the temperature observed is at most equal to the median temperature. This is the same as saying there is a probability of .5 that in a given year the temperatures will be above the median temperature. Each row with some combination of thirty 0s and 1s gives a different set of thirty years during which some annual temperatures

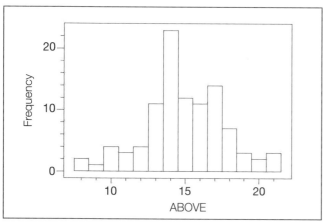

Fig. 2. Histogram of the variable ABOVE

Table 2

Summary Statistics for the Variable ABOVE	
	ABOVE (in years)
n	100
Mean	14.910
Median	15.000
StDev	2.697
Min	8.000
Max	21.000
Q1	13.250
Q3	17.000

are above the median temperature and some are below the median temperature.

Next compute the total number of years with annual temperatures above the median temperature in each thirty-year simulation and save the totals in column 31, labeling it "ABOVE":

Click on Calc → Row Statistics

A window titled Row Statistics will open. Fill in the following information:

Select Sum by clicking on the circle next to it. Input variables: C1–C30 Store results in: C31

Click OK.

The column C31 will be filled with the total of numbers in the first 30 columns. For example, in our simulation the first number in column 31 was 20. That means in the first simulation of thirty years, the annual temperatures were above the median annual temperature during twenty years and below the median annual temperature during the remaining ten years. The variable ABOVE is defined as the "number of years out of thirty with above-median temperature." The possible values that the variable ABOVE can take are

0, 1, 2, . . . , 30. This is a discrete random variable, since the outcomes are integers that can be counted.

The variable ABOVE (the count of 1s in a sample of 30) is a statistic calculated from each simulated sample. A *statistic* is a numerical summary calculated from a sample. The distribution of the values of a statistic obtained from the different possible samples is called a *sampling distribution*. What is the expected behavior of the sampling distribution of the variable ABOVE if $P(0) = P(1) = .5$? To study this, compute graphical and numerical summaries for generated simulations.

Figure 2 shows a histogram of values of the variable ABOVE generated from one such process of 100 simulations. These 100 simulated values for ABOVE give an estimate of the sampling distribution of ABOVE. Note that different sets of 100 simulations will result in different summary statistics and thereby different sampling distributions, but most of them should be similar to the one presented here.

The summary statistics (**table 2**) show that the sampling distribution of ABOVE is fairly symmetric and mound-shaped. There are no outliers or other unusual features. The values of ABOVE range from 8 to 21, with the mean approximately 14.9. The median is 15, and the middle 50 percent of values in the sampling distribution fall between 13.25 and 17.00.

The maximum = 21.00 shows that, in this particular set of 100 simulations, none of the simulations resulted in 23 or more years out of 30 with annual temperatures above the median temperature. This information can be interpreted in the following way:

If for any given year the probability of observing an above-median temperature is .5, then the probability of observing twenty-three or more years out of thirty with above-median temperatures as estimated from this simulation is less than .01 (less than one out of 100).

This simulated probability is called a *p*-value. This probability is very small, indicating that observing twenty-three out of thirty years with temperatures that are above the median is not very likely because of the random variability in the weather process such as temperatures. This set of 100 simulations provides strong evidence that the last thirty years in Central Park are warmer than would be expected; that is, the hypothesis of an unusually large number of warm years is strongly supported.

The student versions of software typically have a limitation on the spreadsheet size. Therefore, this activity is described by generating 100 rows of data. If the software allows, generate a larger number of simulations to better gauge the convergence of the probability in the

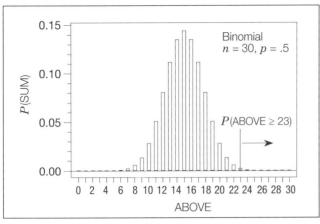

Fig. 3. Mathematical distribution of the variable ABOVE

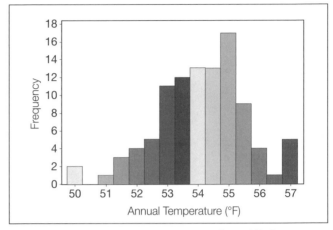

Fig. 4. Histogram of annual temperatures at Central Park

long run. For example, we generated 1000 rows of data (equivalent to 1000 simulations). Out of 1000 simulations, 4 simulations resulted in twenty-three or more years out of thirty with annual temperatures above the median temperature. In other words, the probability of observing twenty-three or more years out of thirty with above-median temperatures as estimated from these simulations is .004. Repeating the process 4 more times (1000 simulations each time) resulted in 1, 2, 3, and 2 simulations with twenty-three or more years out of thirty with annual temperatures above the median temperature. Thus, the probability of observing twenty-three or more years out of thirty with above-median temperatures as estimated from these 4 runs of 1000 simulations is .001, .002, .003, and .002, respectively.

MATHEMATICAL SAMPLING DISTRIBUTION

How close is the simulated probability to the mathematically derived probability? If p = the probability of above-median annual temperature in any given year, then theoretically the variable ABOVE follows a binomial distribution with $n = 30$ and $p = .5$. In other words, the theoretical sampling distribution of the variable ABOVE is binomial with $n = 30$ and $p = .5$ (**fig. 3**). Computation of binomial probability is discussed in almost all introductory statistics books, such as Peck, Olson, and Devore (2005) and Yates, Moore, and Starnes (2003).

We are interested in testing the hypothesis $H_0: p = .5$ (i.e., annual temperature variation in Central Park is due to chance or randomness) versus $H_a: p > .5$ (i.e., higher annual temperatures in Central Park occurred in an unusually large proportion of years). Recall that during the last thirty years of the twentieth century, twenty-three years were observed with temperatures that were above the median.

The exact probability of at least twenty-three out of thirty years with temperatures above the median can be computed using the binomial distribution:

$P(\text{ABOVE} \geq 23 \mid n = 30, p = .5)$

$= P(\text{ABOVE} = 23) + \cdots + P(\text{ABOVE} = 30)$

$= \binom{30}{23} 0.5^{23} (1 - 0.5)^7 + \ldots + \binom{30}{23} 0.5^{30} (1 - 0.5)^0$

$\approx 0.0019 + \cdots + 0.0000$

≈ 0.0026

The p-value, that is, the mathematical probability of observing twenty-three or more years with above-median annual temperatures out of thirty, if there is no trend in annual temperatures, is .0026. The estimated probability of "< .01" found from the earlier simulated sampling distribution (**fig. 2**) and those calculated from 1000 simulations (.004, .001, .002, .003, and .002) are very close to the mathematical probability. Note that different simulated sampling distributions will result in different estimated probabilities, but most of the simulated probabilities should be very close to the mathematical probability.

FREQUENTLY ASKED QUESTIONS

During the activity, students typically raise the following questions:

Will the temperatures in Central Park continue to rise? Looking back at the Central Park data, it seems that the higher-than-expected number of years with warmer temperatures during the last thirty years may not be due to random variation. Does it mean the temperatures in Central Park will keep on rising? This may not necessarily be true. The conclusion about the higher-than-expected number of years with warmer temperatures during the last thirty years is applied only to the time period over which data were collected. The same pattern may or may not continue in the future.

Should we compare annual temperatures with the mean or the median? Prepare a histogram to study distribution of annual temperatures. If the distribution is skewed and/or has outliers, then compare annual temperatures with the median annual temperature. If the distribution is fairly symmetric and there are no outliers, then compare annual temperatures with either the mean or the median annual temperature. This activity was described using the median because the distribution of annual temperatures at Central Park is slightly skewed to the left (see **fig. 4**). But there are no outliers; and as seen from the values of mean and median (**table 1**), the skewness of the distribution is not prominent enough to affect the mean of the distribution significantly. Therefore, either mean or median could be used when doing this activity with Central Park temperatures over the last century.

In general, an advantage of using the median is that the probability of exceeding it is $p = .5$ for the complete dataset. If one uses the sample mean as the critical threshold, one would also need to be convinced that the distribution is symmetric in order to use the mathematical justification that the distribution should be binomial with some n and p, the probability of success being equal to .5.

Why are we using the last 30 years of data? Thirty is not a magic number. We can achieve similar results using twenty-five or thirty-five years of data. We are using data from the later part of the century, since the plot clearly shows a difference in temperatures between the first and the last parts of the century. Also, the last thirty years are closer to the present time than is the early part of twentieth century.

Can we really assume that temperatures are independent? Actually, daily temperatures at any given location are not independent. However, we are looking at the mean temperatures over each year. Averaging over all the days of each year lowers the year-to-year dependence (as measured by autocorrelation) among temperatures considerably, compared to the day-to-day or month-to-month dependence. A more detailed analysis of the data shows that any dependence in the temperatures may be due to the trend itself. Therefore, the assumption of independence, although not exactly true, is fairly reasonable.

What if there are unusually cooler (or low) temperatures? The temperatures in different parts of the United States and in different parts of the world show different patterns. For example, Newnan, Georgia, which is in the southeastern part of the United States, experienced cooler years in the later part of the last century. (The temperature data for Newnan is available at **www.stat.uga.edu/faculty**. Click on Franklin's homepage, then follow the link to the Newnan dataset.) This simulation activity can be easily extended to decide if there are unusually lower temperatures in this part of Georgia. Define hypotheses H_o: $p = .5$ (i.e., variation in annual temperatures is due to random variation) versus H_a: $p > .5$ (i.e., the proportion of years with lower annual temperatures is unusually high), where

p = probability of observing an annual temperature below the median annual temperature in a given year. Then carry out this activity by counting the number of years with below-median annual temperatures, or use the same definition of "heads" and check that the observed number of successes is too low.

CONCLUSIONS

This is a simple activity that requires no special equipment or supplies. It can be conducted with temperatures at any location and takes only one class period to complete, even with discussion. If all students have access to computers, let each student simulate a separate sampling distribution. Students can then see how each simulation leads to a slightly different simulated sampling distribution. All the simulated sampling distributions, however, generally give results similar to the theoretical sampling distribution. If no computer lab is available for students, the teacher may still perform the simulation for students, projecting the simulation for the class to view. A class discussion can follow after students have viewed results of the simulated sampling distribution and the implications of the simulation upon the temperature model.

REFERENCES

National Council of Teachers of Mathematics (NCTM). *Curriculum and Evaluation Standards for School Mathematics*. Reston, Va.: NCTM, 2000.

Peck, Roxy, Chris Olson, and Jay L. Devore. *Introduction to Statistics and Data Analysis*. 2nd ed. Belmont, Calif.: Thomson–Brooks/Cole, 2005.

Quinn, Robert J., and Stephen Tomlinson. "Random Variables: Simulations and Surprising Connections." *Mathematics Teacher* 92 (January 1999): 4–9.

Yates, Daniel S., David S. Moore, and Daren S. Starnes. *The Practice of Statistics*. 2nd ed. New York: W. H. Freeman, 2003.

Is Central Park Warming?

Central Park, located in New York City, is the first urban landscaped park in the United States. It's two and a half miles long and half a mile wide, with an internal loop of approximately six miles. Conceived in the salons of wealthy New Yorkers in the early 1850s, the park project spanned more than a decade and cost the city $10 million.

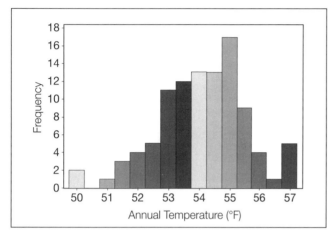

Fig. 1. Annual temperatures (in degrees F) in Central Park, 1901–2000

Table 1	
Numerical Summary of Annual Temperatures	
	Temp (°F)
n	100
Mean	54.097
Median	54.215
StDev	1.447
Min	50.040
Max	57.220
Q1	53.055
Q3	55.113

Figure 1 and **table 1** show, respectively, the distribution of the mean annual temperatures (in °F) recorded at Central Park over the last century and the summary statistics for the data set. In the activity, the mean annual temperatures will be referred to as simply "temperatures."

1. Describe the distribution of temperatures (**fig. 1**), focusing on center, spread, shape, and any unusual characteristics.

2. See the time plot of temperatures in **figure 2** below. Comment on the trend you observe.

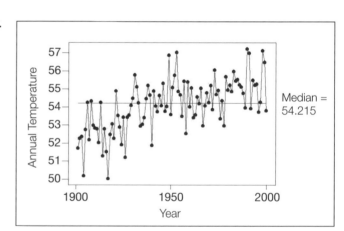

Median = 54.215

Fig. 2. Histogram of the variable ABOVE

3. Compare temperatures over the 100 years with the median temperature (shown by the line in **fig. 2**), and comment on your observations.

4. Of the last 30 years, how many years had recorded temperatures above the median line at 54.215?

Is Central Park Warming?

Sheet 2

5. The following question is raised: "Are we observing an unusually large number of warm years, or is the relatively warm weather of recent years (23 of 30 above the median year) simply due to expected variability in a natural process such as weather?"

A given year has two possible outcomes:

        a. Above-median temperature   *b*. Below-median temperature

Let p = Pr(above-median temperature in any given year)
        $1 - p$ = Pr(below-median temperature in any given year)

What value of p is expected if there is no warming trend and the variability in temperatures is expected variability in weather patterns? _____

_____

_____

Express the above question as a null and alternative hypothesis.

Hint: The null hypothesis is expressed in terms of what's observed being due to expected or typical variability; the alternative hypothesis is expressed in terms of what's observed being due to some effect other than expected variability.

H_o: _____

H_a: _____

6. One way to answer the question in (5) is to explore the temperatures in Central Park for the 20th century. Another approach is to simulate the expected behavior of temperatures on the assumption that there is no trend toward warming—that is, the expected behavior of temperatures over time given typical variability in the temperatures from one year to the next.

Simulation Using a Coin:

Let   Heads = a year with above-the-median annual temperature
        Tails = a year with below-the-median temperature

- Toss a fair coin 30 times. On the following table, note the number of heads in 30 tosses in the column labeled "ABOVE." This completes one simulation. Repeat this procedure 100 times, recording the number of heads each time. Collect data from classmates to complete this table.

- After completing the entire table, count the number of simulations with 23 or more heads. _____

- Use the following formula to estimate the probability of observing at least 23 years of 30 with above-median annual temperatures:

$$\frac{\text{Number of simulations with 23 or more heads out of 30}}{100}$$

Simulation Using Minitab:

- Open a blank worksheet in Minitab.
- Click on Calc → Random Data → Integer.

    A window for Integer Distribution will open.

- Fill in the following information in the Integer Distribution window:

    o   Generate 100 rows of data
    o   Store in column(s): C1–C30
    o   Minimum value: 0
    o   Maximum value: 1

- Click OK. Notice that 100 rows and 30 columns are filled with 0s and 1s.

Is Central Park Warming?

Sim #	ABOVE	Sim #	ABOVE	Sim #	ABOVE	Sim #	ABOVE	Sim #	ABOVE
1		21		41		61		81	
2		22		42		62		82	
3		23		43		63		83	
4		24		44		64		84	
5		25		45		65		85	
6		26		46		66		86	
7		27		47		67		87	
8		28		48		68		88	
9		29		49		69		89	
10		30		50		70		90	
11		31		51		71		91	
12		32		52		72		92	
13		33		53		73		93	
14		34		54		74		94	
15		35		55		75		95	
16		36		56		76		96	
17		37		57		77		97	
18		38		58		78		98	
19		39		59		79		99	
20		40		60		80		100	

Is Central Park Warming?

Sheet 4

Each row represents one simulation of 30 years, and 100 such rows represent 100 simulations of the situation. Next compute the total number of years with annual temperatures above the median temperature in each 30-year simulation as follows and save the totals in column 31, labeling it "ABOVE." This sum will represent the number of years that have above-median temperatures.

- Click on Cal → Row Statistics.

  A window titled Row Statistics will open.

- Fill in the following information in the Row Statistics window:

  o Select Sum by clicking on the circle next to it.
  o Input variables: C1–C30
  o Store results in: C31

- Click OK.

- Count the number of simulations with ABOVE ≥ 23.

- Use the following expression to estimate the probability of observing at least 23 years of 30 with above-median annual temperatures:

$$\frac{\text{Number of simulations with ABOVE} \geq 23}{100} = \frac{}{100} =$$

7. The variable ABOVE is a discrete random variable and will take values 0, 1, 2, . . ., 30. It is the outcome of a simulation, and hence a statistic (summary count from each simulated sample). The distribution of the 100 simulated statistics is a simulated sampling distribution. This sampling distribution represents counts or number of years with temperatures above median under the assumption of random or expected variability. Let us explore how the sampling distribution of the discrete variable ABOVE is expected to behave if there is no warming trend.

- Make a histogram of 100 values of ABOVE. Describe the distribution, focusing on center, spread, shape, and any unusual characteristics.

- Obtain summary statistics for the variable ABOVE.

- Compare the sampling distribution of ABOVE from your simulation with the data distribution in **figure 3**.

Results from one such set of 100 simulations using Minitab are given below for your use.

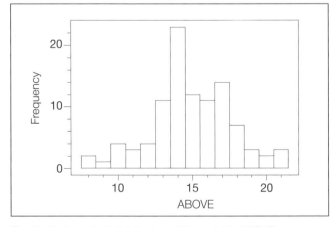

Fig. 3. Mathematical distribution of the variable ABOVE

Table 3	
Summary Statistics for the Variable ABOVE	
	ABOVE (in years)
n	100
Mean	14.910
Median	15.000
StDev	2.697
Min	8.000
Max	21.000
Q1	13.250
Q3	17.000

Is Central Park Warming?

Sheet 5

8. To answer the following questions, use both your simulated results and the results of simulation given in **figure 3** and **table 3**.

   - What would be the expected median value for ABOVE if the behavior of the yearly temperatures in the past 30 years is due to random variation?_____

   - What would be the expected mean value for ABOVE if the behavior of the yearly temperatures in the past 30 years is due to random variation? _____

   - What would be the expected distribution shape for the variable ABOVE? _____

   - Would the value of ABOVE = 23 be considered an outlier? Justify your answer. _____

     _____

     _____

   - What would be the simulated probability (proportion) for ABOVE values 23 or higher?
     [Use Stat → Tables → Tally. Then input C31 → OK] _____

9. On the basis of your answers to (8), do you believe strong evidence exists to support that a warming trend is occurring? Justify your answer. _____

   _____

   _____

10. What probability distribution would we use to model the simulated sampling distribution of ABOVE? _____

    _____

11. Use this probability distribution (10) to find the following:

    The expected value (or mean) of ABOVE:_____

    The standard deviation of ABOVE: _____

    $Pr(\text{ABOVE} \geq 23) =$ _____

12. Compare the values from (11) with the values you obtained from your simulated sampling distribution in (8).

13. What assumptions were necessary to calculate probability by using the distribution described in (10) and (11)?

    _____

    _____

    Are these assumptions reasonable? Discuss your answer._____

    _____

    _____

    _____

    _____

    _____

Using Research Projects to Help Develop High School Students' Statistical Thinking

Randall E. Groth and Nancy N. Powell

Statistics plays a key role in shaping policy in a democratic society, so statistical literacy is essential for all citizens to keep a democratic government strong (Wallman 1993). However, fostering statistical thinking is a complex endeavor. We ultimately need to engage students in all phases of the investigative cycle of statistics, including data gathering, data analysis, and inference.

This article describes two projects in which we attempted to help Advanced Placement (AP) statistics students become proficient at moving through the investigative cycle. Although the projects were implemented in an AP class, they could be included as part of any class in which linear equations and best-fit lines are studied. As we describe the two projects, we also discuss some of our reflections on the extent to which the projects helped our students become more proficient. In the first project, the students focused on the data analysis and inference phases. In the second project, the students had the opportunity to experience all phases of the cycle.

PROJECT 1: ANALYZING AND DRAWING CONCLUSIONS FROM EXISTING DATA FILES

Project 1, shown in **figure 1**, was assigned during the first semester of the school year. At the time, students were using their textbook (Yates, Moore, and McCabe 1999) to study concepts related to correlation. Project 1 helped students develop deeper understandings of some of the concepts that they were studying, and it helped them become comfortable using the computer program Fathom (Finzer, Erickson, and Binker 2001). We saw Fathom as a powerful tool for helping formulate and investigate statistical conjectures, since it allows users to quickly import data, calculate summary statistics, test hypotheses, and build graphs and tables. Students had some freedom to choose the problem that they would study; we encouraged them to explore some of the data files contained in Fathom. Because Fathom comes packaged with existing data files, we were able to move the focus of the project away from the planning and data-collection parts of the investigative cycle and toward the analysis and conclusion parts of the cycle.

In assessing students' work on the first project, we gained several important insights about students' thinking while they engaged in analyzing data. These insights were valuable for guiding future instruction. We found that some students tried to make the inter-cepts of the regression line have meaning even when they did not. Some students tried to make sense of a negative regression-line intercept within the context of plotting SAT scores versus grade-point average.

Perhaps partially because of the stated requirements of the project, students attempted to force meaning on the intercept when it really had no useful physical interpretation. Some students also did not seem to realize that it was dangerous to force meaning on intercepts, or any points, that lie well outside the cluster of points used to produce the regression line. After identifying these gaps in students' understanding, we were able to address them in future instruction.

Although many students identified the slope of the regression line as a rate of change, quite a few had difficulty giving the specific units of the rate of change. In addition, some students confused the slope of the regression line with the correlation coefficient and stated that the slope of the regression line indicated the strength of the linear relationship between the two variables. Finally, we noticed that some students struggled with the idea that two variables can have a correlation close to 1 or –1, yet not cause each other. Identifying these problems in thinking proved to be an important first step in dispelling them.

Since the assignment involved completing a written report, we helped students learn to communicate the results of statistical analysis. In assessing the students' work on the projects, we noticed that several students used the formal term outlier to describe "unusual" data points in a bivariate situation, even though they had only studied formal procedures for identifying outliers in univariate situations. We then had a class discussion of the importance of being careful in using formal terms when reporting statistical results. We also noticed that students tended to say in their reports that they were using the regression line to "find values," when a more precise wording would have stated that the regression line can be used to "predict approximate values." These imperfections in communication allowed us to emphasize that even if statistical analysis is correctly done and the conclusions are appropriate, the study is of limited value if the data-analysis process and the conclusions are not communicated properly.

Project

Choose one pair of variables from the "Data in Depth" folder in Fathom that interests you and that seem to be correlated. Write a short paper organized in the following manner:

Introduction (10 pts.)
Why would someone be interested in investigating the relationship between the two variables you have chosen?

Background (20 pts.)
- Does one of the variables you have chosen actually influence the other? If you think so, make an argument that one does influence the other. If you don't think so, explain why the variables you have chosen are correlated, but yet one does not "cause" the other.
- What are some interesting questions you have that will be answered by analyzing your data set? Include questions that can be answered by finding the equation of the regression line, the slope and intercepts of the regression line, and the strength of the linear relationship between the variables. You will be answering these questions in the "Results" section.

Method (20 pts.)
- What kinds of technology did you use to investigate the relationship between the two variables (e.g., Fathom, TI-83, other)? What did the technology do for you (or what did you make it do for you)?
- What are some other tools or methods (if any) you decided to use to investigate the relationship?

Data Analysis (20 pts.)
- Copy and paste all Fathom output into this section. Once you have your Fathom output pasted into a Word document, you can move the output around in the document—even on a computer that does not have Fathom. So, you might want to copy and paste your output before writing the other sections of this report.
- Does the data set have any "unusual" values? What makes them unusual? Are you choosing to throw out any of the unusual values before you plot a regression line? If so, justify the decision to throw out unusual values.
- What is the equation of the regression line? What are its slope and intercept?
- What is the correlation coefficient, r?
- What graphs did you analyze to investigate the relationship?

Results (20 pts.)
- How strong is the relationship between the two variables? How do you know how strong the relationship is? Are they positively or negatively associated? Now that you know the strength of the relationship, what questions can you answer? What are the answers to these questions, and how did you find them?
- What valuable information do the slope and intercepts of the regression line give you? What questions can you answer now that you know the slope and intercepts of the regression line? What are the answers to these questions, and how did you find them?
- What valuable information does the regression line give you? What questions can you answer now you that know the regression line and its equation? What are the answers to these questions, and how did you find them?
- Discuss the answers to any of the other questions you wrote in the "Background" section.

Conclusion (10 pts.)
What did you learn (e.g., about statistics, technology, baseball, cars, buildings) that you didn't know before? Is there anything you would do differently if you were to write the paper again? What other interesting studies could be done that are related to what you did in this paper?

Fig. 1. Project involving the analysis of existing data

PROJECT 2: EXPERIENCING ALL PHASES OF THE INVESTIGATIVE CYCLE

As a final project for our AP statistics course, students designed their own statistical projects and posters. We instructed students to follow the guidelines given by the American Statistical Association (ASA) at **www.amstat.org/education/posterprojects/index.cfm**, and we allowed them to do joint projects with classmates who had common interests. The ASA guidelines helped students experience all phases of the investigative cycle. We set aside approximately two weeks to allow everyone in class to identify a quantifiable problem of interest, make a plan for investigating it, gather data, analyze the data, and draw conclusions.

As students began to select their problems, several interesting ideas for projects emerged. One student wished to learn whether a relationship existed between a student's grade level and the number of steps that he or she took in a day. A pair of students decided to investigate the relationship between the types of music that students listened to and their grade-point averages. Another pair of students wondered how the distribution of types of cars driven in town compared with the distribution of types of cars driven in the entire United States. Yet another pair became interested in the relationship between brain size and intelligence. A larger group of students wanted to determine whether boys could generally throw a softball faster than girls. Another group decided to compare the blood pressures of athletes to those of nonathletes. Still another group decided to investigate the relationship between the stock price of a company that produced fatty foods and the amount of obesity in the United States. Even though some groups took one or two class periods to settle on a question that interested them, allowing the students to identify their own problems to investigate helped them develop feelings of ownership of the problems at the outset.

Because of the diversity of the questions posed by members of our class, students chose a number of different types of plans for completing the projects. The students who wanted to examine the relationship between the types of music and grade-point averages for students at the high school decided that a survey would be the most efficient way of gathering information for their project. The students who were investigating the number of steps taken in a day, the types of cars driven in town, softball throwing, and blood pressure all chose observational studies as their mode of inquiry. Students investigating the questions about blood pressure, brain size, and stock prices of a company chose to use preexisting data to help answer the questions. The diversity of modes of inquiry used by our students in answering the questions that they had posed reflected some of the methods of data production that they had studied during the AP statistics course.

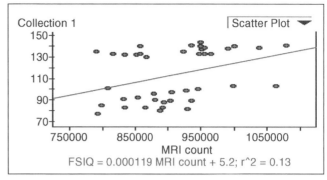

Fig. 2. Fathom output comparing IQ scores to brain size (measured by MRIcount)

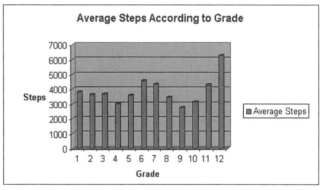

Fig. 3. Excel output comparing number of steps taken in a day to grade level

The students who gathered their own data came to appreciate that gathering data for a statistical study can be a complex endeavor. The pair using a survey to gather information realized that the manner in which they asked the survey questions could easily influence the manner in which people responded. The student who investigated the relationship between the number of steps that a student takes in a day and his or her grade level used an electronic device called a Digiwalker, which we provided. However, he quickly discovered that the device reported inaccurate data. The group investigating the softball-throwing question needed to learn how to operate a radar gun that they borrowed from a policeman. The pair of students gathering information about blood pressure discovered that finding a truly random sample of students was impossible, since some teachers would not let students out of class to have their blood pressure taken. The complicated and often imperfect nature of gathering data was conveyed to students in a manner that was far more memorable than conventional textbook instruction.

Once the data had been gathered, many of our students used both Fathom and Excel to help analyze the data. Using Fathom and Excel enabled students to quickly produce such graphical displays as scatterplots, histograms, and dot plots. Some of the graphs that our students created are shown in **figures 2** and **3**. More student work is displayed online at **www.district87.org/staff/powelln/Statistics /EndofYearProject/Default.htm**. The software-produced graphical displays served as tools to help students think about the validity of the conjectures that they made at the start of the project and sometimes led students to pose new questions that they had not asked at the outset. Fathom also allowed students to produce summary statistics, such as the mean, median, mode, and correlation coefficients, as needed. These summary statistics played a role similar to that played by the graphs, since they led our students to think about their initial conjectures and formulate new conjectures that were based on the data. Project 1's focus on using Fathom during the analysis phase of the investigative cycle certainly paid dividends when our students engaged in this end-of-year project.

To wrap up the project, each group presented its findings to the entire class. We believed that evaluating and discussing the validity of each group's conclusions were important. We designed a written feedback form, shown in **figure 4**, for students to fill out to critique each presentation. We also encouraged students to ask questions at the end of each presentation. Although students were initially hesitant to ask questions about the work of their classmates, several good class discussions did eventually take place as presentations proceeded. For instance, the group that was trying to determine whether boys threw softballs faster than girls did used a two-sample t-test to analyze the data that they had collected. The class then discussed whether that particular test was appropriate for the given situation. At the end of its presentation, another group stated that the increasing consumption of a certain fatty food caused the increasing instances of obesity in the United States. The class challenged this conclusion, since some students argued that several other factors could contribute to increasing obesity. The group investigating the relationship between types of music that students listened to and their grade-point average was also questioned. Some students asked why the group listed the various types of music on the questionnaire instead of letting the students fill in the types of music they enjoyed. The questions and challenges from classmates helped students in each of these groups reflect on and evaluate the various phases of the journeys they had taken through the investigative cycle.

CONCLUSION

As noted at the beginning of this article, helping students develop statistical thinking is extremely challenging. In addition to helping students master the mathematics needed for the analysis and inference phases of the investigative cycle, teachers need to help students master some of the nonmathematical elements of the cycle involved in identifying a problem, creating a plan of attack, and gathering necessary data. Our strategy identified and addressed some of the difficulties students had in the data analysis and inference phases of the cycle in one project and then immersed

them in all phases of the investigative cycle in a later project. We hope that the projects and experiences that we have described will add to the dialogue concerning how best to develop and foster statistical thinking.

REFERENCES

Finzer, William, Tim Erickson, and Jill Binker. Fathom. Emeryville, Calif.: Key Curriculum Press, 2001.

Wallman, Katherine K. "Enhancing Statistical Literacy: Enriching Our Society." *Journal of the American Statistical Association* 88 (March 1993): 1–8.

Yates, Daniel S., David S. Moore, and George P. McCabe. *The Practice of Statistics: TI-83 Graphing Calculator Enhanced.* New York: W. H. Freeman & Co., 1999.

Audience Evaluation Form

Evaluator's first initial and last name: _____

Group topic/Name of presentation: _____

Team members: _____
Write first name, last initial for each member

(Answer the questions below. You may use the back for more room! Then rate each one in the table below.)

1. Did you understand the topic being studied? Was the question that the group studied interesting?

2. *a.* Was the presentation informative/creative?

   b. Was the presentation done professionally?

3. Was the use of visuals/technology (visual aids such as graphs, posters, Web pages, PowerPoint slides)—

   a. correct, and

   b. did they add to the understanding of the material presented?

4. *a.* Were the data-gathering methods appropriate and reliable? Clearly explained?

   b. Did the group's statistical analysis justify and support the group's conclusions?

   c. Did the statistical analysis have any errors or omissions?

5. What questions or comments do you still have for the group after their presentation?

Criterion	Awesome	Really Good	Okay	Good Start	Points Awarded
Question studied understood? #1	4	3	2	1	_____
Presentation: professional, informative, and creative? #2	4	3	2	1	_____
Use of technology/visual aids effective? #3	4	3	2	1	_____
Statistics correct/reliable? #4	4	3	2	1	_____
Overall rating	4	3	2	1	_____
				Total points awarded	_____

Fig. 4 Audience evaluation form for project 2

Area Models and Expected Values

Glenda Lappan, Elizabeth Phillips, William M. Fitzgerald, and Mary J. Winter

TEACHER'S GUIDE

Introduction: In the Activities section of the March 1987 *Mathematics Teacher* the authors presented a model based on area for analyzing games of chance that involved dependent events. In this set of activities we again use the area model but extend the questions to include expected value. *Expected value* is the average payoff in points over a very long run of trials. Students have a natural affinity for games that involve chance. These activities build on this interest to reinforce understanding of dependent probabilities and to introduce the concept of expected value in an experimental way. The activities provide the motivation and concrete foundation for a more formal study of these ideas at a later stage.

The first activity introduces the concept of expected value by analyzing the average points per spin on a variety of spinners over a given number of trials. In the last three activities, the area model is used to determine the probabilities of dependent events, and then the expected value is calculated.

The activities are designed to be used with minimal instruction. However, a richer experience is obtained if for each activity the teacher launches the challenge, discusses some preliminary guesses about the outcomes, and then allows students to explore the problem in small groups with a final whole-class summary of results and generalizations.

Grade levels: 7–12

Materials: Copies of the activity sheets for each student, a set of transparencies for class discussion, and bobby pins or paper clips to use as spinners

Objective: To practice area models in analyzing compound situations and calculating the expected value or long-term average

Prerequisites: Students should have some familiarity with the topic of probability, including the following:

1. The idea that the probability of a particular event *A* occurring is

$$P(A) = \frac{\text{The number of ways } A \text{ can occur}}{\text{The total number of possible events}}.$$

2. The idea that the notation $P(B)$ means "the probability that event B will occur,"

3. Some facility with using the area model to determine probabilities of dependent events (See Lappan et al. 1987.)

DIRECTIONS

Distribute the activity sheets one at a time. Introduce these activities by posing some simple problems involving probabilities, such as flipping a coin, tossing a die, or drawing colored marbles from a bag, to review the definition of probability.

On sheet 1 the probability of spinning and obtaining a particular number is easily determined by subdividing the circular spinner into congruent sectors. If students have never computed experimental probabilities, this is a good time to instruct them to make the spinners, spin each twenty times, record their results, and determine the experimental probabilities.

To calculate the expected value or long-term average, we employ the definition of probability and equivalent fractions. For example, in problem 1, $P(5) = 1/4$, and $P(-2) = 3/4$. Since $1/4 = 25/100$, we would expect the number 5 to occur 25 times in 100 spins. Similarly, since $3/4 = 75/100$, we expect the number -2 to occur 75 times. At the end of 100 spins, the total is

$$25 \times 5 + 75 \times (-2) = -25 \text{ points.}$$

The average points for 100 spins is $-25/100$, or -0.25 points per spin. The average points per spin is called the expected value.

In each problem the number of spins is different. The number of spins is a multiple of the denominations of the probabilities of the events, making it easier to calculate the expected value. The expected value is independent of the number of trials. For example, in problem 1 you should also ask the students to calculate the expected value for 136 spins:

$P(5) = 1/4 = 34/136$, so we expect 5 to occur 34 times for 170 points.

$P(-2) = 3/4 = 102/136$, so we expect -2 to occur 102 times for -204 points.

Thus the total number of points is -34 and the expected value is $-34/136$, or -0.25 points per spin.

In general, for a situation with two events A and B, the expected value is

$$\frac{P(A) \times (\text{Total Trials} \times \text{ pts. for } A) + P(B) \times (\text{Total Trials} \times \text{ pts. for } B))}{\text{Total Trials}}$$
$$= (P(A) \times (\text{pts. for } A) + P(B) \times (\text{pts. for } B).$$

More advanced students may discover this formula, but for the majority of students simply using the definition of probability is the best way to determine expected value.

Before doing the problems on sheet 2 you should review the area model for analyzing dependent events. The spinner from problem 1 on sheet 1 can be used. Change the directions to read "add the numbers you obtain by spinning twice." Ask the questions, "What are all the possible sums?" and "What are the probabilities for obtaining these sums?"

Use the area of the grid that has been subdivided into sixteen congruent squares to represent the probability of 1. Separate the area of the grid vertically to represent the probabilities of 1/4 and 3/4 corresponding to obtaining 5 or –2 on the first spin (**fig. 1**). Separate the area of the grid horizontally to represent the probabilities of 1/4 and 3/4 corresponding to obtaining 5 or –2 on the second spin (**fig. 2**). Analyze the grid to find the probabilities of obtaining the sums –4, 3, and 10: $P(-4) = 9/16$, $P(3) = 3/16 \times 2 = 6/16$, $P(10) = 1/16$.

The sum of –4 is the most likely to occur.

After introducing the game on sheet 2, have the students predict which outcome they think is the most likely to occur. Most students will guess the score of 1 and are surprised to find that the scored of 0 is the most likely and the score of 1 is the least likely. However, the expected value is 0.96. Many students will enjoy knowing that this problem is identical to simulating a one-and-one free throw situation for a player with a 60 percent free throw shooting average. An extension problem would be to analyze the outcome for a two-shot foul for a 60-percent free throw shooter. The analysis would be as follows (**fig. 3**):

$$P(2 \text{ pts.}) = 0.36$$
$$P(1 \text{ pt.}) = 0.48$$
$$P(0 \text{ pts.}) = 0.16$$

The expected value is $P(2 \text{ pts.}) \times 2 + P(1 \text{ pt.}) \times 2 + P(0 \text{ pts.}) \times 2 = 0.72 + 0.48$, or 1.20 points per turn.

On sheet 3 the students will need help determining all the arrangements for placing the coins in two cans. Each arrangement has a different expected value. This problem is easily accessible using the area grids to analyze the dependent events of selecting a can and then selecting a coin from the can. Some students may need help in subdividing the grids. A 6×6 grid works nicely. The students are expected to calculate the expected value for 100 trials. This is the first time the number of trials is not a common multiple of all the denominators.

The game on sheet 4 should be played first to get an idea of the outcomes before analyzing the theoretical possibilities. Note that we are not actually computing an expected value for the game but expected total

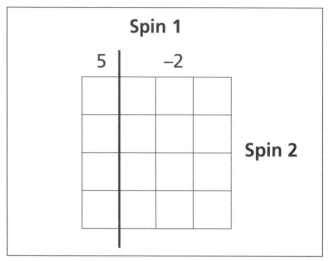

Fig. 1

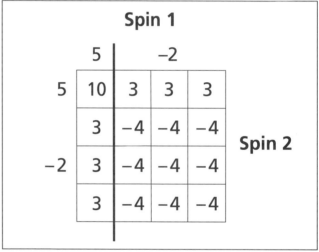

Fig. 2

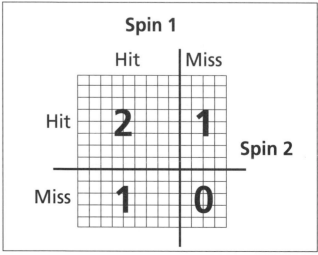

Fig. 3

points for each player. However, the game can be analyzed for an expected value if we ask "What is the average number of points scored in a round?"

In the process of doing these activities, the students may discover that the probability of event *A* followed by event *B* occurring is $P(A) \times P(B)$. If this discovery occurs, use the area model for multiplying fractions to support this conjecture.

SOLUTIONS:

Sheet 1:

1) $P(5) = 1/4$; -2 occurs 75 times for -150 points; the expected total points is -25; the expected value is -0.25 points per spin.

2) $P(5) = 1/3$, $P(10) = 1/3$, $P(-12) = 1/3$; 5 occurs 20 times for 100 points, 10 occurs 20 times for 200 points, -12 occurs 20 times for -240 points; the expected total points is 60; the expected value is 1 point per spin.

3) $P(-1) = 5/12$, $P(2) = 1/6$, $P(3) = 1/4$, $P(-4) = 1/6$; the expected total points in 72 spins is 0 points and the expected value is 0 points per spin.

4) For 240 spins the expected value for each spinner is exactly the same as computed in problems 1, 2, and 3.

Sheet 2:

1) and 2) Answers will vary.

3) See figure 4.

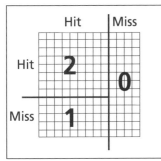

$P(0 \text{ pts. scored}) = 40/100$

$P(1 \text{ pt. scored}) = 24/100$

$P(2 \text{ pts. scored}) = 36/100$

Fig. 4

In 100 turns we expect 0 points to occur 40 times for 0 points, 1 point to occur 24 times for 24 points, and 2 points to occur 36 times for 72 points. The total expected points is 96, and the expected value is 0.96 points per turn.

4) For a 20-percent "hit" area: $P(0 \text{ pts.}) = 80/100$, $P(1 \text{ pt.}) = 16/100$, $P(2 \text{ pts.}) = 4/100$. The expected value is 0.24 points per turn. For a 40-percent "hit" area: $P(0 \text{ pts.}) = 60/100$, $P(1 \text{ pt.}) = 24/100$, $P(2 \text{ pts.}) = 4/100$. The expected value is 0.24 points per turn. For a 40-percent "hit" area: $P(0 \text{ pts.}) = 60/100$, $P(1 \text{ pt.}) = 24/100$, $P(2 \text{ pts.}) = 16/100$. The expected value is 0.56 points per turn. For an 80-percent "hit" area: $P(0 \text{ pts.}) = 20/100$, $P(1 \text{ pt.}) = 16/100$, $P(2 \text{ pts.}) = 64/100$. The expected value is 1.44 points per turn.

Sheet 3:

1)

Can 1:	GG	GS	G	S	
Can 2:	SS	GS	GSS	GGS	GGSS

2) *GG-SS* (**fig. 5a**)
$P(G) = 1/2$
$P(S) = 1/2$
The expected value =
$3.00 per turn

G-GSS (**fig. 5b**)
$P(G) = 2/3$
$P(S) = 1/3$
The expected value =
$3.67 per turn

S-GGS (**fig. 5c**)
$P(G) = 1/3$
$P(S) = 2/3$
The expected value =
$2.33 per turn

-GGSS (**fig. 5d**)
$P(G) = 1/4$
$P(S) = 1/4$
The expected value =
$1.50 per turn

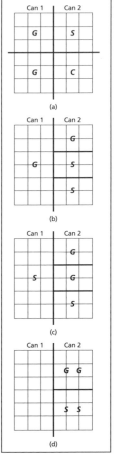

Fig. 5

The best arrangement is to put a gold coin alone in one of the cans and a gold coin and two silver coins in the other can.

3) The best arrangement is to put a gold coin alone in two of the cans and three silver coins in the third can. Then $P(G) = 3/4$, $P(S) = 1/4$. The expected value is $4.00 per turn.

Sheet 4:

1) and 2) Answers will vary

3) $P(A$ scores$) = 3/4$, $P(B$ scores$) = 1/4$ in **figure 6**, The expected number of points for each player on 20 rounds: A gets 15 points and B gets 10 points; for 100 rounds: A gets 75 points and B gets 50 points.

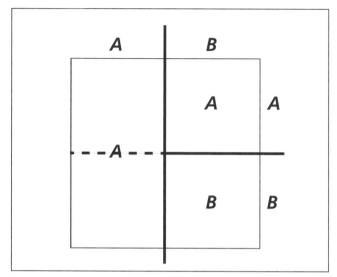

Fig. 6

4) To make the game fair, player B should get 3 points for scoring.

5) Answers will vary.

6) $P(\$20) = 1/16$, $P(\$10) = 9/16$, $P(\$15) = 6/16$ in **figure 7.** The expected value is $12.50 per week. It should make no difference which method you choose for obtaining your allowance; the amount is the same.

		$5		$10
	$10	$10	$10	$15
$5	$10	$10	$10	$15
	$10	$10	$10	$15
$10	$15	$15	$15	$20

Fig. 7

BIBLIOGRAPHY

Armstrong, Richard D. "An Area Model for Solving Probability Problems." In *Teaching Statistics and Probability*, 1981 Yearbook of the National Council of Teachers of Mathematics (NCTM), edited by Albert P. Shulte, pp. 135–42. Reston, Va.: NCTM, 1981.

Dahlke, Richard, and Robert Fakler. "Geometrical Probability." In *Teaching Statistics and Probability*, 1981 Yearbook of the National Council of Teachers of Mathematics (NCTM), edited by Albert P. Shulte, pp. 143–53. Reston, Va.: NCTM, 1981.

Lappan, Glenda, Elizabeth Phillips, Mary J. Winter, and William M. Fitzgerald. "Area Models for Probability." *Mathematics Teacher* 80 (March 1987): 217–23.

Lappan, Glenda, and Mary J. Winter. "Probability Simulation in Middle School." *Mathematics Teacher* 73 (September 1980): 446–49.

Phillips, Elizabeth, Glenda Lappan, Mary J. Winter, and William M. Fitzgerald. *Probabiltity, Middle Grades Mathematics Project.* Menlo Park, Calif.: Addison-Wesley Publishing Co., 1986.

Shulte, Albert P., and Stuart A. Choate. *What Are My Chances?* Books A and B. Palo Alto, Calif.: Creative Publications, 1977.

Analyzing Games of Chance for Expected Value   Sheet 1

In games of chance involving points it is often interesting to know what the *average* points per turn would be over many turns. This long-term average is called the *expected value* for the game. The following games involve spinners. For each turn a player spins and gets the points indicated in the area in which the spinner lands.

1. For the spinner the probability of landing in the area marked -2 is ¾.
    We write $P(-2) = $ ¾.

    Find $P(5) = $ _____ .

    In 100 spins the number of times you would expect

        5 to occur is ___25___ for a total of ___125___ points.

        -2 to occur is _____ for a total of_____ points.

    The expected total points for 100 spins is_____ .

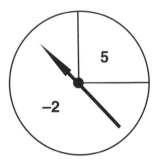

2. Determine the probabilities for the spinner below.

    Find $P(5) = $ _____ $P(10) = $ _____ $P(-12) = $ _____ .

    In 60 spins the number of times you would expect

        5 to occur is _____ for a total of _____ points.

        10 to occur is _____ for a total of _____ points.

        -12 to occur is _____ for a total of _____ points.

    The expected total points for 60 spins is_____ .

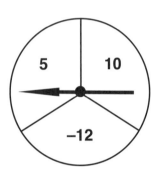

3. Determine the probabilities of this spinner.

    $P(-1) = $_____ $P(2) = $ _____ $P(3) = $_____ $P(-4)$ _____ .

    For 72 turns, the expected total points is _____ .

    The expected value is_____ .

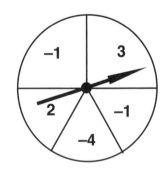

4. Calculate the expected value for 240 spins on each of these spinners. Discuss your results.

Analyzing Games of Chance for Expected Value   Sheet 2

The spinner for this game is divided into two parts representing 60 percent and 40 percent of the total area. A player has a 60-precent chance of landing in the "hit' area. If a player hits on the first spin, he or she takes a second spin on the spinner. Two points are given if both spins are hits; one point is given if the first spin is a hit and the second spin is a miss; and zero points are given if the first spin is a miss. A player spins either once or twice to complete a turn depending on whether a "miss" or a "hit" was obtained on the first spin..

1. Which score do you think is most likely to occur on a turn: zero, one, or two? _____

2. Insert a pencil point into the head of a bobby pin at the center of this spinner. Spin the spinner 50 times. Record your results in the chart below:

Points	Frequency	Total	Experimental Probability
0	_____	_____	_____
1	_____	_____	_____
2	_____	_____	_____
Total Turns		**50**	

3. Analyze this game using this 100 grid to represent total probability of 1. Determine each of the following probabilities.

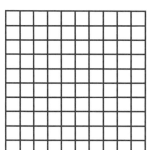

P(0 points scored) =_____ .

P(1 point scored) =_____ .

P(2 points scored) =_____ .

In 100 spins the number of times you would expect

    0 to occur is _____ for a total of _____ points.

    1 to occur is _____ for a total of _____ points.

    2 to occur is _____ for a total of _____ points.

4. Determine the probabilities and expected value for spinner games with 20-percent, 40-percent, and 80-percent probability of landing in the are marked "hit."

Analyzing Games of Chance for Expected Value   Sheet 3

A game of chance involves two gold coins each worth $5 and two silver coins each worth $1. The coins are identical in size and shape. While your partner is blindfolded you place the coins in two identical cans in any way you want (one can could be empty). Your partner then selects a can at random and reaches in and pulls out a coin. The two of you get to keep the coin.

1. What are all the possible ways that the four coins could be placed into the two cans?

2. Use the area model to help analyze each arrangement. For each arrangement calculate the expected value for 100 trials.

    Example: One possible arrangement.

    Can 1   Can 2

G	G
S	S

    Can 1   G, S
    Can 2   G, S

    $P(G) = \frac{2}{4} = \frac{1}{2}$
    $P(S) = \frac{2}{4} = \frac{1}{2}$
    Expected value = $3 per turn

    Can 1 _____

    Can 2 _____

    $P(G) =$ _____

    $P(S) =$ _____

    Expected value = _____

    Can 1 _____

    Can 2 _____

    $P(G) =$ _____

    $P(S) =$ _____

    Expected value = _____

    Can 1 _____

    Can 2 _____

    $P(G) =$ _____

    $P(S) =$ _____

    Expected value = _____

    Can 1 _____

    Can 2 _____

    $P(G) =$ _____

    $P(S) =$ _____

    Expected value = _____

    What arrangement of coins gives the greatest chance of drawing a gold coin? _____ .

3. If three cans are used with three gold coins and three silver coins, what arrangement gives the best chance of drawing a gold coin? Calculate the expected value for this arrangement.

Analyzing Games of Chance for Expected Value   Sheet 4

A spinner is marked with an *A* and a *B* as shown. Each round will consist of either one or two spins. The player with the highest score wins.

On each round player *A* spins first:

- If the spinner lands in the area marked *A*, player *A* scores a point (and this turn ends with player *A* spinning to start the second turn).

- If on the first spin the spinner lands in the area marked *B*, player *B* spins the spinner; player *B* scores 2 points if the spinner lands in *B*, and player *A* scores a point if it lands in *A*.

1. Does this game seem fair? _____ Explain.

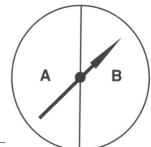

2. With a partner try out the game. Play a total of 20 rounds. Record your results. Calculate the points for each player.

   Who won? _____

Player	Points Scored	Total Score
A		
B		

   Now do you think it is fair?_____ Explain.

3. Find the probability that player *A* will score and the probability that player *B* will score on a given round.

   Use the square grid provided to help.

   Calculate the expected number of points for each player for 20 rounds. For 100 rounds.

   P(*A* will score on this round)
   = _____ .

   P(*B* will score on this round)
   = _____ .

4. How would you modify the awarding of points to make this a fair game?

5. Devise another one- or two-person game with points. Analyze the probabilities and the expected points scored by each player.

6. Your father offers you the following choice for obtaining your weekly allowance: you can receive $12.50 per week, or you can receive the amount of money you get by taking two spins on the spinner shown. Which do you choose?_____ Why? _____
   Analyze how much money you would expect to get per week for 32 weeks if you chose the spinner each time. Compare this with the amount of money you would get if you choose the $12.50 per week.

Investigating the Relationship between Latitude and Temperature

Jean McGivney-Burelle, Raymond J. McGivney, and Katherine G. McGivney

This article describes an engaging data-gathering activity that involves exploring relationships between latitude and average monthly temperatures of cities in the Western Hemisphere. Our extensive work with middle school and high school students has confirmed this activity's attractive features: It is elastic (i.e., it can be adapted for use in either algebra 1 or algebra 2); it is interdisciplinary and can be taught through team teaching; and it requires wise and effective use of technology.

LESSON OVERVIEW

The lesson described here was presented to a class of twenty rising ninth- and tenth-grade students enrolled in a summer mathematics program funded by the State of Connecticut at the University of Hartford in West Hartford. Project TEAM (technology, exploration, application, and modeling), which ran for five consecutive years (1999–2004), was geared for mathematically average students in the city of Hartford as well as some nearby suburbs. Beyond the mathematics, one of the goals of the project was to reduce the racial isolation of students in both city and suburban schools. Teachers from participating middle schools and high schools cotaught with University of Hartford faculty members during this two-week residential mathematics program. Approximately 100 students participated in each session, with about 50 percent of the students coming from the inner city and the other 50 percent coming from the suburbs. There were about an equal number of girls and boys. The students had had at least one year of algebra and were familiar with linear functions and the table and graphing functions on the TI-83+/84 calculator; they also had a general notion of lines of best fit and correlation coefficients. The goal of the summer program was to provide students experience in mathematical modeling—that is, gathering real-world data and finding best-fit models to help formulate and test their conjectures.

We began the activity with a discussion of latitude (see **fig. 1**) and longitude and explained how to determine each. We sketched a picture (**fig. 2**) on the board and explained that the latitude of a city represented by point P is found by drawing a line from P to the center of the Earth, O. The elevation angle, θ, of that point is its latitude. If P is north of the equator, the latitude is northern; if P is south of the equator, the latitude is southern.

Students were asked to discuss possible relationships between latitude and temperature on the basis of their travel experience. Several students conjectured that as

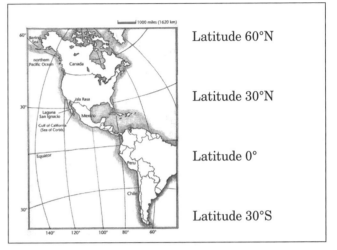

Fig. 1. Latitude lines

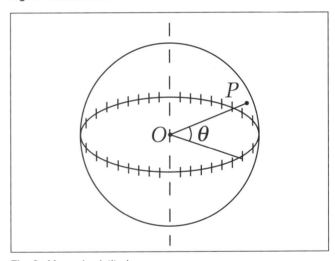

Fig. 2. Measuring latitude

latitude increases, temperature decreases and that as latitude decreases, temperature increases.

To get students to think more deeply about the relationship between latitude and temperature, we posed several follow-up questions, including these:

- At what rate do average monthly temperatures fall as you travel north and rise as you travel south?

- Given the latitude of a city or town, how accurately can we predict its average monthly temperature?

- Is it always true that cities at the same latitude will have the same average monthly temperature?

For this last question, students offered plausible geographic reasons as to why cities at the same latitude might have different temperatures. One student suggested that elevation may play a role, because temperatures are generally warmer at sea level than in the mountains. Another student conjectured that cities with large populations might have higher temperatures than smaller cities.

Because data collection involves many variables and approaches, we agreed on the following restrictions:

- We would search for data only in the geographical area generally called the Americas.

- We would investigate the temperatures only of large cities or cities popular with tourists, because data for these are more readily available.

- Finally, we would focus on the daily average temperatures in April, a heavy travel month.

DATA FOR THE AMERICAS' NORTHERN HEMISPHERE

Building a model

Working in the computer lab, the students collected data from the website **www.worldclimate.com**. We instructed them to identify ten cities with wide-ranging latitudes, collect data regarding latitude and temperature in degrees Fahrenheit, and round both measures to the nearest tenth of a degree. Also, when the website listed temperatures at several locations near a city but not in the city itself, students were told to choose data from that city's major airport.

Students recorded their data in a table; a sample set is shown in **table 1**. To avoid confusion, degree notation (°) was used solely for temperatures, not for latitude.

Using a TI-83+/84 graphing calculator, students found the scatter plot of the ordered pairs (latitude, average April temperature). After examining the scatter plot, students concluded that the data looked "almost linear" with an obvious negative trend. At this point, we instructed students to use their calculators and the technique of least-squares regression to find the line that best fit the data (see **fig. 3**).

The students reported that the linear regression model for these data was $y_N = -0.8x + 85.1$ with a correlation coefficient of about -0.9. This model suggested a strong negative linear relationship between temperature and latitude. The data points appeared to be randomly scattered both above and below the line, providing further evidence that a linear model is appropriate for these data.

Interpreting the model's parameters

At this point, we wanted to be sure that students could make sense of the linear regression model; in particular, we wanted them to understand the significance of the slope and the y-intercept in this problem. When we have asked such questions in the past, students have offered general, seemingly rehearsed responses such as "Slope is rise over run" and "The y-intercept is the

Table 1		
Latitude and Temperature Data for Northern Hemisphere		
City	Latitude	Average April Temperatures (in degrees F)
Recife, Brazil	8.0N	79.0°
Caracas, Venezuela	10.5N	70.3°
Mexico City, Mexico	19.4N	64.4°
Miami, FL	25.8N	74.5°
Charleston, SC	32.9N	64.8°
Phoenix, AZ	33.4N	68.4°
Washington, DC	39.0N	52.9°
Kansas City, MO	39.1N	55.6°
New York, NY	40.7N	50.4°
Chicago, IL	42.0N	48.6°
Caribou, ME	46.9N	37.8°
Vancouver, Canada	49.2N	48.0°
Juneau, AK	58.4N	39.6°

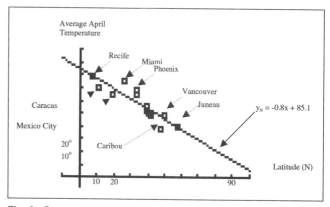

Fig. 3. Scatter plot and least-squares regression line of Northern Hemisphere latitude and temperature data

starting point"—a result, perhaps, of studying these concepts outside any real-world context. Throughout this summer program, we encouraged students to respond to these questions in context. After struggling a bit, several students were able to draw the following conclusions: A slope of -0.8 indicates that "as latitude goes up by one degree, the average temperature [in April] goes down by about 0.8°"; and, further, a y-intercept of 85 means that "at 0 degrees [the equator], the average temperature [in April] is about 85°."

Using the model to make predictions

To facilitate students' reflection on their models, we posed a variety of questions, including these:

- Suppose the class plans to visit New York City (40.7N) during April vacation. What do you expect the average temperature to be? (*Answer:* 52.9°)

- During spring break, Jasmine's family wants to visit a city in the Caribbean whose average April temperature is at least 70°. Would San Juan, Puerto Rico, at 18.4N do? (*Answer:* Barely. Our model predicts the average April temperature to be 70.5°).

These answers can be found by using the Table, Value, or Intersect features on the TI-83+/84 graphing calculator.

Residuals

In addition to the correlation coefficient calculated earlier, residuals provide another goodness-of-fit test. In this case, residuals are the difference between the actual and the predicted average temperatures for each city. These residuals are calculated in **table 2** and graphed versus the corresponding *x*-values in **figure 4**. Because the residuals appear in no systematic pattern, we can surmise that our model is a good fit for the data. (*Note:* Students found these residuals by using the RESID function on a TI-83+/84 calculator.)

Explaining "outliers"

Many students noticed a data point or two (one or two cities) further away from the best-fit line than others. At this point, we informally introduced the term outlier as a way to characterize ordered pairs that stray far off on either side of the graph. For example, in the data we have been describing, Miami and Caracas, among other cities, might be considered outliers (see **fig. 5**).

Using their best-fit model, one pair of students calculated the predicted temperature for Caracas to be 76.8° and saw that it was higher than its actual temperature (70.3°); the students conjectured that this difference may be attributed to the fact that Caracas is at an altitude of 800 meters (2,625 feet). On the other hand, Miami's predicted temperature was much warmer than the model predicted. Students conjectured that this discrepancy was due to the fact that Miami sits at sea level. This discussion underscored the fact that average temperature is a function of many variables, only one of which is altitude.

DATA FOR THE AMERICAS' SOUTHERN HEMISPHERE

After exploring temperature in the Northern Hemisphere of the Americas, we extended our discussion of the relationship between latitude and temperatures into the Southern Hemisphere of the Americas. (*Note:* Data from websites for the Southern Hemisphere are

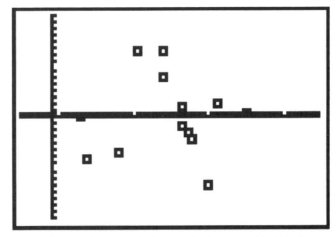

Fig. 4. Scatter plot of residual values for linear model of Northern Hemisphere data

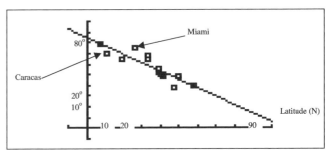

Fig. 5. Outliers in Northern Hemisphere data

less abundant.) One group's first attempt resulted in the data shown in **table 3**.

The linear regression model, $y = -0.5x + 70.1$, has a correlation coefficient of -0.6, which, students noted, indicates a somewhat weaker linear relationship than for temperatures in the Northern Hemisphere. However, they also noted that both Quito and La Paz appear to be outliers (see **fig. 6**). Students were asked to explain why the data points representing Quito and La Paz strayed from the rest of the data. Once again, the elevation of these cities might explain their lower-than-expected temperatures. (See **table 3** for the elevations of both cities.) Although Quito and La Paz fit our criteria as larger or more popular vacation cities, they are atypical in that each is situated about two miles above sea level.

At this stage, we asked students to eliminate several outliers and find a model that might better predict temperatures for cities in less extreme altitudes. We suggested this approach not to imply that the revised model would be "better" but only to show that it may be more useful for this activity. Consequently, we excluded Quito and La Paz from our data set and replaced them with Guayaquil, the largest city in Ecuador, and Brasilia, the capital of Brazil (see **table 4**). Students noted that a case could also be made for replacing Rio de Janeiro. However, as a large, popular tourist city, it fits the criteria established for the activity, and so we left it in the data set.

Table 2

Latitude and Temperature Data for Northern Hemisphere

City	Latitude	Average April Temperatures		Residuals
		Actual	Predicted	
Recife, Brazil	8.0N	79.0°	78.8°	0.2
Caracas, Venezuela	10.5N	70.3°	76.8°	−6.5
Mexico City, Mexico	19.4N	64.4°	69.8°	−5.4
Miami, FL	25.8N	74.5°	64.7°	9.8
Charleston, SC	32.9N	64.8°	59.1°	5.7
Phoenix, AZ	33.4N	68.4°	58.7°	9.7
Washington, DC	39.0N	52.9°	54.3°	−1.4
Kansas City, MO	39.1N	55.6°	54.2°	1.4
New York, NY	40.7N	50.4°	53.0°	−2.6
Chicago, IL	42.0N	48.6°	52.0°	−3.3
Caribou, ME	46.9N	37.8°	48.0°	−10.2
Vancouver, Canada	49.2N	48.0°	46.0°	1.8
Juneau, AK	58.4N	39.6°	39.0°	0.7

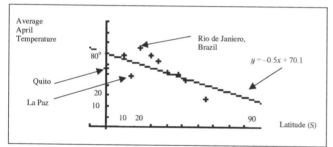

Fig. 6. Scatter plot and line of best fit for Southern Hemisphere data

Although Quito and Guayaquil have essentially identical latitudes and are only 380 miles apart, the difference in their average April temperatures is almost 25°. This difference can be explained by the fact that Quito has an elevation of almost two miles, while Guayaquil is a coastal city.

Using the TI-83+/84 calculator, students found that the linear regression model for these data is

$$y_S = -0.8x + 84.3.$$

The coefficients are almost exactly those of the model for the Northern Hemisphere ($y_N = -0.8x + 85.1$). Consequently, students gave a similar interpretation of the slope and y-intercept of this model as they did for the earlier one. The correlation coefficient is -0.9, which suggests a strong negative trend between latitude and temperature. A graph of these data and the linear model is shown in **figure 7**.

Table 3

Latitude and Temperature Data for Southern Hemisphere

City	Latitude	Average April Temperatures (in degrees F)
Quito, Ecuador (altitude 9,300 ft.)	0.0S	56.3°
Lima, Peru	12.0S	68.8°
La Paz, Bolivia (altitude 12,000 ft.)	16.5S	48.9°
Rio de Janeiro, Brazil	22.9S	75.4°
Porto Alegre, Brazil	30.0S	68.0°
Montevideo, Uruguay	34.8S	62.8°
Valdivia, Chile	39.8S	51.8°
Puerto Deseado, Argentina	47.7S	50.7°
Rio Gallegos, Argentina	51.6S	44.6°
Palmer Station, Antarctica	64.8S	27.7°

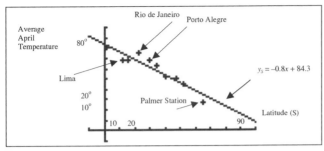

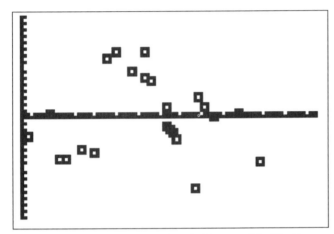

Fig. 7. Scatter plot and line of best fit for revised Southern Hemisphere data with two outliers replaced

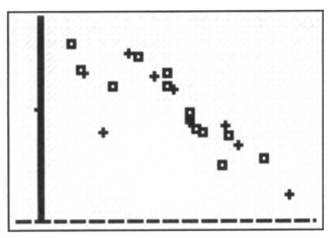

Fig. 9. Scatter plot for combined Northern and Southern Hemispheres data (cities in the Northern Hemisphere are represented by open squares, while cities in the Southern Hemisphere are represented by plus signs).

Fig. 8. Scatter plot of residual values for linear model of Southern Hemisphere data

In **table 5**, we present the residuals for the linear model of the Southern Hemisphere data. Like the residuals for the Northern Hemisphere, the residuals for the Southern Hemisphere appear in no predictable pattern (see **fig. 8**). This fact, taken with the correlation coefficient's value of -0.9, suggests that our model is a good fit for the data.

COMBINED WESTERN HEMISPHERE DATA

As an extension activity, students graphed their combined data for both Northern and Southern Hemispheres. **Figure 9** uses the data from **tables 1** and **4**. The linear model for the combined Western Hemisphere data is $y = -0.8x + 84.6$ with a correlation coefficient of -0.9. Using this model, students predicted the temperature at the equator in April to be about 85° and the temperature at the poles to be about 12.5° degrees.

CONCLUSION

This data-gathering activity covered interesting and important mathematical ground and engaged students from the start. While they searched for their list of cities, we overheard them asking questions and exchanging ideas related to the geographical properties of the cities they chose. The introductory data-gathering portion of this activity piqued students' interest and provided the basis for mathematical relationships they found between latitude and temperature.

This activity can be extended in several ways. For example, the number of variables affecting temperatures could be reduced by limiting data to cities at sea level or at a fixed altitude. As teachers, we appreciated how easy it was to implement this lab. No laboratory equipment or hardware attachments to our graphing calculators were necessary, simply access to the Web and a five-dollar almanac!

Table 4

Revised Southern Hemisphere Data with Two Outliers Replaced

City	Latitude	Average April Temperatures (in degrees F)
Guayaquil, Ecuador (altitude 9 ft.)	2.0S	80.1°
Lima, Peru	12.0S	68.8°
Brasilia, Brazil (altitude 3,500 ft.)	15.8S	67.3°
Rio de Janeiro, Brazil	22.9S	75.4°
Porto Alegre, Brazil	30.0S	68.0°
Montevideo, Uruguay	34.8S	62.8°
Valdivia, Chile	39.8S	51.8°
Puerto Deseado, Argentina	47.7S	50.7°
Rio Gallegos, Argentina	51.6S	44.6°
Palmer Station, Antarctica	64.8S	27.7°

Table 5

Residual Values for Linear Model of Southern Hemisphere Data

City	Latitude	Average April Temperatures		Residuals
		Actual	Predicted	
Guayaquil, Ecuador	2.0S	80.1°	82.8°	−2.7
Lima, Peru	12.0S	68.8°	75.1°	−6.3
Brasilia, Brazil	15.8S	67.3°	72.2°	−4.9
Rio de Janeiro, Brazil	22.9S	75.4°	66.8°	8.6
Porto Alegre, Brazil	30.0S	68.0°	61.4°	6.6
Montevideo, Uruguay	34.8S	62.8°	57.7°	5.1
Valdivia, Chile	39.8S	51.8°	53.9°	−2.1
Puerto Deseado, Argentina	47.7S	50.7°	47.8°	2.9
Rio Gallegos, Argentina	51.6S	44.6°	44.9°	−4.0
Palmer Station, Antarctica	64.8S	27.7°	34.8°	−7.1

Latitude

Sheet 1

Lines of latitude are imaginary lines that run around the earth parallel to the equator. In fact, they are often called parallels. These lines measure the distance north or south of the equator in degrees. The equator is 0°. The North Pole is 90° north (N), and the South Pole is 90° south (S). So as one moves away from the equator, the number of degrees of latitude increases. Kingston, Jamaica, has latitude of 17°N. Atlanta, Georgia, is at 33°N. Hartford, Connecticut, is about 41°N. Anyone who has watched the weather on TV for even a few minutes can tell you that as one moves away from the equator, the temperature changes too. It generally goes down. Kingston, Jamaica, has a mean (average) annual temperature of 80°F. Atlanta, Georgia, is a bit cooler at 61°F. Hartford, Connecticut, averages about 49°F.

We're going to do a lab to discover more about this relationship between latitude and temperature. The easiest way to gather data about this is on the Web.

STEP 1

Go to **http://www.worldclimate.com/**.

STEP 2

Enter a name of a city. You'll probably get several matches. Try Hartford, for example. You'll find that five states have a city named Hartford. If you want "our" Hartford, you'll have to select it. Notice that there are several Connecticut sites to choose from. That is usually because there is more than one weather station collecting data. It usually doesn't matter which one you select.

STEP 3

As soon as you find the weather station you want, select Average 24-Hour Temperature. Record three facts:

Average temperature (in °F)
Latitude (it is toward the top. Careful—don't confuse it with longitude!)
Elevation or height (it is not always listed)

STEP 4

Go back to the home page and repeat this process until you have the necessary data for at least fifteen cities. Choose widely spaced cities. Don't restrict yourself to the United States or to the Northern Hemisphere. Find at least one city on each of the six inhabited continents.

	City	Country	Mean Temp	Latitude	Elevation
1					
2					
3					
4					
5					
6					
7					
8					
9					
10					
11					
12					
13					
14					
15					

Latitude Lab

Sheet 2

1. Create a scatterplot of latitude vs. average yearly temperature.

2. Sketch a line of best fit on top of the scatterplot.

3. Approximate the best-fit equation.

4. What do the slope and the y-intercept measure in this problem?

5. Describe the relationship between latitude and average yearly temperature.

6. Use your model to predict the temperature of the following cities:

City	Latitude	Average Yearly Temp.
Cork, Ireland	51°N	_____
Capetown, South Africa	33.9°S	_____

7. Include the following city on your scatterplot.

City	Latitude	Average Yearly Temp.
Bogotá, Colombia	4.7°N	55.8°F

8. Does this new data point fit on or near your best-fit line?

9. If not, what might explain the reason for this outlier?

Counting Penguins

Mike Perry and Gary Kader

The NCTM's curriculum standards for statistics give a specific objective for students in grades 9–12: to "understand sampling and recognize its role in statistical claims" (NCTM 1989, p. 167). The use of random samples for estimation is a fundamental statistical concept. Random sampling and its consequences can be studied through simulated sampling activities. The nature of sampling variability, the influence of sample size on the quality of estimation, and the role of the underlying population distribution are ideas that can be illustrated with repeated sampling.

Estimating population size is a common type of problem that often requires "spatial" sampling. For instance, estimating the size of an animal population can require selecting samples over a geographic region. Estimating the number of cars in a parking lot requires spatial sampling. Crowd estimation at such events as the Million Man March in Washington, D.C., uses aerial photographs. These types of problems are discussed by Schaeffer, Mendenhall, and Ott (1990).

To obtain estimates of the size of the penguin population of a region in Antarctica, the map of the region is divided into subregions, and aerial photographs are taken of each selected subregion. Penguins appear as dots on the photographs; thus the photographs produce data for the estimation of the size of the total population.

The activity described herein is a simplification of the actual penguin-counting process but employs the same basic ideas and principles. It is intended for grades 9–12. The level of sophistication of the interpretations can be changed to match the level of the students. The full activity requires two or three class periods, but restricted versions can be completed within a single class. The following materials are required.

Sampling board. A region of Antarctica is represented by a sampling board. The 32-inch-by-32-inch sampling board is laid out in a 10 × 10 grid with margins as shown in **figure 1**. Each 3-inch-by-3-inch square represents a subregion. The borders are numbered 0 through 9 so that each square can be located by a pair of coordinates. To consider question 3 of the activity, three boards are needed.

Penguin cards. One hundred penguin cards are required for each sampling board. Each card is slightly smaller than 3 inches by three inches. One side of the card shows a pattern of dots—the penguins. The other side of the card shows a coordinate pair that corresponds to the card's position on the sampling board.

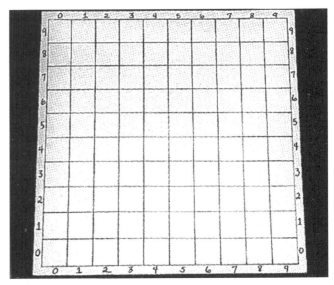

Fig. 1. The sampling board

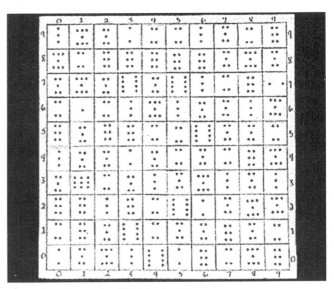

Fig. 2. Penguin cards with dots showing

Figure 2 shows a set of penguin cards on the sampling board with the dots showing. **Figure 3** shows the same cards turned over, with the coordinates face up. To consider question 3, we use three sets of penguin cards, which are described in **table 1**. In other words, board A is set up with no blank cards, one card with a single dot, six cards with two dots, and so forth.

Random-number generator. We prefer to use a die with ten outcomes (for example, an icosahedron). If these dice are not available, a random-digit table or a random-number generator on a calculator could be used.

Table 1

Population Distribution for Number of Penguins on Each Grid

Board A		Board B		Board C	
# of Penguins	Frequency	# of Penguins	Frequency	# of Penguins	Frequency
0	0	0	9	0	20
1	1	1	9	1	14
2	6	2	9	2	9
3	12	3	9	3	5
4	18	4	9	4	2
5	26	5	10	5	0
6	18	6	9	6	2
7	12	7	9	7	5
8	6	8	9	8	9
9	1	9	9	9	14
10	0	10	9	10	20
$\mu = 5$		$\mu = 5$		$\mu = 5$	
$\sigma = 1.65$		$\sigma = 3.15$		$\sigma = 4.07$	

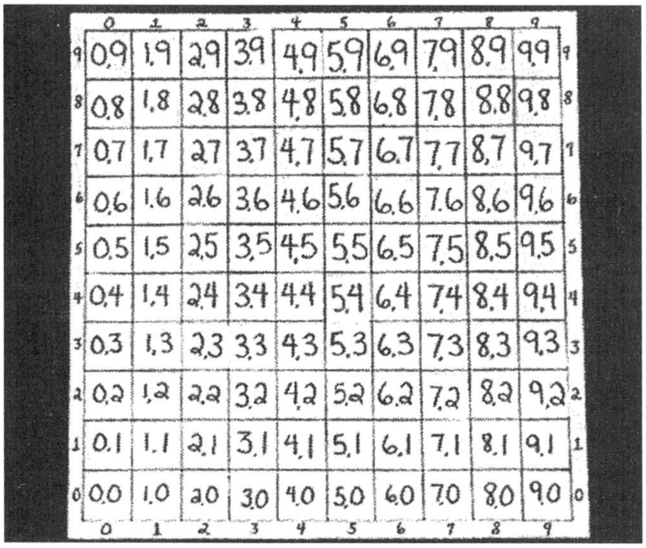

Fig. 3. Penguin cards with coordinates showing

Recording sheets. The data sheets in **figure 4** are designed specifically for the data summaries in this activity.

THE ESTIMATOR

From a random sample of size n, we can estimate the average, or mean, number of dots per square on the entire board with the average, or mean, number of dots per square in our sample. We denote this sample mean by $\overline{X}$. An estimate for the total number of dots N on the board is given by $\hat{N} = 100\overline{X}$.

THE QUESTIONS

1. Is this method a "good" way to estimate the number of dots on the board?

2. How does the sample size affect this estimation procedure?

3. How does the distribution of dots on the board affect the estimation procedure?

To study these questions, two criteria for a good estimation procedure are considered:

- Different samples should give similar results, at least most of the time.

- The estimator should be unbiased. It should neither systematically underestimate nor systematically overestimate the number of dots on the board.

In other words, a large number of independent samples should produce estimates that (*a*) do not vary too much and (*b*) give, on the average, the correct result.

We can study these two properties with repeated sampling, that is, by considering a large number of independent samples. A good rule of thumb for a study of this type is to have about one hundred samples so that reasonable inferences might be made from the data.

THE DATA

Each sampling is set up with the coordinates showing as is shown in **figure 3**. Each student selects two

Data

Board _____   Sample No. _____

Observation	Random Coordinates	Number of Penguins
1	_____	_____
2	_____	_____
3	_____	_____
4	_____	_____
5	_____	_____
6	_____	_____
7	_____	_____
8	_____	_____
9	_____	_____
10	_____	_____

Sum: _____

Mean: _____

Estimate of N ($100 \times$ Mean): _____

Board _____   Sample No. _____

Observation	Random Coordinates	Number of Penguins
1	_____	_____
2	_____	_____
3	_____	_____
4	_____	_____
5	_____	_____
6	_____	_____
7	_____	_____
8	_____	_____
9	_____	_____
10	_____	_____

Sum: _____

Mean: _____

Estimate of N ($100 \times$ Mean): _____

Combine two samples; sample size $n = 20$

Sum 1 + Sum 2: _____

Mean: _____

Estimate of N ($100 \times$ Mean): _____

Fig. 4. Recording sheets

Table 2

		Mean	Min	Q1	M	Q3	Max	Range	IQR
Summaries of Estimates Table									
n = 10									
	A	485	230	450	490	530	670	440	80
	B	491	180	440	500	550	710	530	110
	C	498	170	410	475	590	850	680	180
n = 20									
	A	480	355	445	490	515	565	210	70
	B	481	345	425	485	535	665	320	110
	C	514	270	450	500	565	795	525	115

samples of size *n* = 10 from each of the boards A, B, and C. In a class of about twenty-five students, each student should select 4 samples from each board for a total of twelve samples per student. This quantity can be adjusted according to class size to get about one hundred samples from each board, but the number of samples should always be a multiple of 2 because we later combine samples to get samples of size *n* = 20.

To select a sample of size *n* = 10, roll the die two times to select randomly a coordinate pair. In this way, select ten pairs for each sample. These pairs should be recorded on the data sheet. **Figure 5** shows a sample of size *n* = 10 on the sampling board.

A card can be used more than once within each sample. If a repetition of random coordinate pairs occurs for a particular sample the repetition is used in the sample. This procedure is often referred to sampling "with replacement" because when a card is selected, it is placed back in the board before the next card is chosen.

ANALYSIS

We illustrate several types of analyses, which can be varied to match the needs of a given class. The data used were collected by forty-seven teachers who attended a summer institute during July 1995. Each participant selected two samples from each board, giving a total of ninety-four samples each. These samples were pooled. **Figure 6** gives stem-and-leaf plots for *N* of sample size *n* = 10. Each data sheet contains two samples of size *n* = 10. These samples are combined to give estimates of *N* that are based on samples of size *n* = 20 as outlined at the bottom of each data sheet. In the case of the data from the institute participants, forty-seven estimates were given. **Figure 7** gives stem-and-leafs plots for these results. In **table 2**, means and quartile summaries are presented for each of the six cases, and the box plots are shown for comparison in **figure 8**. In **figures 6** and **7**, the stems are split into *h*, higher, and *l*, lower sets.

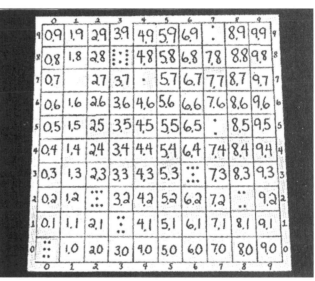

Fig. 5. Penguin cards with sample showing

INTERPRETATION

The cards of each of boards A, B, and C have a total of $N = 500$ dots. The estimator $\hat{N} = 100\overline{X}$ is an unbiased estimator of $N = 500$. The average of the estimates after repeated sampling should be close to 500. Note the means in **table 2**. The averages, or means, of the estimates based on ninety-four samples of size *n* = 10 are 485, 491, and 498. The averages for the forty-seven samples of size *n* = 20 are 480, 481, and 514.

The box plots in **figure 8** suggest some answers to all three questions and are useful for comparing the results of the two sample sizes and comparing the results from the three sampling boards. The corresponding numbers are in **table 2**. Estimates falling either 1.5 (IQR) below the lower quartile or 1.5 (IQR) above the upper quartile have been designated as outliers. Indicated by asterisks in the box plots. A thorough discussion of outliers is given by Landwehr and Watkins (1994).

First consider the effect of sample size. In the first two

Board A

```
1l |
1h |
2l |   30
2h |
3l |
3h |   70, 70, 80, 90
4l |   00, 00, 00, 10, 10, 20, 20, 20, 20, 30, 30, 30, 40, 40, 40, 40
4h |   50, 50, 50, 50, 50, 50, 50, 60, 60, 60, 70, 70, 70, 70, 70, 70, 80, 80, 80, 80, 80, 80, 80, 80, 80, 90, 90, 90, 90, 90
5l |   00, 00, 00, 00, 00, 00, 00, 00, 10, 10, 10, 10, 10, 10, 10, 20, 20, 20, 30, 30, 30, 30, 30, 30, 30, 40, 40, 40, 40
5h |   50, 50, 50, 60, 60, 80, 80, 80, 80, 80
6l |   00, 00
6h |   70
7l |
7h |
8l |
8h |
```

Board B

```
1l |
1h |   80
2l |
2h |   60, 90
3l |   20, 40
3h |   50, 50, 60, 60, 60, 70, 90, 90
4l |   00, 10, 10, 20, 20, 20, 30, 30, 30, 30, 40
4h |   50, 50, 60, 60, 60, 60, 60, 60, 60, 80, 80, 80, 80, 80, 80, 80, 80, 90, 90, 90, 90
5l |   00, 00, 00, 00, 00, 00, 00, 10, 10, 10, 10, 10, 10, 10, 20, 20, 20, 20, 30, 40, 40, 40, 40, 40
5h |   50, 50, 50, 60, 60, 60, 60, 70, 70, 70, 80, 80, 80, 90, 90, 90, 90, 90
6l |   00, 20
6h |   50, 50, 60, 70
7l |   10
7h |
8l |
8h |
```

Board C

```
1l |
1h |   70
2l |   00
2h |   70
3l |   00, 10, 10, 20
3h |   50, 50, 60, 60, 60, 60, 70, 70, 80, 90
4l |   00, 00, 00, 00, 00, 00, 10, 10, 10, 10, 20, 30, 30, 40, 40
4h |   00, 50, 50, 50, 50, 50, 50, 50, 60, 60, 60, 60, 60, 70, 70, 70, 70, 80, 90, 90, 90
5l |   00, 00, 10, 10, 10, 10, 30, 30, 30, 40, 40
5h |   50, 60, 60, 70, 80, 80, 90, 90, 90
6l |   00, 10, 10, 10, 30, 30
6h |   50, 60, 60, 60, 60, 80, 80, 90, 90, 90
7l |   20, 40, 40
7h |   50, 70
8l |
8h |   50
```

Fig. 6. Stem-and-leaf plots of estimates, $n = 10$

Board A

```
1l |
1h |
2l |
2h |
3l |
3h |   55, 70, 80
4l |   25, 30, 30, 30, 40, 40, 45, 45, 45
4h |   50, 55, 55, 60, 60, 70, 75, 75, 75, 75, 85, 90, 95, 95, 95, 95, 95, 95
5l |   00, 05, 10, 10, 10, 15, 20, 20, 25, 25, 30, 30, 30, 35, 40
5h |   50, 65
6l |
6h |
7l |
7h |
8l |
8h |
```

Board B

```
1l |
1h |
2l |
2h |
3l |   45
3h |   50, 75, 80, 90, 90, 95
4l |   00, 05, 05, 15, 25, 30, 35
4h |   50, 50, 50, 55, 55, 60, 70, 70, 70, 85, 85, 85, 90, 95
5l |   00, 00, 00, 10, 20, 30, 30, 35, 40, 40
5h |   50, 60, 60, 70, 70
6l |   00, 05, 25
6h |   65
7l |
7h |
8l |
8h |
```

Board C

```
1l |
1h |
2l |
2h |   70
3l |   35
3h |   50, 60, 70
4l |   00, 05, 10, 10, 35, 40
4h |   50, 50, 70, 70, 75, 75, 80, 80, 85, 90, 95, 95
5l |   00, 00, 00, 00, 05, 10, 15, 30, 30
5h |   50, 50, 50, 65, 70, 70, 80, 95
6l |   10, 20, 35, 45
6h |   50
7l |   10
7h |   95
8l |
8h |
```

Fig. 7. Stem-and-leaf plots of estimates, $n = 20$

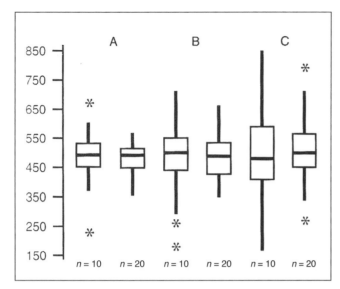

Fig. 8. Estimates from 94 samples, $n = 10$, and 47 samples, $n = 20$

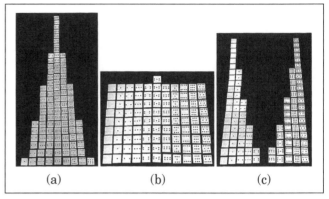

Fig. 9. Distribution of cards

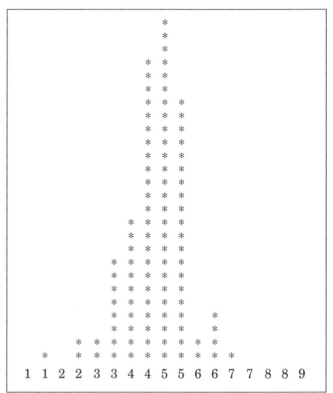

Fig. 10. Estimates of N from board B with $n = 10$

box plots, which summarize estimates from board A, we see an apparent reduction in the variation of estimates when the sample size is increased from $n = 10$ to $n = 20$. The next four box plots show similar comparisons for the results from boards B and C. The reduction is most pronounced for board C, with an interquartile range of 180 for samples of size $n = 10$ and an interquartile range of 115 for samples of size $n = 20$.

Next consider the differences in results among the three boards. In the three box plots for samples of size $n = 10$, estimates from board A show the least variability and those from board C show the most variability. Similar differences are indicated for samples of size $n = 20$.

The simulated data suggest the following:

- Increasing the sample size decreases the variability of estimates.

- Sampling from board A produces the least variability of estimates, and sampling from board C give gives the most.

Can we explain the differences in results among the three boards? The answer lies in the differences among the distribution of dots on each board. The placement of the cards on the board is arbitrary and has no effect on the estimation because of random selection. To demonstrate these differences, penguin cards from each board are arranged as shown in **figures 9a, 9b,** and **9c.** When conducting the activity, do not label these three arrangements but rather allow students to determine which is A, B, and C. Note that the distribution for board A is the most concentrated of the three around the middle. The variation around the mean is the smallest of the three distributions ($\sigma \approx 1.65$), which is why estimates based on samples from board A have the least variation. On the other extreme, the distribution for board C has the greatest variation about the mean

($\sigma \approx 4.07$), which is why estimates based on samples from board C have the most variability.

For some classes, it may be desirable to sample from only one board. In this situation, you would focus on the differences between the sample sizes and not on the effect of the underlying population distribution. In other classes, you might consider only samples of size $n = 10$ and focus on the quality of estimation for one sample size from one particular population.

THE CENTRAL LIMIT THEOREM

It is of interest to note the approximate bell shape of each of the distributions of estimates represented by the stem-and-leaf plots in **figures 6** and **7** with each centered at $N = 500$. For instance, the ninety-four estimates from board B are shown in **figure 10.** This arrangement is a consequence of the central limit theorem: For "large" sample sizes, the distribution of sample means can be approximated by a normal curve with center at the population mean.

In our applications, we are using a multiple of the sample means, but the principle still applies. The normal distribution and central limit theorem are widely discussed in the literature (see, e.g., Moore [1995]). Students may not completely comprehend the concept of the central limit theorem initially. This activity could be extended to do additional simulations with the same sample sizes to see that the replications of the experiment give similar results or to do simulations with different sizes to reinforce the concept.

CLASSROOM DISCUSSION

The activity may raise questions among students and generate discussion. This discussion may lead to further investigations. Potential questions for investigation follow:

- If we do the activity again, will we get the same or similar results?

- In what other situations would this estimation be appropriate?

- What would be some of the constraints or idiosyncrasies of these other situations?

- How many times would you expect to sample in such a simulation activity before you are fairly confident of your results?

- How does sampling without replacement affect results?

REFERENCES

Landwehr, James M., and Ann E. Watkins. *Exploring Data*. Rev. Student ed. Palo Alto, Calif.: Dale Seymour Publications, 1994.

Moore, David S. *The Basic Practice of Statistics*. New York: W. H. Freeman & Co., 1995.

National Council of Teachers of Mathematics (NCTM). *Curriculum and Evaluation Standards for School Mathematics*. Reston, Va.: NCTM, 1989.

Schaeffer, Richard L., William Mendenhall, and Lyman Ott. *Elementary Survey Sampling*. Boston: PWS-Kent Publishing Co., 1990.

This work was supported by the National Science Foundation's Instructional Materials Development Program under grant no. ESI-9150117. The opinions expressed in this article to not necessarily represent the position, policy, or endorsement of the foundation.

Data Collection and Analysis

Sheet 1

Objective

To examine a procedure for estimating the size of a penguin population; N = total number of penguins on an island. The procedure involves the following:

- Partitioning the island into 100 equal-sized (area) grids

- Selecting a random sample of n grids

- Counting the number of penguins on each grid in the sample and finding the average number of penguins on the grids in the sample ($\overline{X}$)

- Estimating the population size with $\hat{N} = 100\overline{X}$

Questions to Address

- Is this estimation procedure a "good" way to estimate the number of penguins?

- How does the sample size affect this estimation procedure?

- How does the distribution of the number of penguins on the grids affect the estimation procedure?

Collecting and Recording the Data

Using a random number generator (e.g., 10-sided dice, a table of random digits, or a calculator), select two random samples of 10 grids (with replacement) from each board.

On the Data Recording Sheet, record the following for each sample:

- The board (A, B, or C) that is being used;

- The random coordinates and the number of penguins on each grid;

- The total number of penguins in the sample (Sum), the mean number of penguins in the sample ($\overline{X}$), and the estimate ($\hat{N} = 100\overline{X}$) of the total number of penguins on the island.

Combine the two samples to form a sample of size 20, and record the results on the Combined Sample Data Recording Sheet.

Analyzing the Estimates from the Class

Samples of Size 10

Combine the estimates based on samples of size 10 from the entire class for each board separately. The analysis of the estimates for each board should include finding numerical summaries (e.g., the mean and the Five-Number Summary) and graphical displays (e.g., stem plots, dot plots, box plots). Record the numerical summaries in the Summaries of Estimates table provided.

Samples of Size 20

Combine the estimates based on samples of size 20 from the entire class for each board separately. The analysis of the estimates for each board should include determining numerical summaries (e.g., the mean and the Five-Number Summary) and graphical displays (e.g., stem plots, dot plots, box plots). Record the numerical summaries in the Summaries of Estimates table provided.

Because the interpretation will include several comparisons, all graphs should have the same scale.

Interpreting the Results

A Good Estimation Procedure Is an Accurate Estimation Procedure

The total number of penguins on each board (A, B, and C) is $N = 500$. We will now examine the accuracy of the estimator . We begin by looking only at the results for samples of size 10 from one board (A).

Look at the graphical displays for the estimates based on samples of size 10 from board A: do the estimates appear to be centered at approximately 500 in each graph? Explain.

Look at the **Summaries of Estimates Table** for samples of size 10 from board A:

- Is the median for the estimates approximately 500? What does this mean?

- Is the mean for the estimates approximately 500? What does this mean?

The estimates of N should generally be on both sides of 500 (the actual for N), and the median and mean should both be close to $N = 500$. This means that about the same number of estimates are below 500 as above 500, and the average of the estimates is close to 500. Thus, $\hat{N}$ does not systematically over- or underestimate N, meaning that $\hat{N}$ is an *unbiased estimator* of N.

In studying the estimation procedure, we are also interested in examining how similar the estimates are from sample to sample. Look at the graphical displays for the estimates based on samples of size 10 from board A:

- Are any of the estimates outliers? Excluding outliers, what are the smallest and largest estimates for N? Excluding outliers, what is the range for the estimates? Between what two values is the middle 50 percent for the estimates? What is the range for the middle 50 percent (the interquartile range [IQR])?

The closer together the values of $\hat{N}$ are from sample to sample, the higher the precision in the estimation procedure. An accurate estimation procedure is one that is unbiased and has high precision.

Effects of Sample Size on the Estimation Procedure

For board A—Compare the box plots and **Summaries of Estimates Table** for $n = 10$ and $n = 20$.

- How do the medians compare? How do the means compare? Does $\hat{N}$ appear to be an unbiased estimator of N for each sample size? Why? Does increasing sample size appear to affect bias? Explain.

- Excluding outliers, how do the ranges compare? How do the IQRs compare? For which sample size are the estimates generally closer together? Does increasing the sample size affect the precision of the estimation procedure? Explain.

Summarize how increasing the sample size affects the accuracy of the estimation procedure.

Interpreting the Results

Sheet 3

Effects of the Distribution for the Number of Penguins on the Grids on the Estimation Procedure

For each board, the population distribution for the number of penguins on each grid is summarized in **table 1**.

- In which distribution does the number of penguins on the 100 grids have the following:

  (1) The least variation from the mean of 5 penguins?

  (2) The most variation from the mean of 5 penguins? Explain.

For samples of size 10—Compare the box plots and **Summaries of Estimates Table** for boards A, B, and C.

- How do the medians compare to 500? How do the means compare to 500? Does $\hat{N}$ appear to be an unbiased estimator of N for each board? Does the distribution for the number of penguins on the grids appear to affect bias?

- Excluding outliers, how do the ranges compare? How do the IQRs compare? For which board are the estimates generally closer together? For which board are the estimates generally farther apart? Does the distribution for the number of penguins on the grids appear to affect the precision of the estimation procedure? Explain.

Summarize the effects of the different population distributions (A, B, and C) on the accuracy of the estimation procedure.

Activities: A-B-C, 1-2-3

Mary Richardson and John Gabrosek

This article describes an activity that revolves around a data set that students help create. The students use data about occurrences of letters in English text to study the relationship between the relative frequency of letters and the percent of Scrabble game tiles for the letter and the relationship between the relative frequency of a letter and the letter's Scrabble game-tile point value.

The activity has two parts. In the first part of the activity, students explore simple linear regression. In the second part of the activity, students explore quadratic regression and cubic regression. The two parts can be completed in sequence, or the teacher may wish to have students complete only the first part. (The regression equations model average point values and may predict values that are not possible for actual data.)

DATA COLLECTION

Each student receives a copy of sheet 1, the data collection sheet, and selects an article in a magazine or newspaper or on the World Wide Web that includes at least 300 letters. As homework, students use their articles to obtain empirical distributions for the relative frequency

of the letters of the alphabet in English text and then complete the table on sheet 1.

During one class period, a class distribution is generated by dividing students into small groups, obtaining group distributions, collecting the data, and tallying the results. During the next class period, students record the class distribution on the "Percent of Scrabble Tiles" table on sheet 2 and on the "Scrabble Tile Point Values" table on sheet 4. **Table 1** is a typical example of a class distribution. The answers given for questions on the worksheets are based on the data in **table 1**. Teachers may prefer to use the data in **table 1** for sheets 2 and 3 instead of data that the students collect.

RELATIVE FREQUENCY OF LETTERS VERSUS THE PERCENT OF SCRABBLE TILES

In this part of the activity, students investigate the relationship between a letter's relative frequency in English text and the percent of Scrabble tiles for the letter. Scrabble is a board game in which players use letter tiles (one letter is on each tile) to construct words. Each letter is associated with a certain number of points; and when a player uses that letter, he or she

Table 1					
Sample Class Distribution					
Letter	Class Total	Class Relative Frequency	Letter	Class Total	Class Relative Frequency
A	631	8.41	O	546	7.28
B	100	1.32	P	170	2.27
C	240	3.2	Q	8	0.11
D	285	3.8	R	457	6.09
E	912	12.16	S	526	7.01
F	148	1.97	T	685	9.13
G	163	2.17	U	216	2.88
H	344	4.59	V	70	0.93
I	553	7.37	W	151	2.01
J	19	0.25	X	20	0.27
K	63	0.84	Y	153	2.04
L	317	4.23	Z	10	0.13
M	179	2.39	Total	7500	≈ 100
N	534	7.12			

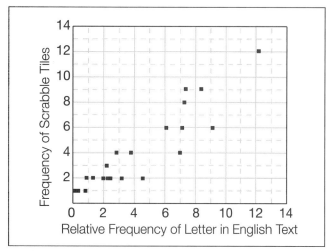

Fig. 1. Scatterplot of percent of Scrabble tiles versus relative frequency of letter in English text

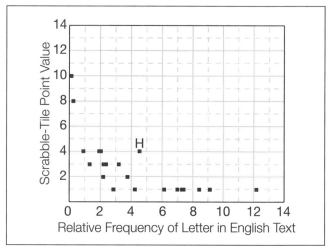

Fig. 2. Scatterplot of Scrabble-tile point value versus relative frequency of letter in English text

receives that number of points. Letters that are harder to use are worth more points. For example, the letter Q is worth ten points for the player, whereas the letter E is worth only one point. A standard Scrabble game set includes one hundred tiles. Two of these tiles are blanks that can be substituted for any letter, and the remaining ninety-eight tiles are distributed among the twenty-six letters of the alphabet. The first table on sheet 2 shows the letter distribution of Scrabble tiles for all letters except L and W.

Each student receives a copy of sheet 2. Students are asked to state whether they think that the association between the percent of Scrabble tiles for a letter and the letter's relative frequency in English text is positive or negative. Students construct a scatterplot showing the percent of Scrabble tiles on the vertical axis and the relative frequency in English text on the horizontal axis. This scatterplot does not include the letters L and W, because the students will predict the Scrabble-tile percents for those letters. There should be a strong, positive, linear association, as shown in **figure 1** for the sample data in **table 1**.

Each student next sketches a "fit-by-eye line" for the points on the scatterplot. The fit is often very poor. Students tend to fit the line to the points with extreme x-coordinates or to outliers rather than to most of the points. After comparing the fit-by-eye lines, students use their calculators to find the least-squares regression line, $y = 0.864x + 0.537$, where y is the Scrabble-tile percent and x is the letter's relative frequency in English text. Each student calculates residuals for both the fit-by-eye line and the least-squares line. The residual for a point is the actual y-coordinate minus the predicted y-coordinate found from the regression equation. In other words, the residual is the vertical distance (magnitude and direction) between the point and the regression line. Each student finds and compares the sum of the squared residuals for the least-squares line and her or his fit-by-eye line. The least-squares line is

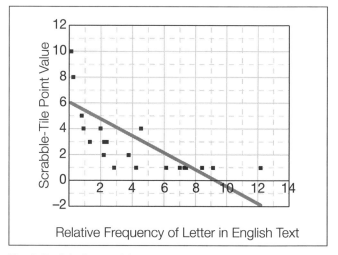

Fig. 3. Straight-line model

likely to have a smaller sum of squared errors than the fit-by-eye-line. The least-squares procedure chooses the line that meets the following two criteria:

- The sum of the residuals is 0.
- The sum of the squared residuals is minimized, subject to the sum of the residuals = 0.

The data used to fit the least-squares line do not include the letters L or W, so the equation can be used to predict the percent of Scrabble tiles for these two letters. Students are given the true percent of Scrabble tiles for the letters L and W and then calculate the residuals.

RELATIVE FREQUENCY OF LETTER VERSUS SCRABBLE-TILE POINT VALUE

In the second part of the activity, students investigate the relationship between a letter's relative frequency in English text and the Scrabble-tile point value for the letter. Each student receives a copy of sheet 4. The table on sheet 4 provides each letter's Scrabble-tile points (for all letters except C and N, whose Scrabble-tile point values will be predicted) and its relative frequency in

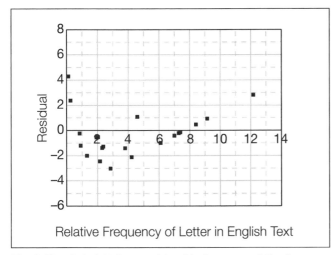

Fig. 4. Plot of straight-line-model residuals versus relative frequency of letter in English text

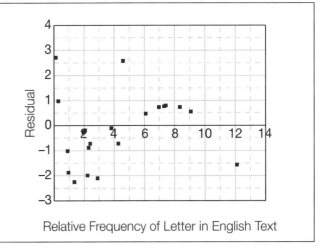

Fig. 6. Plot of quadratic-model residuals versus relative frequency of letter in English text

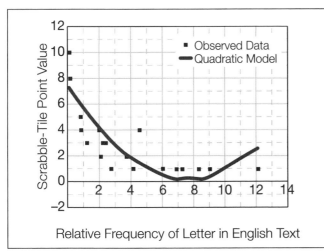

Fig. 5. Quadratic model

English text. Students construct a scatterplot with the Scrabble-tile point value for each letter on the vertical axis and the relative frequency on the horizontal axis. **Figure 2** shows a curved association for the sample data in **table 1**.

Students fit a straight-line model to the data, as shown in **figure 3**, and construct a scatterplot of the residuals (on the vertical axis) versus the relative frequencies of the corresponding letter (on the horizontal axis). If a fitted model is appropriate for a set of data, then we should see a random scatter of points about the horizontal line "residuals = 0." **Figure 4** shows a definite curved pattern for the residual plot of the straight-line model. We ask students to give an intuitive explanation for the pattern. Since a straight-line model is not appropriate for these data, we encourage students to conjecture about other model equations that might be used.

The equation that is most often suggested is that of a quadratic model. We ask students to use their calculators to fit a quadratic model (**fig. 5**) to the data:

$$\hat{y} = -0.121x^2 - 1.844x + 7.52$$

where $\hat{y}$ is the Scrabble-tile point value and where x is the relative frequency in English text, and then construct a residual plot for the model (**fig. 6**). The residual plot of the quadratic model shows a definite curved pattern, although the pattern is not as strong as the pattern for the straight-line model. **Figure 5** shows that when the relative frequency of a letter reaches 6 percent, the Scrabble-tile point values level off at one point.

The quadratic model is not capable of achieving a good fit in the range of letter relative-frequency values from 6 percent to 12 percent because it will not be flat in that region. We encourage students to think of another model that might achieve a good fit for this range. One such model is a cubic model. We ask students to use their calculators to fit a cubic model (**fig. 7**) to the data,

$$\hat{y} = -0.023x^3 + 0.521x^2 - 3.637x + 8.866,$$

where $\hat{y}$ is the Scrabble-tile point value and where x is the relative frequency in English text, and to construct a residual plot for this model, shown in **figure 8**. The residual plot for the cubic model shows a slight curved pattern, but this plot more closely resembles the ideal of a random scatter of points about the horizontal line "residuals = 0," than do the residual plots for either the straight-line model or the quadratic model.

The data used to fit the cubic regression model do not include the letters C or N, so the equation can be used to predict the Scrabble-tile point values for these two letters. Given the true Scrabble-tile point values for C and N, students calculate the residuals.

If students suggest other models, such as exponential or logarithmic functions, they should be encouraged to test those models and base their answers to the questions accordingly. We suggested only quadratic and cubic models because students are probably most familiar with these functions. We certainly would not discourage students from trying other models.

Table 2					
Table for Problem *10a*, Sheet 3					
Letter	Residual	Squared Residual	Letter	Residual	Squared Residual
A	−0.230	10.053	N	−2.000	14
B	−0.280	10.078	O	−0.120	10.014
C	−2.160	14.666	P	−1.230	11.513
D	−0.720	10.518	Q	−0.090	10.008
E	−0.920	10.846	R	−0.970	10.941
F	−0.930	10.865	S	−3.930	15.445
G	−0.110	10.012	T	−4.010	16.08
H	−3.550	12.602	U	−0.200	10.04
I	−0.810	10.656	V	−0.110	10.012
J	−0.230	10.053	W	—	—
K	−0.820	10.672	X	−0.250	10.062
L	—	—	Y	−1.000	111
M	−1.350	11.822	Z	−0.110	10.012

SOLUTIONS

Sheet 2:

1) The association should be positive because a greater number of Scrabble tiles should exist for a letter that occurs more frequently in the English language.

2) The scatterplot, which is based on the data in table 1, is shown in **figure 9**.

3) There do not appear to be any letters whose percent of Scrabble tiles do not follow the pattern for the majority of points. That is, there do not appear to be any outliers.

Sheet 3:

4) The association between a letter's relative frequency and its percent of Scrabble tiles is roughly linear, moderately strong, and positive.

5) Answers will vary. The teacher should be certain that the students do not fit the line to the exceptional points. The line should be fit to most of the points. It is interesting that many students will force the line through the origin.

6) Answers will vary, depending on the student's line. The teacher should make certain that students use two points on the line to calculate the slope. These points are not necessarily two data points.

7) The linear correlation for the data in **table 1** is r = 0.924. The equation of the least-squares line is $\hat{y}$ = 0.864x + 0.537.

8) The linear correlation indicates that a strong, positive, linear association exists between a letter's frequency in English text and its percent of Scrabble tiles.

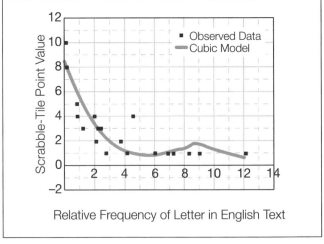

Fig. 7. Cubic model.

9) The least-squares line in **figure 10** is based on data in **table 1.**

10) a. The residuals for the fit-by-eye line will vary from student to student. For this problem, a commonly chosen fit-by-eye line equation is $\hat{y}$ = 1 + 1x. The residuals for the fit-by-eye line are found as residual equals $y - \hat{y}$, where y is the letter's actual percent of Scrabble tiles and $\hat{y}$ is the letter's predicted percent of Scrabble tiles using the fit-by-eye line. For the letter A, which has a relative frequency in English text of 8.41 percent, we have

$$y = 9.18,$$
$$\hat{y} = 1 + 1 \cdot 8.41$$
$$= 9.41$$

for the example fit-by-eye line. So the residual for the

Table 3

Table for Problem *10b*, Sheet 3

Letter	Residual	Squared Residual	Letter	Residual	Squared Residual
A	−1.380	1.904	N	−0.566	0.321
B	−0.363	0.132	O	−1.336	1.784
C	−1.261	1.59	P	−0.458	0.209
D	−0.261	0.068	Q	−0.388	0.15
E	−1.201	1.443	R	−0.323	0.105
F	−0.199	0.039	S	−2.511	6.306
G	−0.649	0.421	T	−2.302	5.299
H	−2.461	6.058	U	−1.056	1.114
I	−2.278	5.189	V	−0.700	0.489
J	−0.267	0.071	W	——	——
K	−0.243	0.059	X	−0.250	0.062
L	——	——	Y	−0.259	0.067
M	−0.561	0.315	Z	−0.370	0.137

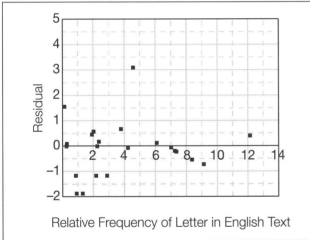

Fig. 8. Plot of residuals for the cubic model versus relative frequency of letter in English text

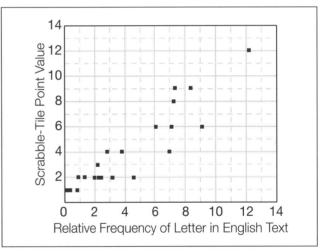

Fig. 9

letter A is 9.18 − 9.41, or −0.23. **Table 2** shows the residuals and squared residuals for the example fit-by-eye line. The SSE is the sum of the squared residuals, or 64.24.

b. The residuals for the least-squares line are found as residual = $y - \hat{y}$, where y is the letter's actual percent of Scrabble tiles and $\hat{y}$ is the percent of Scrabble tiles predicted for the letter by using the least-squares equation. For example, for the letter A, which has a relative frequency in English text of 8.41 percent, y = 9.18, and

$$\hat{y} = 0.864(8.41) + 0.537,$$
$$= 7.8.$$

For the letter A, the residual is

$$= 9.18 - 7.8 = 1.38.$$

Table 3 uses the data in **table 1** to show the residuals and squared residuals for the least-squares line. The sum of the squared residuals, or SSE, is 33.33.

c The least-squares line is likely to have a smaller SSE. The least-squares procedure chooses the line that meets the following two criteria:

- The sum of the residuals is 0.
- The SSE is minimized, subject to the sum of the residuals = 0.

11) The slope of the least-squares line shown in **figure 10** is 0.864. For every 1 percent increase in a letter's relative frequency in English text, we expect its percent of Scrabble tiles to increase by 0.864 percent.

12) For the least-squares line in **figure 10**, the predicted percent calculation for the letter L is

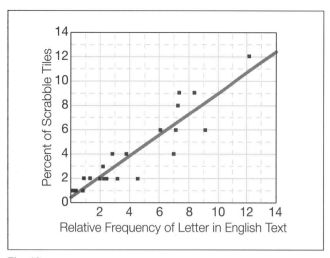

Fig. 10

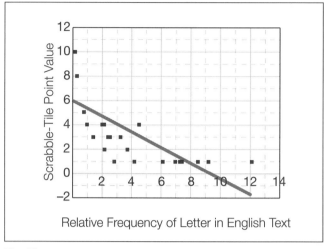

Fig. 12

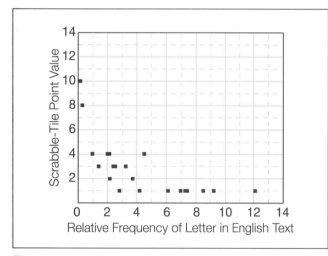

Fig. 11

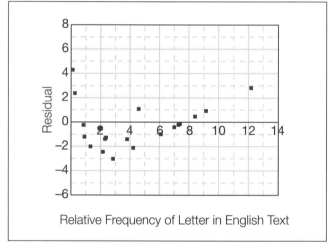

Fig. 13

$$\hat{y} = 0.537 + 0.864 \cdot 4.23,$$
$$= 4.19.$$

See **table 4**.

Sheet 4:

1) The association should be negative because we would expect that the more frequently a letter occurs in the English language, the fewer Scrabble points are given for using that letter in a word. For instance, the letter E is the most frequently used letter in the English language. It has a Scrabble point value of 1 because the letter E is found in many words.

2) The scatterplot in **figure 11** is based on the data in **table 1**.

3) A letter's relative frequency in English text and its Scrabble-tile point value have a curved association. The relationship is fairly strong. As the relative frequency of the letter increases, the Scrabble-tile point value decreases.

4) On the basis of the data in **table 1**, the equation of the least-squares line is $\hat{y} = -6.26x + 5.80$.

5) The least-squares line shown in **figure 12** is based on the data in **table 1**.

Table 4				
Table for Question 12, Sheet 3				
Letter	Relative Frequency in English Text	Percent of Scrabble Tiles	Predicted Percent of Scrabble Tiles	Residual
L	4.23	4.08	4.19	−0.11
W	2.01	2.01	2.27	−0.26

Table 5				
Table for Question 11, Sheet 6				
Letter	Relative Frequency in English Text	Percent of Scrabble Tiles	Predicted Percent of Scrabble Tiles	Residual
C	3.2	3	1.81	−1.19
N	7.12	1	1.08	−0.08

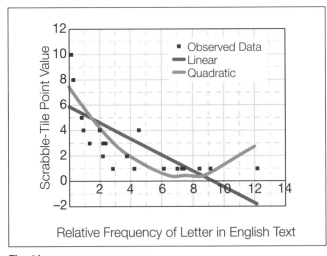

Fig. 14

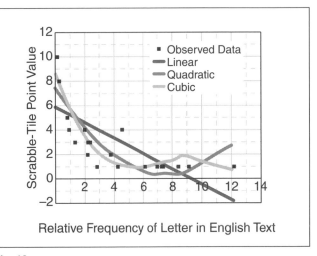

Fig. 16

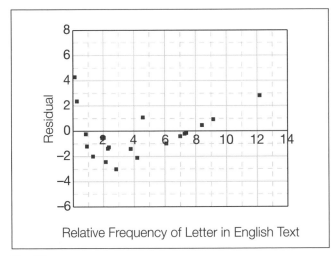

Fig. 15

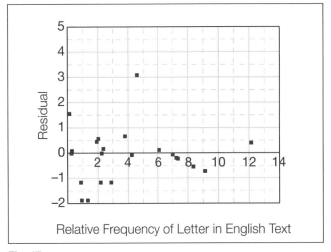

Fig. 17

Sheet 5:

6) The straight-line regression equation is a poor fit, as we can see from **figure 12**. It tends to underestimate the point values for letters at the extremes, that is, letters with very high or very low relative frequencies. The regression equation overestimates the point values for letters with moderate relative frequencies, leading to the parabolic shape found in the residual plot in **figure 13**. (Some calculators provide the residuals in a list.)

7) The fitted model is
$$\hat{y} = 0.121x^2 - 1.884x + 7.52,$$
on the basis of the data in **table 1**. See **figure 14**.

8) The quadratic regression equation is a poor fit, as we can see from the plot in **figure 14**. It tends to underestimate point values for letters with very low relative frequencies. It tends to overestimate point values for letters with moderately low relative frequencies. The regression equation underestimates point values for letters with moderately high relative frequencies. **Figure 15** shows the curved shape found in the residual plot;

9) On the basis of the data in table 1, the fitted model is

$$\hat{y} = -0.023x^3 + 0.521x^2 - 3.637x + 8.866.$$

See **figure 16**.

Sheet 6:

10) The cubic regression equation is a better fit, as we can see from the plot in **figure 16**. It tends to slightly overestimate point values for most letters, but the regression equation closely estimates point values for letters with very low and very high relative frequencies. A slight curved shape is found in the residual plot, as shown in **figure 17**.

11) The predicted point calculation for the letter C is
$$\hat{y} = -0.023(3.2)^3 + 0.521(3.2)^2 - 3.637(3.2) + 8.866$$
$$= 1.809.$$

See **table 5**.

REFERENCE

"Scrabble." Available at **www.hasbro.com/scrabble/en_US/story-cfm**.

The authors gratefully acknowledge the helpful comments and suggestions of the Activities editor and the reviewers during the preparation of this manuscript.

Data Collection

Sheet 1

Instructions

Obtain a newspaper, magazine, or World Wide Web article. Choose a place in the article to begin at random, and tally the next 300 letters one at a time, filling out the individual tally column in the table. Add the tallies, which should total at least 300 letters.

Letter	Individual Tally
A	
B	
C	
D	
E	
F	
G	
H	
I	
J	
K	
L	
M	
N	
O	
P	
Q	
R	
S	
T	
U	
V	
W	
X	
Y	
Z	
Total	

A-B-C, 1-2-3

Sheet 2

Background

"The story of Scrabble is a classic example of American innovation and perseverance. During the Great Depression, an out-of-work architect named Alfred Mosher Butts decided to invent a board game" ("Hasbro Scrabble: History." Available at **www.hasbro.com/scrabble/en_US/story.cfm**.). His goal was to create a game that combined vocabulary skills used in crossword puzzles and anagrams with an element of chance. He originally named the game Lexico, but he later changed the name to Criss-Cross Words, and it was renamed Scrabble, which means "to grope frantically." The name Scrabble was trademarked in 1948.

To create the game, Butts studied the front page of the *New York Times* to calculate the frequency of use in the English language of each of the letters of the roman alphabet. He discovered that vowels occur more frequently than consonants and that the letter E is the most frequently used vowel. After determining how frequently each letter was used, Butts assigned point values to each letter and decided how many of each letter would be included in the game.

The purpose of this activity sheet is to examine the relationship between a letter's relative frequency in English and the percent of Scrabble tiles for the letter.

Use the following table to record the class frequency for the letters of the alphabet in English text. The percent of Scrabble tiles containing the letters are recorded except for the letters L and W. We wish to use this information to examine the relationship between a letter's relative frequency and its percent of Scrabble tiles.

Percent of Scrabble Tiles

Letter	Class Relative Frequency	Percent of Scrabble Tiles	Letter	Class Relative Frequency	Percent of Scrabble Tiles	Letter	Class Relative Frequency	Percent of Scrabble Tiles
A	_____	9.18	J	_____	1.02	S	_____	4.08
B	_____	2.04	K	_____	1.02	T	_____	6.12
C	_____	2.04	L	_____	?	U	_____	4.08
D	_____	4.08	M	_____	2.04	V	_____	2.04
E	_____	12.24	N	_____	6.12	W	_____	?
F	_____	2.04	O	_____	8.16	X	_____	1.02
G	_____	3.06	P	_____	2.04	Y	_____	2.04
H	_____	2.04	Q	_____	1.02	Z	_____	1.02
I	_____	9.18	R	_____	6.12			

1. Is the association between the percent of Scrabble tiles and the relative frequency of a letter positive or negative? Why?

2. Using the graph on sheet 3, make a scatterplot with the percent of Scrabble tiles for each letter on the vertical (*y*-) axis and the relative frequency of the letter on the horizontal (*x*-) axis.

3. Are there any letters whose percent of Scrabble tiles does not follow the pattern for the majority of points? That is, do any outliers exist? If so, which letters are they?

A-B-C, 1-2-3

Sheet 3

4. Use the scatterplot to describe the form, strength, and direction of the association between a letter's relative frequency and its percent of Scrabble tiles.

5. Sketch a fit-by-eye line on your scatterplot.

6. Find the equation of your fit-by-eye line.

7. Use your calculator to find the linear correlation r and the equation of the least-squares regression line. (The regression equations model average point values and may predict values that are not possible for actual data.)

8. Interpret the value of the linear correlation r.

9. Plot the least-squares line on your scatterplot.

10. *a.* Compute the residuals for your fit-by-eye line, then square them and add them together. This quantity is called the sum of squared errors, or SSE.

    b. Use your calculator to compute the residuals for the least-squares regression line, then calculate the SSE.

    c. Which line has a smaller SSE? Are you surprised? Why or why not?

11. Interpret the slope of the least-squares regression line in the context of this problem.

12. Use your least-squares regression line to predict the percent of Scrabble tiles for the letters L and W. Complete the following table.

Letter	Relative Frequency In English Text	Percent of Scrabble Tiles	Predicted Percent of Scrabble Tiles	Residual
L	4.23	4.08		
W	2.01	2.01		

A-B-C, 1-2-3

Sheet 4

The purpose of this activity sheet is to examine the relationship between a letter's relative frequency in English text and the Scrabble-tile point value for the letter.

In the following table, record the class frequency for the letters of the alphabet in English text. The corresponding Scrabble-tile point values are recorded, except for the letters C and N.

Scrabble-Tile Point Values

Letter	Class Relative Frequency	Scrabble Tile Points	Letter	Class Relative Frequency	Scrabble Tile Points	Letter	Class Relative Frequency	Scrabble Tile Points
A	_____	1	J	_____	8	S	_____	1
B	_____	3	K	_____	5	T	_____	1
C	_____	?	L	_____	1	U	_____	1
D	_____	2	M	_____	3	V	_____	4
E	_____	1	N	_____	?	W	_____	4
F	_____	4	O	_____	1	X	_____	8
G	_____	2	P	_____	3	Y	_____	4
H	_____	4	Q	_____	10	Z	_____	10
I	_____	1	R	_____	1			

1. Do you think that the association between the Scrabble-tile point value and the letter's relative frequency will be positive or negative? Why?

2. Make a scatterplot with each letter's Scrabble-tile point value on the vertical (*y*-) axis and each letter's relative frequency on the horizontal (*x*-) axis.

3. Using the scatterplot, describe the form, strength, and direction of the association between a letter's relative frequency and its Scrabble-tile point value.

4. Use your calculator to fit a straight-line model to the data. (The regression equations model average point values and may predict values that are not possible for actual data.)

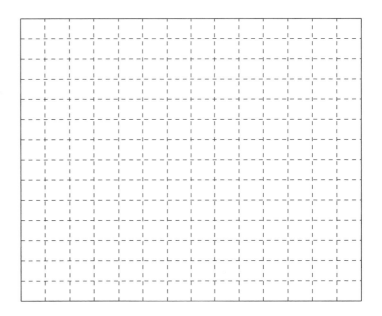

5. Plot the least-squares line on your scatterplot.

A-B-C, 1-2-3

6. Use your calculator to compute the residuals for the straight-line model. Plot the residuals (on the vertical axis) versus the relative frequency of each letter (on the horizontal axis). Give an intuitive explanation for the pattern on this plot, and explain why a straight-line model is not a good model to use to describe the relationship between a letter's relative frequency and its Scrabble-tile point value.

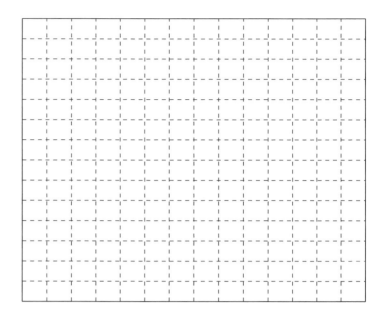

7. Use your calculator to fit the quadratic model $y = a \cdot x^2 + b \cdot x + c$ to the data. Plot your fitted-model equation on the scatterplot in question 2.

8. Use your calculator to compute the residuals for the quadratic model. Plot the residuals (on the vertical axis) versus the relative frequency of each letter (on the horizontal axis). Give an intuitive explanation for the pattern on this plot, and explain why a quadratic model is not a good model to use to describe the relationship between a letter's relative frequency and its Scrabble-tile point value.

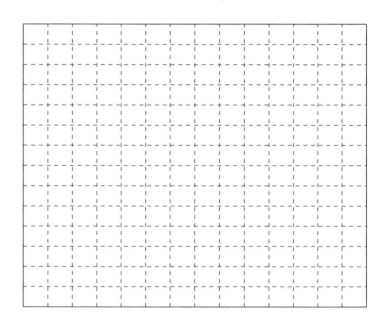

9. Use your calculator to fit the cubic model $y = a \cdot x^3 + b \cdot x^2 + c \cdot x + d$ to the data. Plot your fitted model equation on the scatterplot in question 2.

A-B-C, 1-2-3

Sheet 6

10. Use your calculator to compute the residuals for the cubic model. Plot the residuals (on the vertical axis) versus the letter relative frequency (on the horizontal axis). Explain how this plot shows that a cubic model is a better model to use than a linear or a quadratic model to describe the relationship between a letter's relative frequency and its Scrabble-tile point value.

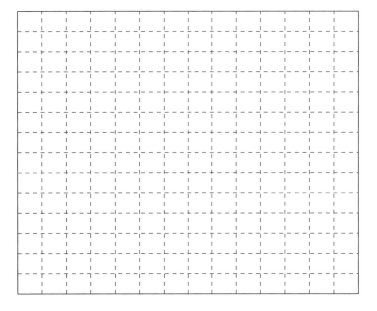

11. Use your cubic least-squares regression equation to predict the Scrabble-tile point value for the letters C and N. Complete the following table.

Letter	Relative Frequency In English Text	Scrabble-Tile Points	Predicted Scrabble-Tile Points	Residual
C	3.20	3	_____	_____
N	7.12	1	_____	_____

How Faithful Is Old Faithful? Statistical Thinking: A Story of Variation and Prediction

J. Michael Shaughnessy and Maxine Pfannkuch

Statistics is a relatively new discipline. Only in the last one hundred years have common methods and common reasoning evolved that can be applied to data from many fields. In the early years, the field of statistics was influenced by the work of Ronald A. Fisher, Karl Pearson, and Jerz Neyman. They focused on developing tools and methods that primarily focused on randomization More recently, exploratory data analysis has been emphasized (Tukey 1977). As statistics continues to mature as a discipline, statistics educators are paying more attention to developing overall models of statistical thinking (Wild and Pfannkuch 1999). This shift in statistics means refocusing the emphasis in teaching from how to do statistics to how to think about statistics. In this next step in the evolution of statistics and statistics teaching, two questions arise: What is statistical thinking? and How can we develop students' statistical thinking?

The authors of this article have found that data sets from the Old Faithful geyser in Yellowstone Park furnish a rich context for introducing such important aspects of statistical thinking as the central role of variation and the importance of asking our students what they would predict. In this article, we first discuss the context of the data, next present a classroom exploration of the data, and then discuss the nature of statistical thinking as it pertains to this Old Faithful data set.

THE CONTEXT

We imagine that we have just arrived at Yellowstone National Park, the home of geyser basins, thermal mud pots, hot springs, acid lakes, and a multitude of fascinating animals and plants. Furthermore, as has actually happened to one of the authors, we have just missed the most recent eruption of Old Faithful Geyser, which has periodically been spewing streams of hot water high into the sky for centuries. How long would we have to wait until the next eruption of Old Faithful? Before reading any further, write down your best estimate.

Readers who have visited Yellowstone Park might have some basis for making an informed estimate of the wait time until the next eruption. However, someone who is unfamiliar with geysers or who has not been to Yellowstone might not have any basis for estimating the wait time. Some geysers are dormant between eruptions for many hours or days. Other geysers erupt almost continuously. How could we obtain better information?

One strategy might be to appeal to a higher authority, by asking a park ranger or by reading the sign that indicates the approximate time of the next eruption. However, a strategy that can help those of us who are not actually at the park is to make a prediction on the basis of past data on Old Faithful. This latter strategy opens the door for an adventure in statistical thinking. We encourage readers to first work through our investigation of the Old Faithful data so that they can experience it in the manner in which we have used it with our own students. We then return to a deeper discussion of the aspects of statistical thinking that can arise while exploring this data set.

THE INVESTIGATION— EXPLORING DATA ON OLD FAITHFUL

Data on wait times between eruptions of Old Faithful are available through such sources as Hand and colleagues (1994). The following is approximately a day's worth of data on wait times for Old Faithful. Old Faithful erupts approximately twenty times each day. The data show the numbers of minutes between the time when Old Faithful stopped erupting to when it first began to erupt again.

Day 1 (minutes between eruptions):

51 82 58 81 49 92 50 88 62
93 56 89 51 79 58 82 52 88

If we had some friends who were planning to visit Yellowstone National Park, how long should we tell them to expect to wait between eruptions of Old Faithful? Before reading further, readers should make an estimate and give some justification for the prediction. They can also construct a graph of the first day's data on Old Faithful's eruptions.

Of course, one day's worth of data does not give much basis for a prediction. Two more days of Old Faithful eruption data, picked at random from a larger data set, follow:

Day 2:

86 78 71 77 76 94 75 50 83
82 72 77 75 65 79 72 78 77

Day 3:

65 89 49 88 51 78 85 65 75
77 69 92 68 87 61 81 55 93

Readers should construct graphical representations of the second and third days' data, similar to the representation of the first day's data. Compare the data for the three days. At this point, what could we predict for our friends? How long should they expect to wait for the next eruption of Old Faithful?

The three days for which data are given are only a small part of a data set, given on sheet 1, for the wait times for 300 consecutive Old Faithful eruptions (Hand et al. 1994). One strategy that we have used is to give each student a strip showing a day's worth of data; have them analyze, graph, and make predictions from it; then have students trade several times with other students; and repeat this process with data for several other days.

TYPICAL STUDENT RESPONSES

Many students first just calculate a mean or determine a median for a day's worth of Old Faithful data and base their initial prediction on a measure of central tendency. The mean of the first day's data is 70.1, and the median is 70.5; the mean of the second day's data is 79.9, and the median is 77; and so forth. Although the mean does furnish a one-number summary of a data set, it can also mask important features in the distribution of the data.

When students begin to create their own graphs, a variety of features of the distribution appear in their graphical representations. **Figures 1** through **4** are examples of students' work depicting data for one or more days of Old Faithful's eruptions. As shown in **figure 1**, some students make stem-and-leaf plots or box plots. Others make histograms, as in **figure 2**, perhaps accompanied by a box plot. Still others create dot plots or bar plots, in which the bar's height represents the length of the wait time between eruptions, as shown in **figure 3**, or plots of the length of the wait versus the number of the wait, as shown in **figure 4**.

Each of these types of representations can highlight or mask particular patterns in the data. Box plots furnish a good visual representation of the range and of the middle 50 percent of the data, as well as allow comparisons of the box size and position for several days. However, stem-and-leaf plots and histograms yield a clearer picture of the data's distribution. Box plots involve data reduction to summarize the data, whereas stem-and-leaf plots and histograms display the actual data. Stem-and-leaf plots or histograms can reveal gaps that are masked in a box plot. Data for the first and third days for Old Faithful appear somewhat bimodal in a histogram, whereas the second day's data are more moundlike, as indicated in **figure 5**.

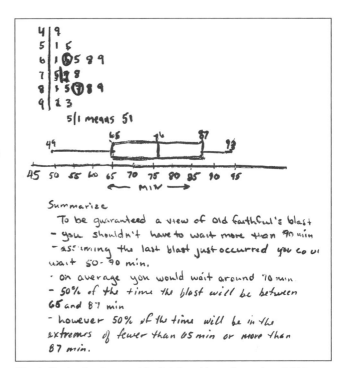

Fig. 1. Student's stem-and-leaf plot and box of one day of Old Faithful data

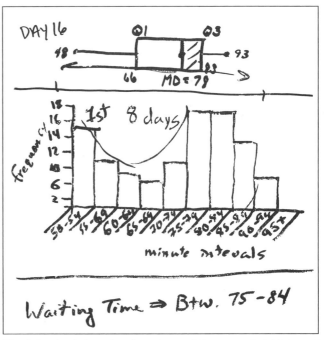

Fig. 2. Student's box plot of one day and histogram of eight days of Old Faithful data

Even more telling, an alternating short-long pattern in Old Faithful's eruptions is visually highlighted by students who create plots of consecutive wait times, dot plots, or bar graphs (**figs. 4** and **5**). This oscillating pattern can be completely missed by students who just calculate a mean or draw a box plot for the data.

Our past teaching, our textbooks, and many state and national assessments have concentrated heavily on measures of central tendency (mean, median, and mode), and we have neglected variation (Shaughnessy

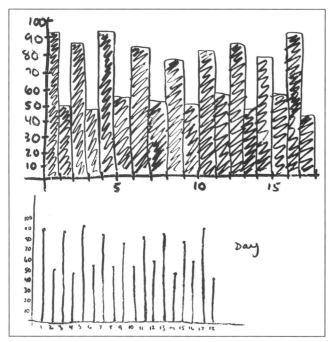

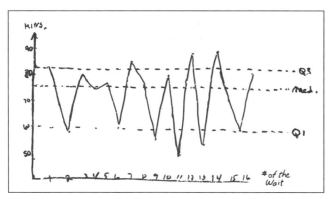

Fig. 3. Student's bar graph and dot plot of wait times for eighteen consecutive eruptions of Old Faithful

Fig. 4. Student's connected plot of sixteen consecutive wait times for Old Faithful eruptions

et al. 1999). However, variation is the essential signature in the Old Faithful data. Variation exists among days and within each day, and patterns in the variation can go completely unnoticed if we, or our students, concentrate only on centers and neglect variation. Students who pay attention to the variability in the data are much more likely to predict a range of outcomes or an interval for the wait time for Old Faithful. Such students make predictions similar to "Most of the time you'll wait between fifty and ninety minutes" rather than a single value of seventy minutes, as shown in the student work in **figure 1**.

The real power in exploring the Old Faithful data set arises when we ask students to share their graphical representations and predictions with one another. Students share a wide variety of graphical representations, and some students express surprise at some of their fellow students' graphs. We have even heard applause for some students who put up a plot of consecutive wait times after students have shared many box plots,

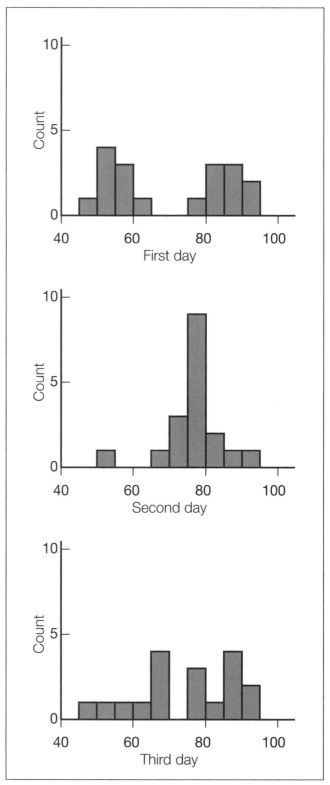

Fig. 5. Minutes between eruptions of Old Faithful; histograms for three different days

stem-and-leaf plots, and histograms. The plot of consecutive wait times is a powerful visual characterization of the alternating pattern in the Old Faithful data. Students who create plots of consecutive wait times often ask for more information. They want to know the length of the previous wait so that they can more accurately predict the next wait interval. They begin

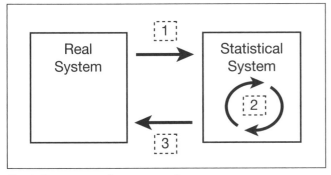

Fig. 6. The process of transnumeration

to make conjectures about the reasons that the data for Old Faithful alternate. They begin to ask what aspect of the geyser system causes this pattern of variation. They are beginning to show statistical thinking.

WHAT IS STATISTICAL THINKING?

The question "What is statistical thinking?" has provoked considerable debate. However, the central element of any definition of statistical thinking is an understanding of variation. According to Moore (1990, p. 135) the core element of statistical thinking is variation: "the omnipresence of variation in processes . . . the design of data production with variation in mind . . . the explanation of variation." Moore believes that students in the future will have a structure of thought that whispers "variation matters." The quality-management field believes that statistical thinking has three key principles: all work occurs in a system of interconnected processes, variation exists in all processes, and understanding and reducing variation are keys to success. Mallows (1998) believes that any definition of statistical thinking that does not include the relevance of the data to the problem is inadequate. Wild and Pfannkuch (1999) believe that statistical thinking is a complex activity, and they have identified five elements that are fundamental to statistical thinking in empirical inquiry in any field:

- Recognition of the need for data
- Transnumeration
- Consideration of variation
- Reasoning with statistical models
- Integrating the statistical and contextual

We are back at the second question posed at the beginning of this article. If these five elements are at the core of statistical thinking, then how can we develop statistical thinking in our students?

OLD FAITHFUL AND DEVELOPING STUDENTS' STATISTICAL THINKING— A DEEPER LOOK

Understanding variation is central to statistical thinking. This Old Faithful activity seeks to promote variation as the "big idea" to which students' attention

should be drawn. An aspect that should be considered in developing statistical thinking is that the reasoning processes are fundamentally different from those of mathematics, since statistical thinking deals with uncertain, empirical data. The student is placed in the role of a data detective. Students must look for patterns, deal with the variation, and make judgments and predictions on the basis of the data. We next discuss the five elements of statistical thinking and further illustrate the use of the Old Faithful data in promoting statistical thinking.

Recognition of the need for data

The foundations of statistical inquiry rest on the assumption that many real situations cannot be judged without gathering and analyzing properly collected data. Anecdotal evidence or one's own experience may be unreliable and misleading for judgments and for decision making. Therefore, data are considered a prime requirement for judgments about real situations. Our initial situation in the Old Faithful activity emphasizes the need for data, since students' predictions of the wait time for the next eruption are probably not within their own experience. A teaching focus should be on the need for data, because much research suggests that students think that their own judgments and beliefs are more reliable than data. In addition, students do not see any purpose in analyzing data, since they already know the "answer." This activity is one in which they are unlikely to know the answer and must therefore look at data to make a judgment.

Transnumeration

Transnumeration is a coined word, meaning "numeracy transformation for facilitating understanding." Transnumeration occurs in three specific phases in a statistical problem and can be viewed from a modeling perspective. A diagram is given in **figure 6**. Transnumeration is a dynamic process of changing representations to engender understanding. If we consider the real system and the statistical system, then trans numeration-type thinking occurs through—

- Capturing measures of the real system that are relevant,

- Constructing multiple statistical representations of the real system, and

- Communicating to others what the statistical system suggests about the real system.

In our Old Faithful activity, some relevant measures have already been "captured," namely, the students receive wait-time data on Old Faithful. The second phase of transnumerative thinking starts with a strip of one day's data, when students need to consider ways to change the data representation to facilitate a prediction. It also occurs when students are asked to draw a graph but are not told what type of graph to draw. Then, when they are asked to share their graphs with the whole class, they soon recognize that different representations convey different types of information about the geyser. The students' dot plots, stem-and-leaf plots, box plots, histograms with varying class intervals, and plots of consecutive wait times all reveal different messages about Old Faithful. Histograms with large class intervals can actually obscure information, and the students must therefore recognize this problem and try several different class-interval widths. To expedite matters, students may want to use Sturges's guideline, which suggests that the ideal number of class intervals is about $1 + \log_2 n$, where n is the number of data values. This sharing of graphs promotes transnumerative thinking and the need to look at multiple representations to detect messages in the data. The third phase of transnumeration begins to occur when the students are asked to communicate their predictions of wait time.

Consideration of variation

Making a judgment from data requires an understanding of variation during the process of statistical inquiry. To make an informed prediction, we first must notice that variation exists, either directly from the data or from the graphs of the data. When students first look at one day's data and then see that a classmate's data for a day look quite different, they begin to see variability from day to day, as well as within a single day. The variation that occurs in the data encourages students to request more data to improve their prediction.

In schools, the emphasis has historically been on descriptive statistics, especially on measures of center, and variation has been neglected. Most students initially report that the mean time of about seventy minutes is the length of time that they would expect to wait before the next eruption. Although it could be argued that this response is appropriate, focusing on one number as the solution addresses neither the variability in the system nor the pattern in the variability. As soon as we ask the question *why*, for example, "Why do patterns appear in the Old Faithful data? we enter the realm of what we call *analytical statistics*. Analytical statistics attempts to find explanations, seeks causes, makes predictions, and looks behind the data. Information about the why questions can be sought in patterns in variation in the data. Since the variation in the Old Faithful data is not random, underlying geological causes or relationships are likely; reasons exist for the variation. Thus, when the students share their graphs, we must encourage them to look at the graphs through a "variation lens" and so must encourage them to search for patterns in variation.

Reasoning with statistical models

According to the research of Konold and others (1997), when dealing with data, students have difficulty making the transition from thinking about and comparing individual cases to thinking about and comparing group propensities. Konold and his colleagues point out that reasoning with statistical models requires the

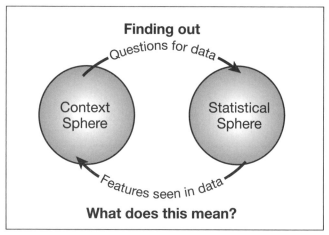

Fig. 7. The integration of statistical and contextual knowledge

ability to carry out both aggregate-based and individual-based reasoning and to recognize the power and limitations of such reasoning across a variety of situations. This aggregate-based reasoning, coupled with recognizing the patterns in the data set, is fundamental to statistical thinking.

When students are asked to draw a graph, they need to see that different days can produce different or similar patterns, but overall they need to see that the pattern or group propensity in Old Faithful is bimodal rather than unimodal. Teaching should focus on the patterns in distributions and patterns in the centers and spreads, not on the individual pieces of data for Old Faithful. Asking students whether knowing one wait time between eruptions would be sufficient for making a prediction may make them aware of the inadequacies of looking at individual examples.

After students have shared their graphs, we have found that they often decide that plots of consecutive wait times are the most appropriate models for depicting the distribution of the Old Faithful data, since such plots emphasize the oscillating character of the data. Thus, part of reasoning with statistical models involves selecting or creating a model that optimally represents and communicates the nature of the real problem and that focuses our reasoning about the data.

Integrating the statistical and contextual

The integration of statistical knowledge and contextual knowledge is a fundamental element of statistical thinking. The statistical model must capture elements of the real situation, and the resultant data carry their own literature base (Cobb and Moore 1997), that is, the data tell a story. Information about the real situation is contained in the statistical summaries, and a synthesis of statistical and contextual knowledge must therefore occur to draw out what can be learned in the context sphere.

At the beginning of the Old Faithful classroom activity, we discuss the context and general behavior of geysers to enable students to understand the meaning of the

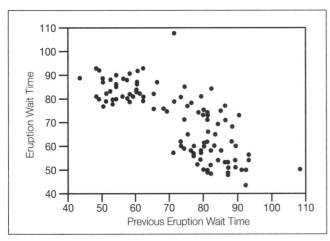

Fig. 8. Comparing adjacent eruption wait times

data. Such contextual knowledge is essential for seeing and interpreting any messages contained in the data. Students play the role of detectives who are looking for patterns to form their predictions; a continuous dialogue should exist between the data and context, as indicated in **figure 7**. For example, the pattern in this particular data set is bimodal and oscillating. Why do the wait times between eruptions oscillate about two mean times? What is the source of this variation? How does this geyser system work?

Another example occurs in a scatterplot of eruption wait times plotted against previous eruption wait time. This plot appears nonlinear, as shown in **figure 8**. This graph was generated using data from the first eight days. The sample correlation coefficient is –0.727. With the high influence point (108, 50) removed, the sample correlation coefficient is –0.722. The evidence of nonlinearity is weak. When a quadratic is fitted, with the high influence point removed, the P-value is 0.08. The nature of this relationship furnishes an opportunity for further investigation in an advanced class. Why do we find more variability in the wait time until the next eruption when the wait time since the previous eruption was a long one? Should other factors about Old Faithful be considered? What is the story contained in the data? Are other geyser systems the same?

ADDITIONAL QUESTIONS TO PROMOTE STUDENTS' STATISTICAL THINKING

Many opportunities occur throughout the Old Faithful investigation to promote further statistical thinking with students. Many of these opportunities can be tapped by asking well-placed questions that prompt students to think about and discuss ideas. A few of those opportunities follow:

Before students are given the data

How much data would you need for a prediction? One wait time? Two wait times? One day's worth of data? Two days' worth of data? A year's worth of data?

When they are given the data

How does your prediction using data compare with your first prediction without the data?

After they draw a graph of the first day's worth of data

What patterns, if any, do you notice?

After they draw graphs of two more days' data

How do your predictions compare with your previous ones?

When they share graphs

Compare and contrast the information revealed by each graph. What is gained or lost with the various graphical representations? How are the graphs related to one another?

When they are ready to interpret the information

What other data would be helpful, for example, duration time of eruptions, to enable you to further understand the wait time between blasts? What does the oscillating pattern tell you about how this geyser works? What other information on this geyser system would help you interpret this pattern?

When they are ready to make a final prediction

Will this prediction be true for the whole year? Will seasonal variation occur? Will this prediction be true over several years? Will yearly variation occur? What, if any, limitations should you put on your prediction? Is your prediction of oscillation valid? In one day, for example, the alternating pattern may be long, long, short, long, long, short, short, long, long, short, and so on. What graphical representation could you use to verify whether the alternating pattern of short and long wait times is generally true?

When they are ready to communicate the information to others

What graphical representation would best communicate your prediction? What other information, besides your prediction, should be communicated?

At the end of the inquiry

What have you learned about making predictions? About variation? How well do your predicted times compare with the actual wait times? If you continued with this problem, what would you investigate next? Can you find an explanation for the pattern? Does a relationship exist between duration times of eruptions and wait times between eruptions? Eruption-duration data for Old Faithful can be found in Foreman and Bennett (1999).

Additional project

Data are available on eruptions of other geysers and volcanoes, such as Kilauea on Hawaii from 1750 to the present (see **www.jason.org** and **hvo.wr.usgs.gov/**

kilauea/history/historytable). Gather information on wait times between eruptions and duration of eruptions for Kilauea or some other volcanoes or geysers. Compare the information with that for Old Faithful, and determine similarities and differences in the patterns of variability.

SO HOW FAITHFUL IS OLD FAITHFUL?

The answer to our original question depends on our statistical thinking. If we measure Old Faithful's wait times for "faithfulness" one eruption at a time, we might conclude that Old Faithful is not faithful at all. We might wait 49 minutes, or we might wait 102 minutes, 58 minutes, or 89 minutes. This pattern does not seem very faithful. However, perhaps wait time is not a good measure of "faithfulness." When we consider the overall pattern in the distribution of Old Faithful's wait times, we find that it is bimodal and oscillating. Using the pattern in the variation, we might be able to predict a time that is close to the wait time for the next eruption. Old Faithful is really very "faithful" to that overall pattern in the distribution of data. We hope that readers find this investigation as helpful for introducing and promoting statistical thinking with their students as we have.

REFERENCES

Cobb, George, and David Moore. "Mathematics, Statistics, and Teaching." *The American Mathematical Monthly* 104 (November 1997): 801–23.

Foreman, Linda, and Al Bennett. *Math Alive.* Course III. Salem, Ore.: The Math Learning Center, 1999.

Hand David J., Fergus Daly, A. Daniel Lunn, Kevin J. McConway, and Elizabeth Ostrowski, eds. *Handbook of Small Data Sets.* London: Chapman & Hall, 1994.

"Jason Project." **www.jason.org**.

Konold, Cliff, Alexander Pollatsek, Andrew Well, and Allen Gagnon. "Students Analyzing Data: Research of Critical Barriers." In *Research on the Role of Technology in Teaching and Learning Statistics,* Proceedings of the 1996 International Association of Statistics Education Round Table Conference, edited by Joan Garfield and Gail Burrill, pp. 151–67. Voorburg, the Netherlands: International Statistics Institute, 1997.

Mallows, Colin. "The Zeroth Problem." *The American Statistician* 52 (February 1998): 1–9.

Moore, David. "Uncertainty." In *On the Shoulders of Giants,* edited by Lynn Arthur Steen, pp. 95–137. Washington, D.C.: National Academy Press, 1990.

Shaughnessy, J. Michael, Jane Watson, Jonathon Moritz, and Christine Reading. "School Mathematics Students' Acknowledgment of Statistical Variation." Paper presented at the research presessions of the 77th annual meeting of the National Council of Teachers of Mathematics, San Francisco, April 1999.

Tukey, John. *Exploratory Data Analysis.* Reading, Mass.: Addison Wesley Longman, 1977.

U.S. Geological Survey, Hawaiian Volcano Observatory. "Summary of Historical Eruptions, 1750–Present." **hvo.wr.usgs.gov/kilauea/history/historytable. html**.

Wild, Chris J., and Maxine Pfannkuch. "Statistical Thinking in Empirical Enquiry. *International Statistical Review* 67 (1999): 223–65.

Yellowstone National Park: National Park Service. **www.nps.gov/yell/**.

Eruptions of the Old Faithful Geyser: Becoming a Data Detective

Sheet 1

Data on wait times between successive eruptions (blasts) of geysers were first collected by the National Park Service and the U.S. Geological Survey in Yellowstone National Park. The data were collected to establish some baseline information that could then be used to track and compare long-term behavior of geysers. If you had some friends who were planning to visit Yellowstone National Park, how long should you tell them to expect to wait between eruptions of Old Faithful? Using the 16 days of data below, estimate your friends' wait time and justify your prediction.

Old Faithful—Minutes Between Blasts
Each row represents about one day's data

1	86	71	57	80	75	77	60	86	77	56	81	50	89	54	90	73	60	83
2	65	82	84	54	85	58	79	57	88	68	76	78	74	85	75	65	76	58
3	91	50	87	48	93	54	86	53	78	52	83	60	87	49	80	60	92	43
4	89	60	84	69	74	71	108	50	77	57	80	61	82	48	81	73	62	79
5	54	80	73	81	62	81	71	79	81	74	59	81	66	87	53	80	50	87
6	51	82	58	81	49	92	50	88	62	93	56	89	51	79	58	82	52	88
7	52	78	69	75	77	53	80	55	87	53	85	61	93	54	76	80	81	59
8	86	78	71	77	76	94	75	50	83	82	72	77	75	65	79	72	78	77
9	79	75	78	64	80	49	88	54	85	51	96	50	80	78	81	72	75	78
10	87	69	55	83	49	82	57	84	57	84	73	78	57	79	57	90	62	87
11	78	52	98	48	78	79	65	84	50	83	60	80	50	88	50	84	74	76
12	65	89	49	88	51	78	85	65	75	77	69	92	68	87	61	81	55	93
13	53	84	70	73	93	50	87	77	74	72	82	74	80	49	91	53	86	49
14	79	89	87	76	59	80	89	45	93	72	71	54	79	74	65	78	57	87
15	72	84	47	84	57	87	68	86	75	73	53	82	93	77	54	96	48	89
16	63	84	76	62	83	50	85	78	78	81	78	76	74	81	66	84	48	93

Task setup

- Either on your own or in pairs, pick any row of these wait times so that you have a sample of a day of Old Faithful wait times.

- Look over the data. Is there anything that you notice, or anything that you wonder about in your sample of data? Jot down some notices and wonders.

Graphical representations

- Create at least one type of visual or graphical representation for that row of data to help to visualize any patterns in the wait times.

- Continue to jot down any additional notices and wonders that occur to you.

- Continue to think about your prediction for the wait times between blasts of Old Faithful.

- Create visual or graphical representations for one additional day of data.

- What do you notice and wonder about?

Share, decide, and compare

- Share notices and wonders with another person or another pair.

- How does the information revealed in your visual or graphical representation help you make a prediction? Would you change your prediction? Why?

- On the basis of your data and graphs, make a group decision about how long you would expect to wait between blasts of Old Faithful.

- Which visual or graphical representation best communicates your prediction? What other information, besides your prediction, is communicated?

- Be prepared to present your graph to the other groups and to defend your group's data-based prediction for the expected wait time.

Connecting Independence and the Chi-Square Statistic

Wes White

The chi-square statistic is used to test goodness of fit, homogeneity of proportions, and independence. Although the first two uses were at least somewhat new to my students, the concept of independence was not. This article begins by reviewing the use of the chi-square statistic for goodness of fit and then looks at how I have helped my students relate their earlier work with independence to the use of the chi-square statistic.

A goodness-of-fit test measures how closely the observed values of a sample fit an expected, ideal ratio. For example, suppose that we want to determine whether a die is balanced, that is, whether each face appears one-sixth of the time when the die is rolled.

We test the die by rolling it sixty times. Ideally, sixty rolls should produce exactly ten rolls of each face: ten 1s, ten 2s, . . . , up to ten 6s. How would we interpret the results? No one would require exactly ten rolls of each face to conclude that the die is balanced. However, if thirty of the sixty rolls came up a 5, for instance, we would conclude that the die is unbalanced. So, just where do we draw the line between a balanced die and an unbalanced die? To help us decide, we need a statistic to quantify how far the sample data have strayed from the expected values.

The chi-square statistic helps with this decision. We suppose that our sixty rolls produced nine, thirteen, thirteen, nine, six, and ten rolls of the faces 1, 2, 3, 4, 5, and 6, respectively. For the goodness-of-fit test, we follow these steps:

1. We find the expected values. We would ideally expect ten rolls for each of the six faces of the die, since one-sixth of sixty is ten.

2. We calculate the chi-square statistic:
$$\sum \frac{(O - E)^2}{E},$$
where O stands for the observed, that is, actual, values and E stands for the expected values that would match the ideal ratio. We expect ten of each face. We calculate as follows:
$$\sum \frac{(O - E)^2}{E} = \frac{(9 - 10)^2}{10} + \frac{(13 - 10)^2}{10} + \frac{(13 - 10)^2}{10}$$
$$+ \frac{(9 - 10)^2}{10} + \frac{(6 - 10)^2}{10} + \frac{(10 - 10)^2}{10}$$
$$\approx 3.6.$$

Each numerator is positive after squaring, and the further that the observed values stray from their expected value of 10, the greater the fraction.

3. We next compare our calculated 3.6 with some threshold, or critical, value in the chi-square table. See **table 1**. These critical values depend on the number of fractions summed. Since a die has six faces, we add six fractions. Our degree of freedom (d.f.) is one less than this number:
$$\text{d.f.} = n - 1 = 6 - 1 = 5.$$

Table 1					
The Chi-Square Table (Partial)					
Degrees of Freedom	*p*-Values				
	0.700	0.500	0.050	0.020	0.010
1	0.148	0.455	3.841	5.412	6.635
2	0.713	1.386	5.991	7.824	9.210
3	1.424	2.366	7.815	9.837	11.340
4	2.195	3.357	9.488	11.670	13.280
5	3.000	4.351	11.070	13.390	15.090

4. We find the critical chi-square range limits that bracket 3.6. In the row with 5 degrees of freedom, we see that our calculated 3.6 falls between 3.000 and 4.351. The top row then tells us that our *p*-value is between 0.50 and 0.70.

5. We interpret this result as follows: We suppose that a die was perfectly balanced. It would give results as extreme as these, or more extreme, in varying from the expected values somewhere between 50 percent and 70 percent of the time. This *p*-value is very high, that is, our results are common for a balanced die. We can conclude that our die is balanced.

But if this probability had been less than 5 percent, we might reject the notion that the die is balanced. The reader may want to repeat the same five steps using the following set of sixty rolls from a different die: five 1s, eleven 2s, four 3s, nineteen 4s, nine 5s, and twelve 6s. The resulting chi-square statistic is 14.8. The corresponding *p*-value is between 0.01 and 0.02. See **table 1**. We infer from the new data that these results were very unlikely—with less than 2 percent probability—to have occurred with a balanced die. Therefore, the data seem to indicate that the die is unbalanced.

With my Advanced Placement and college-level statistics classes, I covered the use of the chi-square test to measure goodness of fit and then looked at how to use the chi-square statistic in connection with independence.

Two events are considered independent if the occurrence of one of them does not alter the probability of the other. My students knew the following fundamental property of independence: if two events are independent, then the joint probability—the probability that both will happen—equals the product of their individual probabilities.

I began the connection between independence and chi square with the following example:

"Suppose," I told my class, "that we want to know if a basketball team's performance is independent of whether it plays at home or away. See **table 2** for its season record so far.

Table 2			
The Team's Record This Season At Home and Away			
	Wins	Losses	Row Totals
Home	22	8	30
Away	13	7	20
Column totals	**35**	**15**	**50**

"We will treat these results as a sample. To gain greater insight from them, we will include row percents. See **table 3**.

Table 3			
The Team's Record This Season At Home and Away (*Including Row Percents*)			
	Wins	Losses	Row Totals
Home	22 (73%)	8 (27%)	30 (100%)
Away	13 (65%)	7 (35%)	20 (100%)
Column totals	**35**	**15**	**50**

"As you can see, the team won 73 percent of its home games, compared with 65 percent of its away games. So it appears that the team does better at home than away—that its performance does depend on the site. Why is that conclusion not the final answer to the question?

"This table is merely a sample. As with all hypothesis tests, we want to know whether the difference suggested by these results would hold up over many games or seasons or whether they can be attributed to sample variability.

"If the team's performance were truly independent of the site over this fifty-game sample, what numbers would we expect in each cell of the matrix?"

I had students start with the upper-left cell—the number of home wins. We let event W mean that the team wins and event H mean that the team plays at home, and I asked students to find the probability of event (W and H), meaning that the team wins at home, assuming that the events are independent.

They saw that they could use $P(W$ and $H) = P(W) \times P(H)$. To get the values for $P(W)$ and $P(H)$, they were helped by thinking of these probabilities in terms of relative frequency. The team plays thirty of its fifty games at home, so $P(H) = 30/50 = 0.6$. Similarly, since the team won thirty-five of fifty games, $P(W) = 35/50 = 0.7$. Thus, if winning and playing at home were independent, we would have $P(W$ and $H) = P(W) \times P(H) = 0.7 \times 0.6 = 0.42$.

I made sure that my students realized the meaning of this result: assuming independence, the team would have won twenty-one home games out of its fifty total games, since $0.42 \times 50 = 21$. So they put twenty-one in the upper-left cell of the table. **Table 4** shows these expected values in parentheses.

Table 4			
The Team's Record This Season At Home and Away (*Including Expected Values*)			
	Wins	Losses	Row Totals
Home	22 (21)	8 (9)	30
Away	13 (14)	7 (6)	20
Column totals	**35**	**15**	**50**

By a similar calculation, the students found the number of home losses that we would expect if independence is assumed:

$$P(H) \times P(L) = 30/50 \times 15/50$$
$$= 0.6 \times 0.3$$
$$= 0.18$$

Then $0.18 \times 50 = 9$. They wrote 9 in the upper-right cell, as shown in **table 4**. They continued in this manner, finding the expected values for the other cells of the matrix.

As they filled in these values, arriving at the "observed-expected chi-square table" (**table 4**), I helped them focus on this essential observation: The definition of independence from probability can, by itself, create the expected-value matrix.

I asked the students whether any of them were convinced that the team plays differently at home than they do away. Only two students in the high school class and one in the college class were convinced.

I pointed out that each cell differed from its expected value by 1 and asked the students how they could evaluate the significance of this difference. This question led to the following computation using the chi-square statistic:

$$\sum \frac{(O-E)^2}{E} = \frac{(22-21)^2}{21} + \frac{(8-9)^2}{9} + \frac{(13-14)^2}{14} + \frac{(7-6)^2}{6}$$
$$= 3.97.$$

To interpret the resulting value of 0.397, we needed to find the degree of freedom. I explained that in a table like this one, we use $(r-1)(c-1)$, where r is the number of rows and c is the number of columns. Thus, we have d.f. $= (r-1)(c-1) = (2-1)(2-1) = 1$.

Referring to the first row of the chi-square table (**table 1**), the students saw that 0.397 was between 0.148 and 0.455, so it corresponds to a p-value of between 50 percent and 70 percent. We later worked this problem on the TI-83 calculator and obtained the precise value, $p = 0.5287333251$, or roughly 53 percent.

This result led to the following conclusion: If we assume independence between performance and site, the chance is 53 percent that we would get these—or more extreme—results purely by sample variability, that is, by luck. Because the p-value is so high, we accept that the team's winning or losing is independent of site.

Before summarizing the discussion, I wanted students to better understand why we use $(O-E)^2/E$—the expression defining the chi-square statistic. I reviewed that we had just considered a season record in which a team won 73 percent of its home games compared with only 65 percent of its away games and concluded that the outcomes of the games were independent of the site.

I next had the students consider what would happen if we multiplied every observed value by 10. See **table 5**. I helped them see that all the row percents would be preserved, so the team would again be winning 73 percent of its home games compared with only 65 percent of its away games. And I asked whether they would still reach the conclusion of independence.

Table 5

The Team's Record (Including Expected Values)

	Wins	Losses	Row Totals
Home	220 (210)	80 (90)	300
Away	130 (140)	70 (60)	200
Column totals	**350**	**150**	**500**

Again they voted. This time, healthy minorities in both classes believed that independence might be rejected. I asked one student why he reached a different conclusion even though the proportions were the same. He said, "The high proportion of winning at home has continued longer, and that changes things." He was right. We did the analogous computation after creating **table 5** and found that the p-value had dropped to 0.046, giving significant evidence and allowing us to reject independence at the 0.05 level.

Finally, I asked the students how the calculation of

the chi-square statistic accounted for the intuitive fact that when unequal proportions persist over a longer time and with a larger sample, independence becomes less tenable. The answer is that the chi-square statistic uses squares in the numerator but not in the denominator. The new, larger denominators do not keep pace with the new, larger, squared numerators, so the fraction total increases.

My presentations to both high school and college statistics classes have evolved somewhat. On the first day, I cover the goodness-of-fit method and give a homework assignment. On the second day, I do what has been described in this article. On both teaching levels, these two lessons have been well received: the goodness-of-fit presentation because it is somewhat intuitive to students, and the independence presentation because it is strongly linked with the previously studied concept of independence from probability.

SOLUTIONS

Part One: The Goodness-of-Fit Test— *Sheet 1*:

1)–2)

Type	O	A	B	AB
Observed	35	42	16	7
Expected	(45)	(40)	(11)	(4)

3)
$(35-45)^2 / 45 + (42-40)^2 / 40 + (16-11)^2 / 11 + (7-4)^2 / 4 = 6.844949495$

4) d. f. $= c - 1 = 4$ categories $- 1 = 3$

5) Hit 2nd DISTR. Arrow down to X^2cdf. Input X^2cdf(minimum, maximum, d.f.) $= X^2$cdf(6.844949495, E99, 3) = 0.077. So the p-value is 0.077

6) Because p-value $= 0.077 > 0.05$ (the commonly used threshold for a significantly rare event), we cannot reject the blood type population proportions being tested. Our data are consistent with these proportions—but not by much.

Part Two: Independence — *Sheet 2*:

1)

Game Type	Wins	Losses	Total
Home	25 (19.5)	5 (10.5)	30
Away (road)	14 (19.5)	16 (10.5)	30
Total	**39**	**21**	**60**

2) $(25-19.5)^2 / 19.5 + (5-10.5)^2 / 10.5 + (14-19.5)^2 / 19.5 + (16-10.5)^2 / 10.5 = 8.864468864$

3) d.f. $= (r-1) \times (c-1) = 1 \times 1 = 1$

4) Hit 2nd DISTR. Arrow down to X^2cdf. Input X^2cdf(minimum, maximum, d.f.) = X^2cdf(8.864468864, E99, 1) = 0.00290775. So the p-value is 0.0029

5) Since p-value = 0.0029 < 0.05, we reject that the Denver Nuggets' performance is independent of whether they are at home or on the road. We have statistically significant evidence that their performance does depend on where they are playing.

Part Two: Independence — *Sheet 3:*

1)

Game Type	Wins	Losses	Total
Home	18 (15.5)	13 (*15.5*)	31
Away (road)	12 (*14.5*)	17 (*14.5*)	29
Total	**30**	**30**	**60**

2) $(18 - 15.5)^2 / 15.5 + (13 - 15.5)^2 / 15.5 + (12 - 14.5)^2 / 14.5 + (17 - 14.5)^2 / 14.5 = 1.668520578$

3) d.f. $= (r - 1) \times (c - 1) = 1 \times 1 = 1$

4) Hit 2nd DISTR. Arrow down to X^2cdf. Input X^2cdf(minimum, maximum, d.f.) = X^2cdf(1.668520578, E99, 1) = 0.1964568043. So the p-value is 0.196.

5) Since p-value = 0.196 > 0.05, we fail to reject that the Memphis Grizzlies' performance is independent of whether they are at home or on the road. We do not have statistically significant evidence that their performance does depend on where they are playing. In other words, the Memphis Grizzlies' performance, through March 2, 2010, is consistent with their performance being independent of whether they are playing a home or a road game.

Part One: *The Goodness-of-Fit Test*

Medical records for people in the United States show the following percentages of the four blood types:

Type	O	A	B	AB
Population Proportion	45%	40%	11%	4%

You believe that these population proportions may be wrong. So you perform a random sample of $n = 100$ people and find these sample results:

Type	O	A	B	AB
Observed	35	42	16	7

Do these results provide statistically significant evidence that at least one of the four population proportions is wrong?

Steps:

1. Find the expected values. For each blood type, calculate the percentage of the sample size $n = 100$.

2. Write these in parentheses and enter them into the table.

Type	O	A	B	AB
Observed	35	42	16	7
Expected				

3. Now calculate the chi-square statistic: $\sum(O - E)^2/E$

4. Determine the degrees of freedom. For goodness of fit, d.f. $= c - 1$, where c is the number of categories.

5. Now, use the X^2cdf function on a TI-83 or TI-84 graphing calculator, or use a chi-square table to determine the p-value.

6. Draw a conclusion.

Part Two: *Independence*

Let's compare the current home and away records of two NBA basketball teams: the Denver Nuggets and the Memphis Grizzlies. The question is which team, if either, has the same ability to win its away (road) games as its home games. When a team has this ability, we say that its performance is independent of home vs. road venue.

The Denver Nuggets' Record as of March 2, 2010

Game Type	Wins	Losses	Total
Home	25	5	30
Away (road)	14	16	30
Total	**39**	**21**	**60**

Is the Denver Nuggets' performance independent of home vs. road venue?

Steps:

1. Find the expected number of home wins, home losses, road wins, and road losses, assuming independence. Place it in parentheses in the table.

   Example: Expected number of home wins = total games × Prob(home game) × Prob(win) = 60 × 30/60 × 39/60 = 19.5.

Game Type	Wins	Losses	Total
Home	25 (19.5)	5 (___)	30
Away (road)	14 (___)	16 (___)	30
Total	**39**	**21**	**60**

2. Now calculate the chi-square statistic: $\Sigma(O - E)^2/E$.

3. Determine the degrees of freedom. For independence, d.f. $= (r - 1) \times (c - 1)$, where r is the number of columns and c is now the number of categories.

4. Use the X^2cdf function on a TI-83 or TI-84 graphing calculator, or use a chi-square table to determine the p-value.

5. Draw a conclusion.

Part Two: *Independence*

Sheet 3

Is the Memphis Grizzlies' performance independent of home vs. road venue?

The Memphis Grizzlies' Record as of March 2, 2010			
Game Type	*Wins*	*Losses*	*Total*
Home	18	13	31
Away (road)	12	17	29
Total	**30**	**30**	**60**

Steps:

1. Find the expected number of home wins, home losses, road wins, and road losses, assuming independence. Place it in parentheses in the table.

   Example: Expected number of home wins = total games × Prob(home game) × Prob(win) = 60 × 31/60 × 30/60 = 15.5.

2. Now calculate the chi-square statistic: $\Box(O - E)^2/E$.

3. Determine the degrees of freedom. For independence, d.f. = $(r - 1) \times (c - 1)$, where r is the number of columns and c is now the number of categories.

4. Use the X^2cdf function on a TI-83 or TI-84 graphing calculator, or use a chi-square table to determine the *p*-value.

5. Draw a conclusion.

Conclusion to Part Two: Independence

The Denver Nuggets' record through March 2, 2010, shows significantly that the team's performance does depend on whether the games are at home or away. In contrast, the results for the Memphis Grizzlies over the same period are consistent with the irrelevance of home vs. away venue.

Appendix A

Relationship of Mathematics Teacher Activities to the Common Core State Standards for Mathematics (CCSSM)

In instances when an appropriate content standard from CCSSM was not available, such as with Willcutt (1973) below, we included the mathematical practice standard that could be represented by the activity. The lack of appearance of the standards for mathematical practice with other activities does not indicate that those activities do not address a mathematical practice standard.

Author and Title	Corresponding Topic Clusters from CCSSM
Chapter 1 Number and Measurement	
Albrecht (2001) *The Volume of a Pyramid: Low-Tech and High-Tech Approaches*	Geometry—Geometric Measurement and Dimension • Explain volume formulas and use them to solve problems
Çağlayan (2006) *Visualizing Summation Formulas*	Functions—Linear, Quadratic, and Exponential Models • Construct and compare linear, quadratic, and exponential models and solve problems
Hansen and Lewis (2007) *Finding a Parking Spot for the Binomial Theorem*	Algebra—Arithmetic with Polynomials and Rational Expressions • Use polynomial identities to solve problems
Herman, Milou, and Schiffman (2004) *Unit Fractions and Their "Basimal" Representations: Exploring Patterns*	Grade 8—The Number System • Know that there are numbers that are not rational, and approximate them by rational numbers
Hill (2002) *Print-Shop Paper Cutting: Ratios in Algebra*	Number and Quantity—Quantities • Reason quantitatively and use units to solve problems
Olson (1991) *A Geometric Look at Greatest Common Divisor*	Grade 6—The Number System • Compute fluently with multidigit numbers and find common factors and multiples
Slowbe (2007) *Pi Filling, Archimedes Style*	Geometry—Circles • Understand and apply theorems about circles
Wagner (2003) *We Have a Problem Here: 5 + 20 = 45?*	Grade 8—Expressions and Equations • Work with radicals and integer exponents
Willcutt (1973) *Paths on a Grid*	Mathematical Practice • Look for and make use of structure
Chapter 2 Algebraic Symbols	
Gamble (2005) *Teaching Logarithms: Day One*	Functions—Building Functions • Build new functions from existing functions

Author and Title	Corresponding Topic Clusters from CCSSM
House (1987) *An Electrifying Introduction to Algebra*	Algebra—Seeing Structure in Expressions • Interpret the structure of expressions • Write expressions in equivalent forms to solve problems
Johnson (1986) *Making –x Meaningful*	Algebra—Seeing Structure in Expressions • Interpret the structure of expressions
Kinach (1985) *Solving Linear Equations Physically*	Algebra—Reasoning with Equations and Inequalities • Understand solving equations as a process of reasoning and explain the reasoning • Solve equations and inequalities in one variable
Kobayashi (2006) *Relations among Powers of 2,* *Combinations, and Symbolic Algebra*	Algebra—Seeing Structure in Expressions • Interpret the structure of expressions • Write expressions in equivalent forms to solve problems
Leiva (1980) *Math Magic*	Algebra—Seeing Structure in Expressions • Interpret the structure of expressions
Uth (1955) *Teaching Aid for Developing* *(a + b)(a—b)*	Algebra—Arithmetic with Polynomials and Rational Expressions • Use polynomial identities to solve problems
Vandyk (1990) *Expressions, Equations, and Inequalities*	Algebra—Reasoning with Equations and Inequalities • Solve equations and inequalities in one variable Functions—Interpreting Functions • Understand the concept of a function and use function notation

Chapter 3 Functions

Andersen (1973) *Griefless Graphing for the Novice*	Functions—Interpreting Functions • Analyze functions using different representations Functions—Building Functions • Build new functions from existing functions
Davidenko (1997) *Building the Concept of Function from* *Students' Everyday Activities*	Functions—Interpreting Functions • Interpret functions that arise in applications in terms of the context • Analyze functions using different representations
Day (1993) *Solution Revolution*	Algebra—Reasoning with Equations and Inequalities • Represent and solve equations and inequalities graphically Functions—Interpreting Functions • Analyze functions using different representations

Author and Title	Corresponding Topic Clusters from CCSSM
Edwards and Chelst (1999) *Promote Systems of Linear Inequalities with Real-World Problems*	Algebra—Reasoning with Equations and Inequalities • Represent and solve equations and inequalities graphically Functions—Interpreting Functions • Interpret functions that arise in applications in terms of the context • Analyze functions using different representations Functions—Building Functions • Build a function that models a relationship between two quantities Functions—Linear, Quadratic, and Exponential Models • Interpret expressions for functions in terms of the situation they model
Hershkowitz, Arcavi, and Eisenberg (1987) *Geometrical Adventures in Functionland*	Functions—Interpreting Functions • Understand the concept of a function and use function notation • Interpret functions that arise in applications in terms of the context • Analyze functions using different representations Functions—Building Functions • Build a function that models a relationship between two quantities Functions—Linear, Quadratic, and Exponential Models • Construct and compare linear, quadratic, and exponential models and solve problems
Moore-Russo and Golzy (2005) *Helping Students Connect Functions and their Representations*	Functions—Interpreting Functions • Analyze functions using different representations Functions—Building Functions • Build new functions from existing functions Functions—Linear, Quadratic, and Exponential Models • Construct and compare linear, quadratic, and exponential models and solve problems
Moyer (2006) *Non-Geometry Mathematics and The Geometer's Sketchpad*	Functions—Interpreting Functions • Analyze functions using different representations Functions—Building Functions • Build new functions from existing functions Functions—Linear, Quadratic, and Exponential Models • Construct and compare linear, quadratic, and exponential models and solve problems

Author and Title	Corresponding Topic Clusters from CCSSM
Peterson (2006) *Linear and Quadratic Change:* *A Problem from Japan*	Functions—Interpreting Functions • Interpret functions that arise in applications in terms of the context • Analyze functions using different representations Functions—Building Functions • Build a function that models a relationship between two quantities Functions—Linear, Quadratic, and Exponential Models • Construct and compare linear, quadratic, and exponential models and solve problems
Van Dyke (2003) *Using Graphs to Introduce Functions*	Functions—Interpreting Functions • Interpret functions that arise in applications in terms of the context • Analyze functions using different representations Functions—Building Functions • Build a function that models a relationship between two quantities Functions—Linear, Quadratic, and Exponential Models • Construct and compare linear, quadratic, and exponential models and solve problems

Chapter 4 Geometry

Edwards (2005) *Using Overhead Projectors to Explore* *Size Change Transformations*	Geometry—Similarity, Right Triangles, and Trigonometry • Understand similarity in terms of similarity transformations
Froelich (2000) *Modeling Soft Drink Packaging*	Geometry—Modeling with Geometry • Apply geometric concepts in modeling situations Geometry—Geometric Measurement and Dimension • Explain volume formulas and use them to solve problems
Gernes (1999) *The Rules of the Game*	Mathematical Practice • Construct viable arguments and critique the reasoning of others
Hirsch (1974) *Pick's Rule*	Functions—Building Functions • Build a function that models a relationship between two quantities
Nelson and Williams (2007) *Sprinklers and Amusement Parks:* *What Do They Have to Do with* *Geometry?*	Geometry—Modeling with Geometry • Apply geometric concepts in modeling situations Geometry—Circles • Understand and apply theorems about circles

Author and Title	Corresponding Topic Clusters from CCSSM
Palmer (1946) *Discovering the Tangent*	Geometry—Similarity, Right Triangles, and Trigonometry • Define trigonometric ratios and solve problems involving right triangles
Quinn and Ball (2007) *Explore, Conjecture, Connect, Prove: The Versatility of a Rich Geometry Problem*	Geometry—Expressing Geometric Properties with Equations • Use coordinates to prove simple geometric theorems algebraically
Reys (1988) *Discovery with Cubes*	Grade 7—Geometry • Solve real-life and mathematical problems involving angle measure, area, surface area, and volume Functions—Building Functions • Build a function that models a relationship between two quantities
Toumasis (1992) *The Toothpick Problem and Beyond*	Functions—Building Functions • Build a function that models a relationship between two quantities Grade 7—Geometry • Draw, construct, and describe geometrical figures and describe the relationships between them

Chapter 5 Statistics and Probability

Bryan (1988) *Exploring Data with Box Plots*	Statistics and Probability—Interpreting Categorical and Quantitative Data • Summarize, represent, and interpret data on a single count or measurement variable
Franklin and Mulekar (2006) *Is Central Park Warming?*	Statistics and Probability—Interpreting Categorical and Quantitative Data • Summarize, represent, and interpret data on a single count or measurement variable Statistics and Probability—Making Inferences and Justifying Conclusions • Understand and evaluate random processes underlying statistical experiments Statistics and Probability—Using Probability to Make Decisions • Calculate expected values and use them to solve problems • Use probability to evaluate outcomes of decisions
Groth and Powell (2004) *Using Research Projects to Help Develop High School Students' Statistical Thinking*	Statistics and Probability—Interpreting Categorical and Quantitative Data • Summarize, represent, and interpret data on two categorical and quantitative variables • Interpret linear models

Author and Title	Corresponding Topic Clusters from CCSSM
Lappan, Phillips, Fitzgerald, and Winter (1987) *Area Models and Expected Values*	Statistics and Probability—Making Inferences and Justifying Conclusions • Make inferences and justify conclusions from sample surveys, experiments, and observational studies Statistics and Probability—Conditional Probability and the Rules of Probability • Understand independence and conditional probability and use them to interpret data • Use the rules of probability to compute probabilities of compound events in a uniform probability model Statistics and Probability—Using Probability to Make Decisions • Calculate expected values and use them to solve problems • Use probability to evaluate outcomes of decisions
McGivney-Burelle, McGivney, and McGivney (2008) *Investigating the Relationship between Latitude and Temperature*	Statistics and Probability—Interpreting Categorical and Quantitative Data • Summarize, represent, and interpret data on two categorical and quantitative variables • Interpret linear models
Perry and Kader (1998) *Counting Penguins*	Statistics and Probability—Making Inferences and Justifying Conclusions • Understand and evaluate random processes underlying statistical experiments • Make inferences and justify conclusions from sample surveys, experiments, and observational studies Statistics and Probability—Using Probability to Make Decisions • Use probability to evaluate outcomes of decisions
Richardson and Gabrosek (2004) *Activities: A-B-C, 1-2-3*	Statistics and Probability—Interpreting Categorical and Quantitative Data • Summarize, represent, and interpret data on two categorical and quantitative variables • Interpret linear models
Shaugnessy and Pfannkuch (2002) *How Faithful Is Old Faithful?* *Statistical Thinking: A Story of Variation and Prediction*	Statistics and Probability—Interpreting Categorical and Quantitative Data • Summarize, represent, and interpret data on two categorical and quantitative variables Statistics and Probability—Making Inferences and Justifying Conclusions • Understand and evaluate random processes underlying statistical experiments • Make inferences and justify conclusions from sample surveys, experiments, and observational studies

Author and Title	Corresponding Topic Clusters from CCSSM
White (2001) *Connecting Independence and the* *Chi-Square Statistic*	Statistics and Probability—Making Inferences and Justifying Conclusions • Understand and evaluate random processes underlying statistical experiments • Make inferences and justify conclusions from sample surveys, experiments, and observational studies Statistics and Probability—Using Probability to Make Decisions • Calculate expected values and use them to solve problems • Use probability to evaluate outcomes of decisions

Appendix B

Additional Articles on *www.nctm.org/more4u*

1. **Number and Measurement**

   • Foster, Drew W. "Diving In Head First: Finding the Volume of Norris Lake" *Mathematics Teacher* 102 (September 2008): 90–97. Finding the volume of a lake using technology

   • Hall, Randy. "Get the Most Pop for Your Buck!" *Mathematics Teacher* 101 (April 2008): 609–13. Dimensional analysis

   • Utley, Juliana, and John Wolfe. "Geoboard Areas: Students' Remarkable Ideas." *Mathematics Teacher* 97 (January 2004): 18–26. Geoboards and area

2. **Algebraic Symbols**

   • Margulies, Susan. "Algebra A_ _ _ _ _ _ _ _ s." *Mathematics Teacher* 86 (January 1993): 40-41. Truth value of state relationships

3. **Functions**

   • Blubaugh, William L., and Kristin Emmons. "Algebra for All: Graphing for All Students." *Mathematics Teacher* 92 (April 1999): 323–26. Popcorn graphs

   • Holliday, Berchie W., and Lauren R. Duff. "Using Graphing Calculators to Model Real-World Data." *Mathematics Teacher* 97 (May 2004): 328–41.—Correlation

   • Horak, Virginia M. "Biology as a Source for Algebra Equations: Insects." *Mathematics Teacher* 99 (November 2005): 55–9.Direct variation

   • Kasprzak, Edward M. "Design a Window." *Mathematics Teacher* 95 (May 2002): 346–59. Problem solving

   • Rauff, James V. "A Millennium Prize Problem for Students." *Mathematics Teacher* 96 (January 2003): 26–39. Decision problems

4. **Geometry**

   • Brown, Betty. "Exponential Growth through Pattern Exploration." *Mathematics Teacher* 98 (February 2005): 434–42. Using iterations to form a Sierpinski triangle

   • Ebert, Dave. "Using Statistical Testing to Approximate π." *Mathematics Teacher* 100 (October 2006): 216–19. Approximating π

   • Gould, S. Louise. "The Tellem Weavers Meet the Graphing Calculator." *Mathematics Teacher* 99 (November 2005): 230–36. Patterns in weaving

   • Jung, Inchul, and Yunghwan Kim. "Using Geometry Software to Revisit the Ellipse." *Mathematics Teacher* 97 (March 2004): 184–91. Ellipses

   • Lege, Steve. "Why Not Three Dimensions?" *Mathematics Teacher* 92 (October 1999): 560–63. Modeling three-dimensional objects

   • Madden, Sean P., James P. Downing, and Jocelyn M. Comstock. "Paper Moon: Simulating a Total Solar Eclipse." *Mathematics Teacher* 99 (January 2006): 312–20. Modeling a solar eclipse

   • Offerman, Theresa Reardon. "Foam Images." *Mathematics Teacher* 92 (May 1999): 391–99. Modeling cross-sections

   • Piatek-Jimenez, Katrina. "Building Intuitive Arguments for the Triangle Congruence Conditions." *Mathematics Teacher* 101 (February 2008): 463–66. Triangle congruence

5. Statistics and Probability

- Beseler, Susan. "The Three-Point Shoot-Out: The Logic of Hypothesis Testing." *Mathematics Teacher* 99 (April 2006): 582–87. Hypothesis testing

- Kader, Gary D., and Christine A. Franklin. "The Evolutions of Pearson's Correlation Coefficient." *Mathematics Teacher* 102 (November 2008); 292–99. Correlation

- Keller, Brian A., and Heather A. Thompson. "Whelk-Come to Mathematics." *Mathematics Teacher* 92 (September 1999): 475–81, 485–89. Modeling a biological phenomenon

- Kranendonk, Henry A. "Country Data Project." *Mathematics Teacher* 100 (November 2006): 284–90. Population distribution

- Lanier, Susie, and Sharon Barrs. "Let's Play Plinko: A Lesson in Simulations and Experimental Probabilities." *Mathematics Teacher* 96 (December 2003): 626–33. Probability and games

- Nord, Gail, Eric J. Malm, and John Nord. "Counting Pizzas: A Discovery Lesson Using Combinatorics." *Mathematics Teacher* 95 (January 2002): 8–14. Combinatorics

- Smith, Richard J. "Equal Arcs, Triangles, and Probability." *Mathematics Teacher* 96 (December 2003): 618–21. Points on a circle, probability

- Teppo, Anne R., and Ted Hodgson. "Dinosaurs, Dinosaur Eggs, and Probability." *Mathematics Teacher* 94 (February 2001): 86–92. Probability